Customer-Driven Project Management

Gill Tulloch

Customer-Driven Project Management

A New Paradigm in Total Quality Implementation

Bruce T. Barkley

James H. Saylor

Boston, Massachusetts   Burr Ridge, Illinois
Dubuque, Iowa   Madison, Wisconsin   New York, New York
San Francisco, California   St. Louis, Missouri

Library of Congress Cataloging-in-Publication Data

Barkley, Bruce.
Customer-driven project management : a new paradigm in total quality implementation / Bruce T. Barkley and James H. Saylor.
p. cm.
Includes bibliographical references and index.
ISBN 0-07-003739-6
1. Industrial project management. 2. Total quality management.
I. Saylor, James H. II. Title.
HD69.P75B38 1993
658.5′62—dc20 93-34343
CIP

McGraw-Hill

A Division of The ***McGraw-Hill*** *Companies*

Copyright © 1994 by McGraw-Hill, Inc. All rights reserved. Printed in the United States of America. Except as permitted under the United States Copyright Act of 1976, no part of this publication may be reproduced or distributed in any form or by any means, or stored in a data base or retrieval system, without the prior written permission of the publisher.

4 5 6 7 8 9 0 BKM BKM 9 0 9

ISBN 0-07-003739-6

The sponsoring editor for this book was Larry Hager, the editing supervisor was Ruth W. Mannino, and the production supervisor was Donald Schmidt. This book was set in Century Schoolbook. It was composed by McGraw-Hill's Professional Book Group composition unit.

Information contained in this work has been obtained by McGraw-Hill, Inc., from sources believed to be reliable. However, neither McGraw-Hill nor its authors guarantees the accuracy or completeness of any information published herein, and neither McGraw-Hill nor its authors shall be responsible for any errors, omissions, or damages arising out of use of this information. This work is published with the understanding that McGraw-Hill and its authors are supplying information but are not attempting to render engineering or other professional services. If such services are required, the assistance of an appropriate professional should be sought.

To Cathy—Your love, friendship, and insights made this book possible.

Bruce

To Rosann—Your love and support throughout the years contributed to our many VICTORIES

Jim

Contents

Preface

Customer-driven project management (CDPM) provides a new paradigm in total quality and project management. CDPM focuses today's organization on total customer satisfaction by integrating two very effective, contemporary management approaches: total quality management and project management. The result is a business or public enterprise which builds real value into customer-driven programs and projects, from beginning to end.

CDPM blends and broadens total quality management and project management into a new way of doing business. The approach uses the customer or customer's voice as the driver of the project deliverable. It stresses total customer satisfaction, continuous improvement, people involvement, and measurements from total quality management. It uses project management approaches to emphasize the provision of a successful deliverable. It targets new patterns of relationships and new processes between and within organizations in order to meet the challenges of today's world.

The farm business provides an illustration. For most enterprises, whether private or public, there are two ways to approach farming. The first kind of farmer is the organic farmer. The organic farmer tends to the soil, enriches it, nurtures it, feeds it, repairs it, prepares it, and lives or dies with it. The soil is the sustaining force of the farming activity. The organic farmer sees the soil as the nesting womb of the crop, the environment for growth, and the basic indicator of success. If the soil is good, the crop will follow. The organic farmer believes that healthy soil over the long term is worth building, even at the expense of short-term crop productivity. Quality begins and ends in the soil for the organic farmer.

The second kind of farmer is the commercial farmer. The commercial farmer tends to the crop, waters it, fertilizes it, feeds it, repairs it, and lives or dies with it. The crop is the product, and the issue is the amount of crop that can be produced. The commercial farmer believes that crop production is the indicator of success. Therefore, the commercial farmer focuses on the annual crop, not the soil. Quality begins and ends in the crop for the commercial farmer.

Now this story is not exactly accurate, since we took some liberties both with commercial and organic farming. But it helps to make a point, namely, that if the organization is the soil and its deliverables are the crop, managers have two options: to tend either to the soil or to the crop. Tending to the soil means working to establish a cultivated and enriched environment in which people can grow and produce to meet customer needs. Tending to the crop means working to design and produce outputs and deliverables as a first priority, focusing on product or service design and delivery, and placing emphasis on things, resources, specifications, and requirements. Some managers tend the soil; others the crop. If too much attention is paid to the soil, the crop can fail; if too much attention is paid to the crop, the soil suffers because its mineral base is not sustained.

To stay with the story, total quality management is the organic farmer, focusing on the organizational soil and the long-term capacity of the organization to satisfy customers; project management is the commercial farmer, focusing on delivering the short-term crop deliverable. This book addresses the challenge of taking the strengths of each to produce a new kind of management, called customer-driven project management. For an organization that is preparing for the future, CDPM is the process of combining continuous improvement of the soil with effective project management techniques for the production of the crop, all driven by the customer from beginning to end.

In gearing up for the challenges of the future, organizations, both public and private, face a swirl of forces and pressures from within and without. There seem to be many ways to farm the organization and produce good crops. From within comes increased awareness of the immense potential of employee involvement and empowerment. This potential comes from a reservoir of energy and innovation that once unleashed by an enlightened leader can reassert the leadership of U.S. industry and government in the global economy. From without, however, comes the realization that this energy must be focused to continually connect with customer's needs and expectations. When it is not, scheduling and costing issues tend to drive the work at the expense of customer satisfaction. It's not that customers are not interested in timely delivery within budget. It is that they typically feel that a product or service that meets their need is more important than time or cost, and they show it by paying premium prices for quality work. It does little good to put bad work out on schedule and within budget. Supplier firms and agencies that do project work for other organizations as customers, so-called contracting project management suppliers, often lose sight of the changing needs of the customer as the project progresses. It does little good to place the power of a supplying organization in gear if its direction is so locked in place by command and control systems and legal contracting and procurement requirements that it cannot respond. This tightly controlled approach cannot react to the dynamic changes that occur as suppliers find new ways to improve and customers identify new needs during the project work.

Two tracks of thought prevail in today's organization about how to respond to this dilemma, both reflected in customer-driven project management. The first, the quality improvement movement, focuses on the continuous improvement of internal processes targeting customer satisfaction. The customer initiates the right projects and products through a quality improvement process. These projects and products will enhance and improve quality for their customers. This process empowers teams to make recommendations for improvements. These can be operational improvements or major system and structural changes. Unfortunately, quality improvement teams are not typically seen as project implementation teams and are not often equipped or empowered to produce the reengineered system enhancements that would fundamentally improve quality.

The second track, project management, targets production and implementation. It gets project suppliers into the customer's process early so that up-front decisions can benefit from anticipated delivery problems. This can be done by nesting project management in a broader, more sweeping improvement process in the customer's organization. The Department of Defense is taking this approach on major acquisitions, and it is hinting at going a step further. Soon we will see a movement toward customer-driven teams that fully integrate quality improvement with project management. This new trend toward customer-driven project management is what this book is about.

It should not be surprising that these two tracks of thought and practice—quality improvement and project management—are not integrated in practice, and they remain separated in the literature as well. In effect, the quality people do quality and the project people do projects. In most organizations, these quality and project initiatives never meet, much less become integrated into an organizationwide focus on the customer.

Let's step back and explore why an organization pursues quality improvement. The determination to delivery quality in an organization results from many forces. The two major ones are

1. The organization faces stiff competition in a tight economy. Leaders and managers seek quality as a road to survival in these

uncertain times. They must continuously improve by focusing on the customer's needs to assure a productive role for the organization in the future.

2. The organization has internal barriers. It then needs to identify where its internal strengths are and play to them.

In the first case, a satisfactory response to the customer is the highest priority for everyone in the organization, every day in every way. This is the process of alignment now so popular in total quality literature. In the second case, in essence, the organization places its people—especially its core project design and product delivery people—in direct touch with the challenges of real customer needs. This removes the internal barriers to teamwork and empowerment. But the missing link here is the lack of flow and continuity from the quality effort to the production effort, from finding quality opportunities to making the structural and system changes necessary to exploit them. There are few opportunities for empowered teams to move through the whole cycle of improvement, including major system enhancements, from beginning to end. Thus, quality improvement is typically at risk of underestimating the potential for system redesign, because it does not encourage fundamental reinvention of the system and the deliverable necessary to produce the new system.

Let's use the fictitious example of Atlas Industries, a typical supply and project management firm. Atlas began its process with a vision statement written by the President, stating the key significance of customer service, continuous improvement, and teamwork. Atlas's top management followed up with a mission statement of what turned out to be an extraordinary expression of the need for change in the way Atlas did business. Atlas had always been a traditional project and public works product development firm, responding to invitations to bid by public and corporate customers. But its management had looked at its project management process through the eyes of its new quality improvement process and found major problems. Atlas had wonderful performance in meeting specifications, on time and within budget, but had less-than-exciting endorsements from its previous customers. Its customers found the firm inflexible and distant—sometimes describing it as technically arrogant—once the work started, and the firm rarely developed lasting customer relationships. Somehow, Atlas had developed a capacity to deliver projects very efficiently through sophisticated project management systems but could not break through the customer's early decision making in order to become part of its quality improvement and needs assessment effort. If productivity were measured in output per dollar invested, Atlas was doing well. If productivity were measured as value added to customer, Atlas was a drag on the economy, despite the fact that it made its cost and schedule objectives as well as its profit margin goals.

To resolve its difficulties, Atlas management decided to go through the eight phases of the customer-driven project management process described in this book. It first looked at its project management system within the framework of quality improvement principles. It found that once customer specifications and requirements were written and the project entered the concept stage, the teams were left with little understanding of the underlying issues and root causes that determined the need for the customer's project in the first place.

It also found that despite its matrix organization, once it was into the project process itself, the handoffs from concept into design, definition, production, and operation were difficult and incomplete. Tension inevitably resulted inside the team. The underlying root cause for this tension was twofold. The first was an implicit misunderstanding on the part of all team members about the meaning of critical path planning and interdependence. They thought that they could not start a task until the previous one was finished and handed off, since that was what the critical path plan suggested. Little communication occurred unless tasks intersected on the critical path. Second, because of the separation of the project team from the customer's quality improvement process and developing sense of needs, and because of this hand-off issue, projects were driven more by schedule and cost controls than by customer issues and needs downstream in the project development cycle. Most of the upstream learning about the customer's quality problems was lost to Atlas project team members.

Recognizing that it had to do something about these issues of disconnection and isolation, Atlas articulated a new vision, mission, and value statement. It fostered the use of customer-driven project management. The major components of its new approach were

1. Seek out business with customers willing to use customer-driven project management. With the customer as the project driver, empowered teams would lead a project through performance and quality improvement in one unbroken process.
2. Support the new arrangement with a telecommunication network to make the customer as accessible as any team member.
3. Identify the customer's representative as the customer-driven project manager.
4. Empower customer-driven teams by cooperative agreement to complete the project while continuously making improvements focusing on customer satisfaction.

The story does not end there, since through this process Atlas was equipping itself for a lasting relationship with its new customers. It soon began to work more upstream with its selected customers and was able to stabilize its project relationships to the point that it was constantly improving the feedback from customers on finished projects.

This book expands on the experience of many Atlas-type organizations needing to broaden their application of quality improvement and project management. It is about integrating project management techniques, tools, and organizational relationships within a broad, customer-driven quality management process. The resulting paradigm will serve as a model for organizations seeking to go beyond project management and total quality management to achieve success with future projects for decades to come.

Bruce T. Barkley
James H. Saylor

Acknowledgments

Special acknowledgments are due to GenCorp Aerojet and the University of Maryland for their cooperation. Specifically, the Electronics Systems Division of GenCorp Aerojet and the University of Maryland University College and the University of Maryland, College Park, College of Business and Management, Maryland Center for Quality and Productivity, were helpful in providing us with unique experiences and insights that generated the ideas contained in this book. We would also like to thank the many team members and students who provided the wisdom and practical experience that allowed us to persist in the continuous improvement of the many concepts included in this book.

In addition, this book is the result of efforts of many others. Contributions and comments were provided by Scott Armstrong, Frank Herrmann, Greg Just, Margo Schachter-Karstadt, Shirley Kondek-Noel, and Richard Rawcliffe. Richard Rawcliffe deserves recognition for his adaptation of the quality function deployment example for a "cup of coffee." Our appreciation also goes to the developers of the many Department of Defense (DoD) Total Quality Management and Project Management documents. The Federal Quality Institute is another valuable source of ideas. They deserve recognition for a job well done. Their efforts provide much of the inspiration for this book. In addition, we would like to express our special thanks to Peter Angiola of the Office of the Under Secretary of Defense for Acquisition, Total Quality Management, for an overflow of information from the Department of Defense. Our thanks are also extended to William Cleland for his several reviews of the manuscript.

We would also like to thank the several project management software companies that provided the use of their software for review. Symantec Corporation's Time Line and Scitor Corporation's Project Scheduler 5 were both used to produce figures in this book:

Symantec Corporation for Time Line and Time Line for Windows
Jayme Kelly, Public Relations
10201 Torre Avenue
Cupertino, Ca. 95014-2132

Scitor Corporation for Project Scheduler 5
Lenore F. Dowling, Director Marketing and Sales
Productivity Software Division
393 Vintage Park Drive
Suite 140
Foster City, Ca. 94404

Primavera Systems, Inc. for SureTrak Project Scheduler
Mr. Eric S. David, SureTrak Marketing Manager
Two Bala Plaza
Bala Cynwyd, Pa. 19004

Microsoft Corporation for Microsoft Project for Windows 3.0
One Microsoft Way
Redmond, Wa. 98052-6399

Computer Associates International, Inc. for CA Super Project
Laura Wood, Public Relations Dept., 6th Floor
One Computer Associates Plaza
Islandia, N.Y. 11788

Special credit is due to the following sources of public domain or copyright materials. Public domain information was used from the following sources:

United States Department of Commerce, National Institute of Standards and Technology for information contained in the 1991 Application Guidelines for the Malcolm Baldrige National Quality Award.

Office of Deputy Assistant Secretary of Defense for Total Quality Management OSUSDA(A) TQM for many sources of information within the Department of Defense.

Federal Quality Institute for many sources of information on total quality management.

The following should receive distinguished credit as source for copyright materials:

Warren Bennis for the quote from *The Unconscious Conspiracy*.

David I. Cleland for excerpts from *Project Management Strategic Designs and Implementation*.

Harold Kerzner for the definitions of project success from his second and fourth editions of *Project Management*.

Massachusetts Institute of Technology, Center for Advanced Engineering Study for Deming's Fourteen Points. This information originally contained in *Out of Crisis*. The Fourteen Points are an updated version from the book.

Special recognition goes to Rosann Saylor for her continuous assistance in the composition of the book. Her efforts greatly enhanced its quality.

As with all worthwhile efforts today, this was truly a “team” effort. All members of the team deserve special recognition for their specific contributions to customer satisfaction.

Chapter

1

Introduction to Customer-Driven Project Management

Focus: This chapter describes why the customer-driven project management approach is needed today. It also outlines the customer-driven project management approach.

Introduction

An organization cannot survive and prosper in today's world without customers. Customers allow an organization to exist. Many modern organizations have lost sight of this fundamental principle. This failure to meet customer needs and expectations has contributed to the United States' lack of competitiveness in the global economy. In many cases, U.S. industry is providing the wrong product and services for domestic and global markets because these organizations have taken the customer out of design and development activities. Such organizations focus on pushing the product or service onto the customer, rather than on the customer pulling a product or service out of the organization. This is particularly true of organizations that provide a product or service as a result of project management efforts.

Many project management organizations have not developed an ability to respond rapidly to changing customer needs and expectations. Many excellent organizations have perfected the ability to "lock in" on specifications and produce a product within schedule and budget; yet they have not developed the ability to listen to their customers. They seek to define the customers' project requirements rather than determining customers' needs and expectations. Project managers tend to lock in too early on the specifications. This frequently results in an isolation from the customer, with the ultimate consequence of leaving the deliverable on the doorstep for the customer. Such organizations do not keep close to their customers. They avoid the possibility of customer changes. Besides putting up the obvious barriers to satisfying customers, they deny the dynamics of the learning process. If change is allowed, it is usually at a high cost in terms of both dollars and relationships. All too often the project ends up as a disappointment, not adding any value for the customer. This alienates the customer for any future deliverables.

In addition, project teams are usually encouraged to produce too quickly. They are rewarded by output. Therefore, they are not motivated to listen for customer needs and expectations. Frequently, they are completely secluded from any customer contact beyond status presentations. These problems stem from the failure in both the private and public sectors to realize that a project team should be empowered to continuously improve the project. The project team must represent both supplier and customer. It should have the responsibility, authority, and resources (empowerment) to satisfy the customer. This type of project team stands a better chance of producing a deliverable (product and/or service) that delights and adds value for the customer. Project teams can perform and make improvements better if they understand both the total customer and supplier environments. With this type of project team, the organization's potential for keeping and gaining customers increases dramatically because it does not only produce a deliverable but focuses on total customer satisfaction.

In addition to project management organizations, many companies and agencies use total quality management processes to strive for success. They have embarked on the road to quality improvement through quality-improvement teams. These quality-improvement teams are frequently limited in their implementation of quality-improvement recommendations. In many cases, they are empowered to take action only on quality issues. In some cases, the quality-improvement recommendation is given to a process owner for implementation. In other cases, the quality-improvement team lacks project management skills to transition from the improvement recommendation to a project deliverable.

In many of today's organizations, the total quality management and project management approaches are separated. This restrains the organization's ability to achieve total customer satisfaction.

Customer-driven project management (CDPM) is a management approach that focuses on producing deliverables that achieve total customer satisfaction. CDPM builds on the strengths of both total quality management (TQM) and project management. It uses the total quality management emphasis on continuous process improvement, people involvement through teams, quantitative methods, and customer focus, and it stresses the project management methods for planning, controlling, and delivering successful deliverables. The target is to gain and keep customers for the long-term prosperity of the organization. Customer-driven project management applies in any activity where

- A project deliverable can be defined.
- A customer or customer's voice can be identified.

In CDPM, a *project deliverable* includes any product, service, or combination provided to a specified customer to satisfy a particular need at a certain time. *Projects* include three kinds of planned, short-term activities: (1) those producing complex technical products, capital, or information systems and requiring traditional project management approaches, (2) those producing quality improvements of any kind from teamwork, and (3) those resulting from natural work teams. In any of these project categories, the approaches and techniques addressed here will be useful. One of the collateral benefits of broadening this definition is that it introduces project management techniques to a much wider universe of users than in the past, building on its wide application in engineering and construction firms performing contract work.

Customer-driven project management works for large and small businesses, manufacturing and service industries, and public and private organizations.

Why Customer-Driven Project Management?

Customer-driven project management expands both project management and total quality management approaches to meet the challenges of the global economic environment. Customer-driven project management provides a management approach adaptable to the new world of rapid change, rising complexity, and rabid competition. Today, political, technological, social, and economic changes are swift. In the short period since the end of the 1980s, the world has been turned around. For instance, the former USSR empire has crumbled. The Berlin wall has been torn down. Japan has become the world's number one economic power. The United States is just one of many players in the global marketplace. The technological advances, especially in computers and telecommunication, of the information age have brought about rising complexities in the processes used to perform our work. Competition on a global scale is a fact of life. Everyone is competing for the new global markets. With competition fierce in all aspects—technology, cost, product quality, and service quality—new approaches must be sought in order to be competitive and share in global economic growth. Customer-driven project management provides an approach to confront these challenges today and in the future.

Today's world

Today's world is radically different from that of the recent past. It is a new environment where old solutions no longer work. The "same old way" simply does not bring about the necessary results. Technology is not the prospective cure-all. Throwing resources at a project for short-term progress does not foster long-term customer satisfaction. Our paradigm (mind-set) must change to reflect the reality of today's world to achieve success.

Some of the major considerations in today's world compared with yesterday's issues are shown in Table 1.1. These conditions require—no, demand—change.

There are many players in the competitive global economy. Japan has replaced the United States as the new world economic leader. In addition, there are many other formidable players. This fresh economic playing field requires everyone to transform their management philosophies, principles, methods, tools, and techniques into a management system that allows everyone to work smarter to respond rapidly to satisfy customers.

Rabid competition is the way of the world economy. Just because a product or service is available does not mean it will sell. Customers are more selective in buying goods and services. In fact, as customers have more and more options, they become increasingly discriminating, demanding added value. This makes keeping and getting new customers more important than ever. Customer satisfaction is the focus of all competing organizations. The organizations that can answer constantly changing customers' demands will succeed in this new environment of rabid competition.

Uncertainty is now always a concern. With a rapidly changing world order, certainty can no longer be taken for granted. No organi-

TABLE 1.1 Considerations: Yesterday's and Today's World

Yesterday's world	Today's world
United States as top economic power	United States as one of players
Make it, it sells	Rabid competition
Certainty	Uncertainty
Reasonable cost	Lowest possible cost
Large budgets	Optimizing budgets
Stable technology	Rapidly changing technology
Waste: many resources	Conservation: limited resources
Quality is supplier-driven	Quality is customer-driven
Reasonable time to market	Accelerated time to market
Hard copy/space/distance controlled	Telecommunications/information controlled

zation is safe from some sort of distress. External factors affecting the organization are progressively out of control. The organization's systems are constantly requiring updating to optimize productivity, quality, and costs. New or changed products and services are perpetually being introduced. Stronger competition is increasingly the norm. In addition, customer needs are continuously changing. Continuous vigilance of all factors affecting the organization, the product/service, the competition, and customers is a necessity.

Economic pressures are a fact. This makes costs and budget a factor today and tomorrow. Lowest-possible cost is the aim of all internal processes. It is no longer good enough to strive for reasonable cost. Everyone has the same technological advantages to make use of economies of scale, automation, and other production and service techniques to reduce costs. Customer satisfaction and profits in today's world depend on providing a product and/or service at the optimal, lowest-possible cost.

In addition, economic pressure makes optimizing budgets an everyday reality. Currently, budgets are shrinking in most organizations, causing a reexamination of priorities to stress more than ever "more for the buck." Economic pressures will continue to dominate choices and decisions in public and private organizations. The demand for increasing value for less cost will continue into the next century.

Rapidly changing technology makes stability impossible. Failing to keep pace with the latest technology can bring obsolescence within a short period of time. Many products today have a very short life cycle. The impact of new technology, especially in information processing and communications, may determine supremacy.

Conservation of limited resources is a necessity. Global competition for scarce resources will only increase in the new global marketplace. Compounding competition, however, is the need to protect and preserve the environment. With many nations competing for few resources, coupled with concern for the environment, waste and loss are everyone's enemies. Organizations must learn new techniques of quality, productivity, and project delivery, focusing on elimination of variation to optimize all resources.

Reasonable production times no longer meet customers' needs. Accelerated production times are essential in many industries. The organization first to the marketplace is usually the winner. In today's world, speed is a competitive advantage.

Customer-driven quality is critical to long-term growth. Since customers define quality by their satisfaction, the supplier forcing a deliverable on a customer does not foster customer satisfaction. Customer satisfaction pulls quality from the supplier. Today, the customer or customer's voice must direct every aspect of the deliverable. This is the only way to ensure quality.

Today's world is controlled by telecommunications and information. Gone are the days when hard copies, space, and distance were the dominating factors of competition. The organization that can speed the right information to the right place at the right time is ahead of its rivals.

Today's world demands change

Adapting to today's economic world with an eye to the future requires an organization to be totally responsive to customers. Specifically, the successful organization will be the one that can change to apply the new paradigm for prosperity in today's global environment. Table 1.2 lists the required changes.

Continuous improvement of processes, people, and products aimed at customer satisfaction is essential. The "if it's not broke, don't fix it" attitude does not promote the critical thinking necessary for growth. Continuous improvement is the only way to survive. This is the proactive approach to change. This new view of "everything can be made better through process improvement" stimulates the creativity and innovation needed for constant growth. In addition to continuous improvement of processes, people must constantly upgrade their knowledge and skills through a lifelong learning system. Further, products and/or services require progressive enhancement to meet or exceed changing customer needs and expectations.

Systems thinking must replace functional orientation. In today's world, everyone's horizon must be expanded beyond such narrow-

TABLE 1.2 The World Demands Change

From	To
If it's not broke, don't fix it	Continuous improvement
Functional orientation	Systems view
Sequential design and production	Concurrent design and production
Inspection of defects	Prevention of defects
Quality not important	Quality critical
Accept current processes	Reengineer processes
Development	Innovation
Rigid organizational structure	Flexible organizational structure
Many organizational layers	Few organizational layers
Compete	Cooperate
Individual performance	Team performance
People specialized, controlled, eliminated	People add value, flexible, empowered
Strong management	Strong management and leadership
Leadership only at the top	Leadership everywhere in the organization
Short-term outlook	Long-term vision
Individual merit reward system	Team performance reward system
Education for management	Education and training for everyone
Focus on profit	Driven by total customer satisfaction

minded occupational disciplines as engineering, manufacturing, accounting, education, training, logistics, and the like. Organizations struggling for success in the world economic environment cannot afford to subsidize functional "fiefdoms" that suboptimize resources. Progressive organizations must view the combination of all their processes as a system focused on customer satisfaction. Thus, everyone in the organization must have a systems outlook geared to achieving organizationwide excellence. This means optimizing quality and performance.

Concurrent design is necessary especially in industries where time to market is a competitive advantage. Time to market is increasingly a differentiator in the marketplace for both products and services. Concurrent design of products and services reduces the time to market significantly over traditional sequential design methods.

Inspection-based quality assurance needs to be supplanted by prevention techniques. Again, the industrial mind-set expects defects. This inspection-based viewpoint adds excessive cost to the product or service—a cost that most organizations can no longer afford and customers do not need to support. By shifting the emphasis to prevention techniques, the right thing is done right the first time. Thus cost is reduced while product and/or service quality is increased. Prevention techniques focus on improvement of all the processes in an organization to maximize the capabilities of processes.

Quality focused on customer satisfaction is required. Disregard for quality or concern for just conformance to requirements is no longer good enough to keep and maintain customers. Today's knowledgeable customers demand satisfaction. This is how they define quality. Hence, organizations must determine what will satisfy their customers and then focus on striving to meet customers' needs and expectations. To accomplish this goal the organization must identify the elements of quality that are of vital importance to their particular customers. The key elements of quality for a customer might include, among others, the following items: perceived product and/or service quality, performance, reliability, supportability, durability, features, availability, esthetics, serviceability, maintainability, usability, environmental impact, conformance to requirements, customer service, logistics, training, warranty, and life-cycle cost. For instance, some customers demand reliability as a key element of customer satisfaction. In other cases, customers might value performance as the critical element. For still other customers, the key element of quality may be availability. For such a customer, it is of primary importance that the product or service be available in the right quantity, at the right time, and at the right place.

Innovation through constant incremental improvement must be pursued by many organizations and industries. Building on the old and creating new uses are critical for future survival. Economic considerations dictate making do with what already is available. Such an approach targets innovative enhancements of existing systems and deliverables as a major method of satisfying requirements.

Flexible organizational structures with as few layers as possible are best able to respond rapidly to customers' changing demands. Rigid structures cannot react fast enough to keep pace with a formidable competitor, whereas an organization with only a few layers has the "lean and mean" structure needed to contend in today's world. Customers will no longer subsidize the cost of the huge waste created by large bureaucratic organizations. Organizations must be trimmed to the absolute core, with decentralization of empowerment to the people closest to the customers and the processes. The organizational structure of an achieving organization has fewer managers.

Cooperation among governments, industries, companies, organizations, teams, and individuals is essential for survival. Cooperation among government, industry, labor, and education is critical to a high-growth, high-wage economy. In addition, management and labor must learn to cooperate for a prosperous economy. Further, departments and functional organizations must break down barriers to optimize organizational productivity. Also, individuals need to work together in teams to respond rapidly to customers. In addition, organizations must develop supplier partnerships and customer relationships. All cooperative efforts aim at win/win solutions instead of the win/lose situation fostered by competition. Only through cooperative relationships can global success be realized.

Groups, especially teams, are the organizational structures of choice. Although individuality is important, teams multiply the capabilities of each team member. In today's complex workplace, teams are the only structure capable of providing the high level of performance, flexibility, and adaptability necessary to respond rapidly to customers and to provide deliverables that delight them.

People are the most important, flexible, and versatile resource capable of adding customer value to a product or service. Empowered people are the only resource with the ability to respond quickly to customers by optimizing the output of a process based on a thorough analysis of the customer requirements and the process. Therefore, specializing or eliminating people greatly reduces an organization's ability to keep or gain customers, and that significantly decreases its chances for survival. People are the most important resource for gaining an advantage over competition. To optimize this essential resource, forward-thinking organizations must strive to provide a high-quality work environment where both the people's and the organization's needs are satisfied while striving to delight customers.

Strong leadership at the top and at all levels is needed, instead of strong management. Guiding people to achieve a common goal is the focus of improved performance in any organization. Strong management is still required to ensure that a project is completed as required, but leadership is essential to maximize the human potential to care about and satisfy customers. Managers simply ordering accomplishment will not make an excellent organization. Leadership involves the sustained, active, hands-on participation of all leaders, continuously setting the example, coaching, training, mentoring, and facilitating empowered people.

A long-term view needs to replace the emphasis on short-term results. Frequently, the emphasis on a short-term outcome has a long-term consequence. For example, the takeover of one organization by another may bring short-term financial rewards. In the long term, the takeover can result in many people losing their jobs. In another case, a short-term focus on stocks causes an organization to reduce investment in capital equipment and/or training of its people, possibly having an adverse effect on the long-term survival of the business. The viewpoint of the organization must be targeted on the long-term future to stay in business. The advanced organization has a vision for the future with a strategic plan for achieving that vision.

The reward and recognition system must shift from individual merit to team-performance systems. A team-performance reward and recognition system provides the incentive for optimizing the results of teamwork to accomplish a mission. The team-performance system should credit the individual's contributions to the success of the team while at the same time providing a reward for effective teamwork. All reward and recognition systems should be geared to each individual team member wanting to contribute to the best of his or her abilities for the ultimate successful outcome of the team.

An education and training investment for everyone in the organization must be the cornerstone of any organization using a continuous improvement system to persevere in today's environment. People must be improved continuously through education and training, with an eye toward the future. High-growth labor markets demand people with specific higher-level education and skills. These conditions require organizations to adopt a viewpoint that people truly make the difference in the competitive world economic environment, and such organizations must make an investment to create a lifelong learning system.

Customer satisfaction through its processes and deliverables must drive an organization. A focus on profit as the primary purpose of an organization is obsolete. As shown in Fig. 1.1, the successful organization strives to meet customer expectations through continuous improvement of its processes, people, and products focusing on total

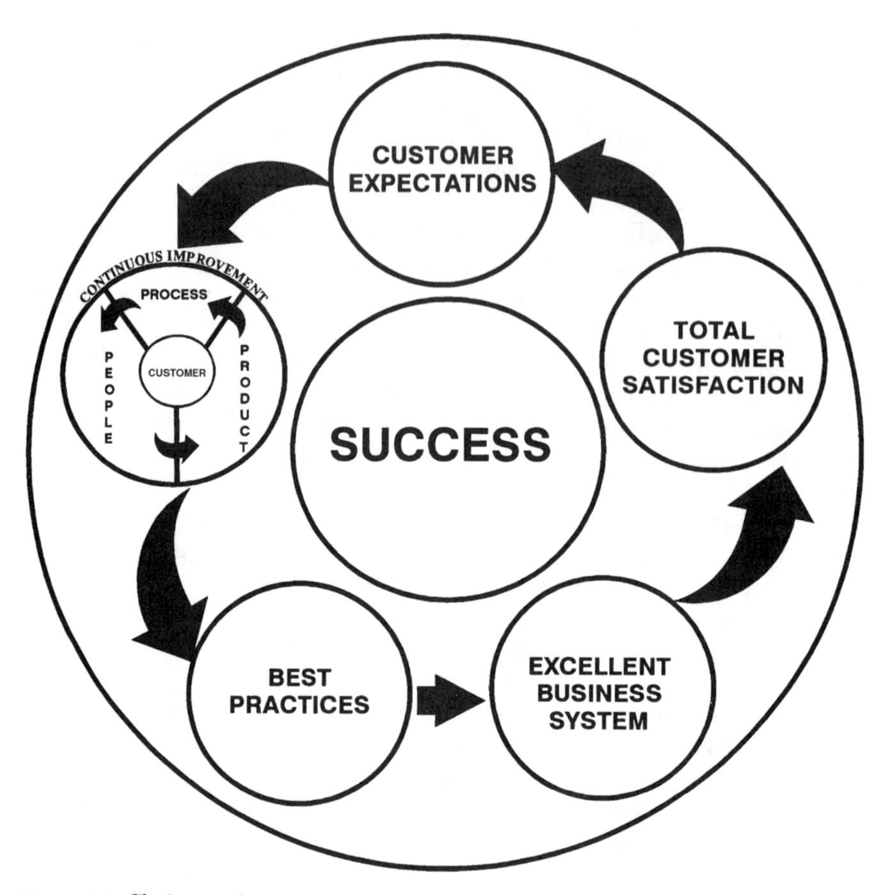

Figure 1.1 Chain reaction.

customer satisfaction. This promotes use of the best business practices, which leads to excellent business systems and provides total customer satisfaction. The chain reaction starts with the customer expectations, and total customer satisfaction is the focus of all efforts. The result is a successful organization. The specific definition of success varies by organization, but this is the vision. It can mean survival or growth, profits, returns on investment, and stockholder dividends. The view of total customer satisfaction applies to both external and internal customers. The entire organization must be aimed at satisfying the ultimate user of the product or service. Within the organization, each process pleases the next process in the chain. Everyone works with an internal supplier and customer relationship to improve the process for total customer satisfaction of both internal and external customers.

Focus on the future

Customer-driven project management focuses on the future. Although every organization is unique, there are certain objectives, strategies, tactics, and operations that will work for organizations striving for a vision of the future. Figure 1.2 shows the vision, overall mission/objective, strategy, tactics, and operations necessary for any organization to prosper today with a view toward the future.

The vision is always aimed toward excellence. The objective is total customer satisfaction, both internal and external. This means meeting or exceeding customer expectations, and requires establishing and maintaining a customer-driven organizational culture focused on doing whatever it takes to add value for the customer. Creating a customer-driven organizational culture requires vision and the leadership to make that vision a reality; involvement of everyone and everything in a systems approach; continuous improvement of people, processes, and products; training and education for everyone; ownership with empowerment; an appropriate rewards and recognition system; and years of commitment and support—all aimed at total customer satisfaction.

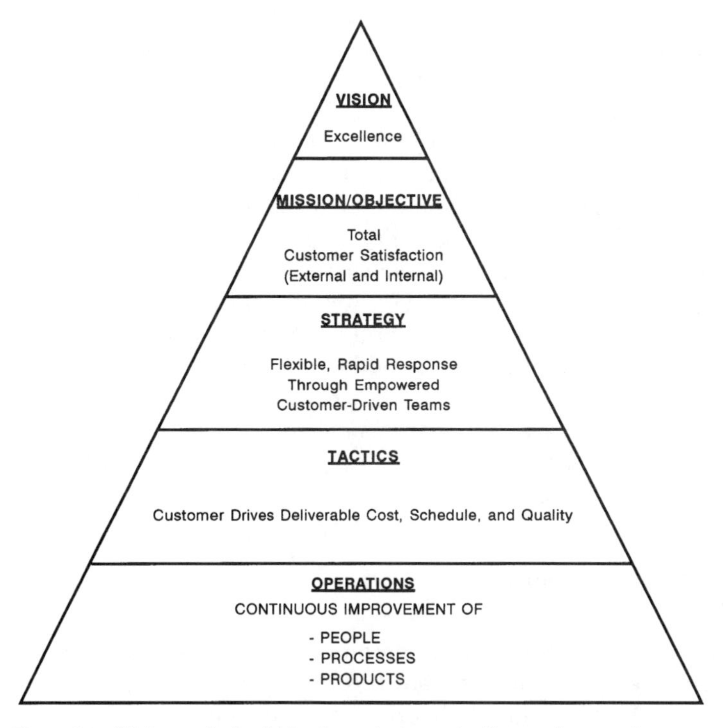

Figure 1.2 Vision, mission/objective, strategy, tactics, and operations.

The strategy revolves around a flexible, rapid response through empowered customer-driven teams. This strategy requires an organizational structure featuring customer-driven teams with leadership and empowered people. The organization and its teams must completely understand customers and their processes. The teams must be organized so as to enable them to develop a partnership with suppliers and a relationship with customers that allows them to adapt their processes immediately to any customer's desire.

The tactics include the customer's voice driving cost, scheduling, and quality of the deliverable. Such an approach includes providing the best possible deliverable at an optimal life-cycle cost. The deliverable, whether a product, service, or both, must be world class to compete in the global economy.

The operations include continuous improvement of processes, people, and products. Continuous improvement involves the elimination of loss in all processes, investment in people, and constant upgrade of products.

Customer-Driven Project Management

Customer-driven project management (CDPM) provides a flexible and responsive management approach that is able to act in the face of or react to all the forces of today's economic world. This customer-driven approach focuses an organization on determining and acting on the internal and external forces that influence the customer. CDPM gears an organization to improving quality, increasing productivity, and reducing costs to satisfy its customers. It focuses on total customer satisfaction through the highest-quality deliverables, whether products or services, at the lowest life-cycle costs to compete in the global environment. Customer-driven project management stresses constant training and education, right-sized and team-based organizations, and use of people's full capabilities to add value while maximizing the human resource. Customer-driven project management advocates the optimal use of technology along with people to achieve a competitive advantage. Customer-driven project management is a management approach for today and tomorrow.

Customer-driven project management background

Full exploration of the concept of customer-driven project management requires an examination of what has happened over the years to change some of the traditional theories of management, the classic approaches to quality, and the prevailing methods of project management. Customer-driven project management borrows heavily from a diverse set of many "old school" management, quality, and project management concepts.

Historically, management concepts have focused on the functions of organizational control, including planning, organizing, staffing, directing, tasking, structuring, coordinating, budgeting, evaluating, inspecting, and reporting. These concepts were derived primarily from industrial and corporate firm models. However, many other organizations were influenced by these theories, including government and nonprofit organizations. This has led to the wide acceptance of these management principles.

The basic assumptions were that managers were paid to exercise control over the organization's resources, including people, to ensure that they met the objectives of management. Scientific principles such as unity of command, uniformity, centralization, delegation, discipline, and work flow were emphasized. This scientific approach to management was derived and promoted during the early twentieth century by Frederick Taylor, Henry Fayol, and Max Weber to support the concept of efficiency and control. Production was more important than people, and people were "slotted" into job functions. The *management* of things was the emphasis. In this environment, people were specialized or eliminated. The production output was the major focus. Project management grew out of this kind of thinking.

Then the psychologists, including Lillean Gilbreath, following on the abuses of the industrial revolution, began to focus on the human element in the workplace. They concluded that it was not the monotony of the work that caused employee dissatisfaction, but rather management's lack of interest in the workers. Then Elton Mayo (1927) defined the organization as a social system and concluded that work groups determined attitudes and behaviors, not top management, and that productivity was directly related to employee satisfaction and particularly to the amount of attention paid to employees (the "Hawthorne effect"). Many others then tried to integrate the productivity and the human elements, including Peter Drucker (1954) through management by objectives and Blake and Mouton (1969) through the managerial grid. Douglas MacGregor (1960) defined theory *X* managers as authoritarians who believe that people are inherently lazy and do not want to perform and theory *Y* managers as nurturers who believe that people are innately responsible and capable of exercising initiative and making worthwhile contributions. Other behaviorists worked to explain motivation in the workplace.

As the management approaches transformed from the turn of the century, quality techniques began to evolve after World War I. The concepts that were to crystallize into total quality management began with the statistical quality concepts of Dr. Walter Shewhart, as published in 1931 in his book, *Economic Control of Quality of Manufactured Products.* His statistical approach, called *statistical quality/process control,* is the foundation for the *quality* management approach. His approach was refined by many practitioners during World War II to improve the quality and productivity of America's war products.

This quality approach has evolved continuously to today's total quality management concept. Total quality management (TQM) includes a wide range of management practices, methods, tools, and techniques. Since TQM is a collection of the best of many disciplines focusing on the actions required to survive and prosper in the environment of global competition, there are many contributors. Some of the major contributors include W. Edwards Deming, Joseph M. Juran, Armand V. Feigenbaum, Kaoru Ishikawa, Genichi Taguchi, Philip B. Crosby, Peter Drucker, Tom Peters, H. James Harrington, A. Richard Shores, and many others in organizations such as the U.S. Department of Defense.

The total quality management approach evolved to meet the needs of today's global environment, the result of world economics since World War II. After World War II, in a United States–dominated industrial world that had little competition from foreign manufacturers and service providers, quality was not seen as very important. Everything that was made was sold, almost regardless of quality, since the United States was in most cases the only producer. Quality in the post-World War II years continued to be a "second-class citizen." First priority was placed on quantity and production—getting the products out the door. Companies employed inspectors at the end of the process to find defects and rework them.

Then in Japan, in the 1950s, W. Edwards Deming introduced the Japanese to his 14 "points," including use of statistical techniques to measure continuous improvement, constancy of purpose, elimination of inspection, massive training and education of the work force, removal of fear as a motivator, and the development of leadership. Deming revealed that the cost of poor quality was a lower market share and lower profitability and higher and costlier inventories. J. M. Juran published his *Quality Control Handbook,* which outlined the application of basic statistical control to business practices. The message was that our organizations were smothering people in command and control systems and that the workers did not feel empowered to think and act on the basis of their insights about how business was practiced and customers served. From these early teachings, many new concepts were fostered and continuously improved to make Japan an economic world power.

During the late 1970s and early 1980s, the threat of competition started to awaken some U.S. organizations to the need to improve their quality so as to be competitive. Many organizations in the United States turned to the original quality masters who helped Japan. In addition, they adopted ideas from many new experts. Tom Peters, Robert Waterman, Jr., and Philip Crosby were influential during this early period of the rebirth of American quality. Tom Peters and Robert Waterman, Jr., provided many ideas about the elements of success of the top companies in the United States in their book, *In Search of Excellence* (1982). Philip Crosby's zero defects program was implemented in many organizations.

As the 1980s progressed, many organizations developed their own management approaches based on previous concepts to meet their particular needs. These approaches have been consolidated under total quality management. The key significance of TQM is the emphasis on customer satisfaction, continuous improvement, measurement, and employee involvement.

To turn to project management, its early history involves major systems and capital investments. Project management approaches became essential for certain activities in the United States after World War II. Project management originated to manage large, complex projects. The foundation of modern project management can be traced to the Manhattan Project. This project focused attention on the development of project management approaches that are continually being refined today.

Project management became necessary because traditional organizations structured around functional activities, such as engineering, manufacturing, support, finance, and human resources, could not meet the demands of building an aircraft, developing a computer system, constructing a building, or producing other complex projects. These types of projects required optimizing and integrating all the functions of the project. From its beginnings in mostly government and construction work, project management moved toward consumer products as well as one-time systems requiring a high degree of technical capability, such as computer systems. As technology increasingly became the basis for economic development, project management prospered as the dominant mode of organization in technology-oriented organizations for producing complex deliverables. This led to the establishment of a Project Management Institute to establish professional standards.

Project management techniques and processes are, to one degree or another, applicable in any organization faced with the need to ensure reliable scheduling, budgeting, and quality of measurable outputs. Project management is currently employed in large industries, particularly construction, oil, power, and chemicals, as well as weapons systems development. Applications also include development of information management, computer systems, research and development, and a wide variety of new hardware systems that require the cooperative work of specialists or specialized contractors. Project management is now typically associated with the integrating function that is so essential in fitting together the work of diverse functions and jobs into a total system. This is the primary reason that this type of management is often identified as a "systems approach."

According to project management standards, a successful project is one that is completed on time and within cost and meets performance criteria. The primary focus of project management has always been on completion of the project on schedule and within budget. Further, performance standards were determined by specifications early in the project. As the project progressed over a long period of time, the project management team often lost sight of the customer. There was little concern for customer needs and expectations. Quality simply meant meeting the specifications. There was no action to focus on customer satisfaction. Simply, there was no interest in making any quality improvements. Project management techniques typically produced unsatisfactory results for customers because project managers were driven by factors unrelated to total customer satisfaction.

Each one of these management approaches developed in response to specific requirements. The traditional management approach was designed for the industrial age of mass production. The quality management approach evolved to cure some of the ills of traditional management. The project management approach was formulated to handle complex development activities. Today, successful organizations have new requirements. These requirements focus on a flexible, rapid response to customers, pointing to the need to integrate the best from the traditional management, quality, and project-oriented approaches to optimize the organization of today. This requires formulating a management approach that ensures that customers drive the product and/or service, while the supplier continuously makes improvements to the deliverables—a fundamentally different way of thinking about how customers are treated in designing and producing deliverables. This approach links total quality management with the project management system (see Fig. 1.3). The link starts with customer expectations that focus the organization approach. It uses the customer to drive both total quality management process improvement and project management. Using the customer to drive process improvement and project management leads to enhanced process capability and project deliverables. This process provides total customer satisfaction, ensuring the product and/or service will be delivered on time, at reasonable cost; will meet technical performance standards; and will sat-

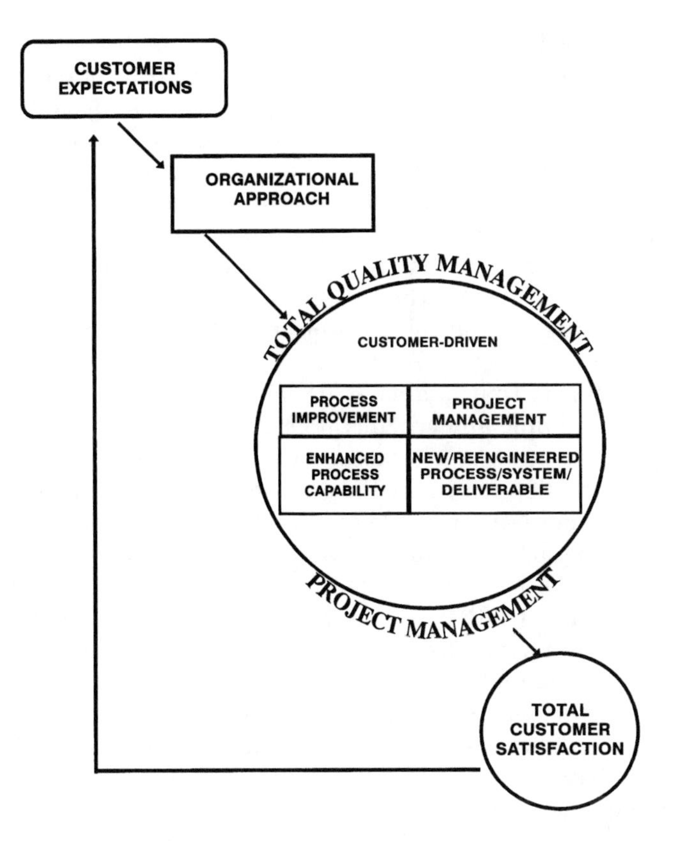

Figure 1.3 Linking TQM with project management.

isfy the customer. It builds a "project learning" process that emphasizes developing the job efficiently and effectively, and it ensures the right output for the customer. This results in a system that optimizes resources and saves hundreds of public and private dollars in producing products and services that add value to the economy.

Definition of customer-driven project management

Customer-driven project management is a management philosophy, a set of guiding principles, a methodology, and a set of tools and techniques that stress customer-driven deliverables, including products and services. It focuses on the performance and improvement of a project and the design and delivery of a product or service. It applies the proven techniques of project management, continuous improvement, measurement, people involvement, and technology. It integrates project management, total quality management, and a customer-driven structure.

Customer-driven project management (CDPM) uses a methodology that is entered into jointly by customers (whether inside or outside the organization) and project management suppliers. Customer-driven project management determines which projects to undertake in order to maximize customer satisfaction. It then takes those projects through concept, definition, production, operations, and closeout using proven project management and total quality management tools and techniques within a customer-driven structure. CDPM is a process that is wholly driven by the customer at every turn and which places the customer in the role of project leader from start to finish.

The essence of CDPM is that quality is defined by the customer's total satisfaction. This simple and succinct definition of quality forms and shapes both the way projects are chosen and the way projects are implemented. Quality is defined in CDPM through a process that is controlled by the customer (either inside or outside the organization), who selects, plans, designs, and implements projects with the help of the CDPM team. The CDPM team is often a combination of several organizations. In essence, CDPM is an enhancement of the project management concept because it offers a new way of ensuring that the project deliverable is a quality deliverable. This occurs because the customer's drive pulls the deliverable from the supplier's organization through a quality-improvement process. In addition, the process through which a deliverable is actually produced—the customer-driven project management process—is itself assessed and improved continuously as the project progresses through its life cycle.

Customer-driven project management involves the following:

1. The project is determined by cooperation between customer and supplier through a structured process.
2. The customer drives the project through customer-driven teams.
3. Customer-driven teams link the customer, process owners, and suppliers. Teams consist of the customer or customer's voice as leader, a project facilitator, a program manager, process owners, and suppliers as appropriate.
4. Customer-driven teams are fully empowered to perform and improve the project.
5. A disciplined customer-driven project management methodology is used.

A further understanding of customer-driven project management comes from looking at the terms.

Customer-driven means the customer or customer's voice is the primary focus. The customer leads the way. Customer satisfaction becomes the focus of all efforts, providing the constancy of purpose vital to success.

Project is any series of activities that has a specific end or objective. Almost all activities in an organization can be defined in terms of a project.

Management involves optimizing resources, that is, getting the most out of both technology and people. The target is on managing the project and leading the people to a deliverable that achieves total customer satisfaction.

Customer-driven project management philosophy

The customer-driven management philosophy requires a fundamental belief in the customer as the focus of all efforts in an organization. It requires a confidence in the development of a joint understanding of customer needs and expectations along with internal processes of the supplier's organization. The philosophy stresses a systematic, integrated, consistent, disciplined approach involving customers, process owners, and suppliers through all the phases of the project. Telecommunications and information systems make possible this integration. Teams must be the organizational structure of choice, with the customer as team leader. The doctrine must empower each team to "own" its project enough to continuously improve it. The philosophy must stress metrics as the means to focus attention on meaningful outcomes for the customer. There must be a basic belief in cooperation as the primary means to success. Also, rewards and recognition must be acknowledged as essential elements. In addition, there must be a belief that designing in quality and long-term prevention is important. Relationships are vital. Involvement of everyone and everything needed in the project in a focused effort is essential. Vision and leadership are predominant values. Further, creativity and innovation must be encouraged. The intense desire of everyone in the organization to nurture supplier partnerships and customer relationships is a critical point of view. This philosophy must be adopted by everyone and requires a passionate, dedicated, hard-working team to be successful.

Customer-driven project management principles

The customer-driven project management principles are the essential rules required for success. Customer-driven project management requires the creation and maintenance of an environment of integrity, ethics, trust, open communications, teamwork, empowerment, pride of accomplishment, and commitment. Actions that satisfy the customer must be recognized and rewarded. Everyone must be oriented to perform and improve the project. Customers, process owners, program managers, and suppliers are joined on the same customer-driven team. Cooperation and teamwork are encouraged. A continuous improvement system is created and maintained. There is continuous training and education of everyone. The customer or customer's voice drives the project. Everyone is empowered to perform and improve the project. Supplier partnerships and customer relationships are actively developed. The customer-driven project management approach is fostered by the example of leadership. Everyone is focused on prevention of defects, reduction of variability, and elimination of waste and losses. Quality in design is emphasized. Process measurements stressing metrics focus on continuous improvement to satisfy the customer. Optimal life-cycle cost is the goal. Leadership at all levels nurtures customer-driven teams. Teams are the primary structure for customer-driven project management.

The uniqueness of customer-driven project management

Customer-driven project management is a unique approach to management. Table 1.3 shows the unique aspects of customer-driven project management. CDPM includes all the aspects of project management and total quality management. Its major contributions are detailed below:

1. *The way projects are selected.* In current project management processes, projects are selected by a customer and then typically

TABLE 1.3 Unique Aspects of Customer-Driven Project Management

Traditional project management	Customer-driven project management
Selected by customer and bid out	Selected by structured cooperation between customer and project supplier
Customer is an outsider to the project team	Customer is driver-leader of project team
Concept phase defines project details	Concept phase performs quality improvement analysis; project details specified in design phase
Project teams perform tasks in functional organizations and "hand off"	Customer-driven teams fully empowered to perform tasks and improve processes to satisfy customers
Project manager has role of controller	Customer has role of leader
Improvements target one-shot modifications of deliverables	Continuous improvement of processes and the deliverable by customer-driven teams

bid out, with the winning project firm taking on the job with little background on the project. In CDPM, projects are selected out of a structured quality-improvement process conducted through a cooperative arrangement between the sponsoring customer and a selected supplier project firm. The contract is signed before the project is selected so that both customer and supplier enter the quality-improvement process free of biases about the deliverable and free of uncertainties about who will do the work.

2. *The role of the customer in the project management process.* In traditional project management, the customer is an outsider to the project team, the client. The team is largely run by a project manager, who directs the team toward the deliverable and checks with the client or customer for periodic approvals or signoffs. In CDPM, the ultimate customer for the deliverable actually drives the team as leader from beginning to end and makes the key decisions along the way.

3. *The role of the initial concept phase.* The typical concept phase in current project management process involves assessment of the customer's needs and project objectives and development of a scope of work, resource needs, and other project details. In CDPM, the concept phase is basically a quality-improvement process that goes through four defined steps of quality improvement before identifying the project:

   a. Definition of the customer's quality issue.

   b. Understanding and definition of the key process involved and its steps.

   c. Identification and selection of key improvement opportunities.

   d. Analysis of improvement opportunities to identify root causes and set priorities.

   This is the way the project is selected in CDPM. The team is established early to go the full cycle from quality (customer) analysis to system development to project management.

4. *The boundaries of team empowerment.* Project teams are organized around work packages or tasks, and each team member is made accountable for a separate work package. The boundary of empowerment is basically set at the work-package level, and the evaluation of performance is typically on the basis of the quality, cost, and schedule of the work-package output produced. Integration of the work packages is assumed to occur in the development of the end deliverable. In CDPM, each team member is fully empowered to contribute beyond the work-package level, to participate in broader decisions about the customer's needs and the progress of the project. Each team member has unlimited access to the customer—and internal team members—through

regular contacts so that feedback can be ensured and to the whole team if appropriate.

5. *The role of the project manager.* In traditional project teams, the project manager is the primary controller of the project. The project manager is basically responsible for producing the deliverable, with the team in support. In CDPM, the leader of the project is the customer. The project manager becomes a team member and project facilitator, with primary responsibility to help the team meet the customer's needs. In CDPM, the team, with the customer as leader, performs many of the controlling functions now performed by the traditional project manager.
6. *The way the project management process is improved.* In the current lexicon of project management, there are four or five sequential phases, depending on the nature of the project. Improvements in the process are typically addressed by focusing on whether certain functions are completed in any given phase. In CDPM, continuous improvement in the project management process is treated as in any other work process, proceeding from isolating the work processes themselves, how each customer need is met from upstream to downstream, what opportunities there are for quality improvement, where the root causes are, how priorities are set, and how corrective actions and elimination of variation are recommended and implemented.

In sum, customer-driven project management is an enhancement of traditional project management. It requires the complete and continuous integration of customer requirements—as they develop during the project—into every aspect of project planning and control. If project management can be explained as the phased planning, control, and development of a deliverable by a project team for a customer, then CDPM can be described as the definition, design, development, and delivery of a project deliverable by the customer. Customer-driven project management strives to cultivate a long-term relationship between customer and project supplier, so the project does not end with delivery but extends into a partnership of continuous improvement.

The customer-driven project management process

The customer-driven project management process involves the following:

- A total quality management environment
- A project management system
- A customer-driven management team structure

The total quality management environment, project management system, and the customer-driven management team structure are integrated to form the customer-driven project management process. The elements of the customer-driven project management are shown in Fig. 1.4. First, the customer-driven project management process demands the creation and maintenance of a total quality management environment, an environment in which the entire organization is focused on an objective to create total customer satisfaction through a systematic, integrated, consistent, organizationwide perspective.

The second element of the customer-driven project management process involves the use of a project management system. The project management system provides a set of processes and the support systems for analyzing, planning, implementing, and evaluating a project through the entire project life cycle.

The third part of the customer-driven project management process requires the establishment of a customer-driven project management team structure. This structure provides the framework for the customer to drive the project. It also links the project supplier and customer through customer-driven teams. Further, it integrates all the processes needed in the project supplier's organization to bring the project to a common focus.

Figure 1.4 Elements of the customer-driven project management process.

The customer-driven project management life cycle

Customer-driven project management begins with a customer's expectations for a deliverable product and/or service and a supplier's willingness to provide that deliverable. The life cycle continues as long as the customer desires. Customer-driven project management follows the typical project management life cycle of concept, definition, production, operations, continuous improvement, and eventually closeout. The customer-driven project management cycle is shown in Fig. 1.5. The cycle is analysis/concept, definition, production, operations, and improvement/closeout. With customer-driven project management, the project starts with quality improvement, the analysis part of the concept stage. It can continue indefinitely as long as the customer and supplier choose. This is the improvement part. The cycle may repeat indefinitely until eventually the project must be closed out. In customer-driven project management, the aim is to avoid closeout as long as possible through continuous improvement. In fact, the original project may evolve many times, with the deliverable constantly changing to respond to the customer.

The customer-driven project management improvement methodology

The customer-driven project management improvement methodology forms the disciplined, structured process for ensuring that a deliverable totally satisfies the customer. The customer-driven project management improvement methodology contains eight steps, as shown on the perimeter circle of Fig. 1.6. The eight steps are as follows:

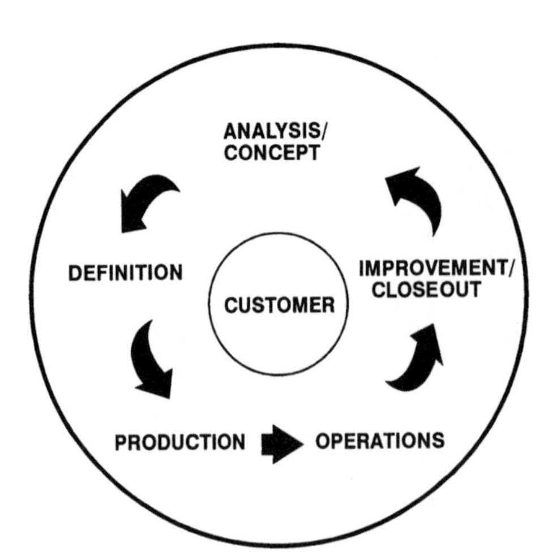

Figure 1.5 Customer-driven project management cycle.

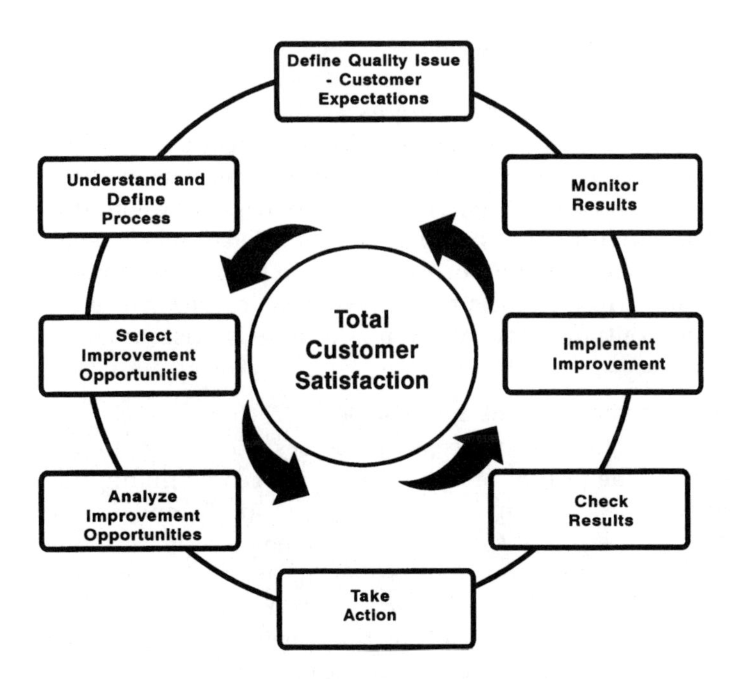

Figure 1.6 Customer-driven project management continuous improvement methodology.

Step 1: Define the quality issue. This first step defines the customer, brings the customer into the team to guide it, and produces a statement of the ultimate customer requirements. In this initial step, the customer assumes the role of customer-driven project leader. This system requires a shared focus of the outcome of the project between supplier and customer. This vision transcends the individual interests of both organizations involved. Under this system, the customer agrees to become the leader of the customer-driven team. The customer-driven team is made up of a project team facilitator and team members. The team members are process owners, the program manager, functional representatives, and supplier(s), as appropriate. The team members represent the supplier of the deliverable. The team's output of this step is a project mission statement.

Step 2: Understand and define the process. This step defines the target customer processes. This phase develops an understanding of the project's critical processes. It focuses on defining continuous improvement opportunities of mutual benefit to both supplier and customer. This phase is designed to outline the basic project concepts. It also gives the team a full view of all the processes involved in the project. The team's output for this step is improvement opportunities.

Step 3: Select improvement opportunities. During this step, the team selects high-priority improvement opportunities. These improvement opportunities enhance the project's processes to meet the customer's needs and expectations. This step involves looking for the critical project process for satisfying the customer's needs and expectations. High-priority processes are selected for further analysis. Two kinds of improvement opportunities are emphasized: (1) project process actions that will increase the performance of the process and (2) enhancements that will make the project deliverables satisfy the customer. The team's output from this step is a list of critical improvement opportunities.

Step 4: Analyze the improvement opportunities. In this step, the customer-driven team examines the project processes to enhance the project's performance. The project processes are analyzed by finding root causes of quality issues, identifying the vital few causes, and/or looking for variation in the process. This is the diagnostic step, involving collecting, sorting, and examining information about the processes. The team asks "why" and then asks "why" again and again until it gets to fundamental issues that prevent the project's processes from performing in an optimal way. The team's output from this step is the project objectives.

Step 5: Take action. This step delivers the project and relates most closely to the traditional project management phases of concept, definition, and production. During this phase, actions are taken to explore alternatives, state the project concept, define the project deliverable, demonstrate and validate the project, and develop and produce the project deliverable. During this phase, the team actually prepares plans, organizes, staffs, controls, and coordinates the project deliverable. In addition to the project management actions, the process-improvement actions are taken during this step. The team's output from this step is the project deliverable for testing.

Step 6: Check results. This is the test and measure step. The team measures the project and the project processes. The performance of the project deliverable and the project processes is evaluated based on the metrics of customer satisfaction. The customer-driven project team's output from this step is a validated project deliverable with capable project processes.

Step 7: Implement the improvement. Once the project and process output has been proven to be effective, the team ensures that the new process steps or project deliverables added are made a permanent part of the system.

Step 8: Monitor results for continuous improvement. The customer-driven team stays in business over several cycles of the target process to monitor performance, address unanticipated quality and productivity problems, and work for continuous improvement. This phase assumes a long-term relationship between customers and project management suppliers. Eventually, the project may be closed out.

Main Points

Customer-driven project management borrows from a diverse set of many management, quality, and project management concepts.

Traditional management principles come from scientific and human relations theorists.

Quality management has its roots in many quality and productivity techniques developed in the United States during World War II.

The early quality management techniques emphasized statistics.

Quality management concepts of continuous improvement, people involvement, measurement, and customer satisfaction evolved into today's total quality management approaches.

Project management originated as a method to manage large, complex programs.

Typical project management focuses on completing a project on or before schedule, within cost, and at the specified level of quality.

Project management aims at a project that has a definite completion date.

Project management relies on resources on an as-needed basis.

Customer-driven project management is a management approach that stresses customer satisfaction as a primary driver.

Customer-driven project management uses an eight-step process while applying both project management and total quality management tools and techniques.

Customer-driven project management addresses both the performance and improvement of the project.

Customer-driven project management involves the following:

1. The project is determined by cooperation between customer and supplier through a structured process.
2. The customer drives the project through customer-driven teams.
3. Customer-driven teams link the customer, process owners, and suppliers. Teams consist of the customer or the customer's voice

as leader, project facilitator, program manager, process owners, and suppliers as appropriate.

4. Customer-driven teams are fully empowered to perform and improve the project.
5. A disciplined customer-driven project management methodology is used.

The customer-driven project management philosophy is summarized as follows:

- *C*ustomers are the focus of all efforts.
- *U*nderstanding the customer and all processes is a target.
- *S*ystems approach must be pursued.
- *T*eams are the organizational structure of choice.
- *O*wnership with empowerment is essential.
- *M*etrics must be used to gain attention.
- *E*ncouragement of teams and individuals helps cooperation.
- *R*eward and recognition are keys to success.
- *D*esigned-in quality and prevention are important.
- *R*elationships are vital.
- *I*nvolvement of everyone and everything is critical.
- *V*ision and leadership are needed values.
- *E*ncouragement of creativity and innovation is necessary.
- *N*urturing of suppliers and customers is desired.

The principles of customer-driven project management are

- *P*ursue a TQM environment for project performance.
- *R*eward and recognize appropriate performance.
- *O*rient everyone to perform and improve.
- *J*oin all key players on the team.
- *E*ncourage cooperation and teamwork.
- *C*reate and maintain a continuous improvement system.
- *T*rain and educate everyone.
- *M*ake the customer the driver.
- *A*ct to ensure that everyone can perform and improve.
- *N*urture supplier and customer relationships.
- *A*ct to set the example.
- *G*et everyone focused on prevention.
- *E*mphasize quality in design.
- *M*easure processes meaningfully.
- *E*ncourage an optimal life-cycle cost.

- *N*urture leadership at all levels.
- *T*eam the complete project.

Customer-driven project management is unique in the following areas:

- Projects are selected through a structured, cooperative process.
- The customer is the driver-leader of the project team.
- A quality analysis focusing on customer satisfaction is performed during the concept phase.
- Customer-driven teams are fully empowered to perform and improve the project to satisfy customers.
- The customer-driven project manager's main role is as leader.
- Customer-driven project management stresses continuous improvement of the project and the deliverable.

The customer-driven management process involves the following:

- Total quality management environment
- Project management system
- Customer-driven management structure

The customer-driven project management life cycle includes

- Concept
- Definition
- Production
- Operations
- Continuous improvement/closeout

The customer-driven project management improvement methodology consists of the following eight steps:

1. Define quality issues.
2. Understand the process.
3. Select improvement opportunities.
4. Analyze the improvement opportunities.
5. Take action.
6. Check results.
7. Implement the improvement.
8. Monitor the results for continuous improvement.

Chapter

2

The Foundation of Customer-Driven Project Management

Focus: This chapter describes the foundation of customer-driven project management in project management and total quality management.

Introduction

Customer-driven project management's foundation is in the integration of the project management and total quality management approaches. Customer-driven management merges the proven methodologies, tools, and techniques of project management and total quality management under a single customer-driven management approach. Customer-driven project management expands the boundaries of both total quality management and project management by using the customer (or customer's voice) to drive an organization to complete a project, focusing on total customer satisfaction.

Historically, a project manager's primary purpose was to use the organization's resources to meet the objectives set by the organization's management. Production was normally the most important objective of the organization. This naturally places management's emphasis on completing projects, focusing on internal operations. Managers emphasized the management functions of planning, organizing, staffing, coordinating, directing, and controlling. In most organizations, directing and controlling were the primary functions of management. This traditional management approach stresses strong task-oriented management, especially at the top of the organization, to meet organizational goals. Often these goals are driven by schedule and cost rather than quality.

With CDPM, the customer leads the project, requiring the customer to use the organization's resources to achieve customer satisfaction. The customer-driven project leader's purpose is to optimize the use of all resources through the use of people in customer-driven teams to meet objectives set by the customer. Total customer satisfaction is the most important objective of the CDPM organization. Hence, management's emphasis is on internal operations, focusing on the customer. This requires a greater concentration on all the management functions. In addition, this makes leadership essential to guide the teams. In customer-driven project management, strong people-oriented leadership and effective task-oriented management throughout the organization are both necessary to satisfy the customer.

In traditional organizations, production usually was more important than people. People were viewed as just another resource. They were just "slotted" into job functions as part of the organizing and staffing function of management. In the day-to-day operations of projects, the human resource, like all other resources, was to be minimized to maximize profit. In fact, as most organizations concentrated on the directing and controlling functions of management, they viewed people as just another commodity to be controlled by structuring, eliminating, and specializing.

In customer-driven project management, people are the most important resource. People are the primary means to add value to a deliverable that is necessary when striving for total customer satisfaction. People are used on customer-driven teams where they can best contribute. People not only need to perform their process, but they are also expected to continuously improve it. People are viewed as a valuable asset that adds value to the product. This people resource must be developed by coaching, facilitating, training, and supporting.

These basic changes to traditional project management, which form the foundation of customer-driven project management, evolved from a wide range of earlier management practices, manufacturing productivity enhancement efforts, quality-improvement efforts, and project management methodologies. Customer-driven project management uses concepts that provide an organization with the means to meet the many challenges of today while ultimately moving toward the future.

In summary, customer-driven project management has its origin in project management and total quality management. Project management was formulated out of the need for a management approach that would meet the demands of managing complex projects. Total quality management has evolved to meet the demands of survival in today's global economic environment. Customer-driven management is designed to focus on striving for success in project delivery through total customer satisfaction.

Foundation of Project Management

Project management has its roots in the experience of managing complex technological and system developments during World War II. During World War II, traditional management approaches proved deficient in integrating the many aspects of the development and production of complex weapon systems. After World War II, the need to manage large, complex undertakings increased the interest in project management approaches. This was fostered by successful efforts, such as the Manhattan Project. In the early 1950s, project management started to evolve into a more systematic approach to completing programs. Project management became necessary as industries took on specific jobs, usually defense- or civil engineering–related. These programs were typically for the management of major space, weapons, and construction projects through the stages of design, development, manufacturing, testing, and production. In the 1960s, project management began to be implemented in many organizations besides those in defense, space, and construction industries. Project management became essential in the computer industry. By the 1970s, project management was recognized as an established management approach for many organizations involved in government, education, and private endeavors. Today, project management continues to progress into a management approach essential to producing many deliverables. Further, project management software helps perform many of today's project management tasks.

Since it evolved from the management of complex projects, project management usually involved the management of defined, nonroutine activities aimed at distinct time, financial, and performance goals for a systems development project. Through the years, project management has been refined through application in a wide range of industrial and service organizations. The most well-known use of project management is within the Department of Defense industries to develop weapon systems. Weapon systems such as the B-2 aircraft, with its state-of-the-art design, would not be possible without highly sophisticated project management techniques. Modern construction projects could not be built without using project management. Today, computer companies, the movie studios, small businesses, and even the music industry use project management.

The basic project management techniques have remained fairly standard over the years. However, the greatest impact on project management has been with the use of technology. Technology, especially automation and telecommunications, has allowed project management techniques to expand in breadth and scope.

What project management is

Project management is the management of an activity that has a defined start and finish. Because project management is usually viewed as having a definite finish, the focus of project management is usually completion of the project as scheduled. The objective in project management is to complete the project before or on time, at or below cost, and within technical performance specifications.

Project management can be called *program management, product management,* and *construction management* in relation to the major areas where it is used. *Program management* is usually the term employed in the Department of Defense. *Product management* is the term used to manage a product in a commercial industry. *Construction management* is used in a building industry.

The uniqueness of project management

Project management is unique because of the following:

- It has a defined specification, deliverable, and end point.
- It borrows and integrates resources.

Both these unique factors present challenges to an organization. First, since the focus of project management is on completing the project at its defined end, there is always a focus on the time element. This focus frequently results in a constant battle between the three basic competing elements of project management: time, cost, and performance. Balancing these three parameters while striving to complete the project requires constant attention to ensure that the priority is not just on getting the deliverable out the door. This constant consideration for time, cost, and performance tradeoffs requires developing positive internal coalitions. It also makes the relationship with the customer critical to success.

Second, the borrowing and integration of resources from functional departments constantly create the potential for conflict between functional resources and project resources. Collaborative unions between functional managers and project managers are essential to resolve issues relating to the dual responsibilities of project team members.

Time, cost, and performance tradeoffs

Traditionally, there are three factors that are key to the success of project management. Harold Kerzner, in *Project Management,* states:

"[W]e define project success as the completion of a project within the constraints of time, cost, and performance." Each of these factors is fundamental to successful project management because together they represent the most important project management characteristics, as follows:

1. Completion of the project within allocated resources. This is the *cost factor* of project management.
2. Completion of the project within allocated schedule. This is the *time factor* of project management.
3. Completion of the project within explicit criteria, standards, and specifications. This is the *performance factor* of project management. This is also sometimes called the *quality factor*.

These factors are not considered equal in every project. In some projects, it may be critical to have the product on time. For instance, a weapon system may be required to perform a certain military mission, or a computer program may be required to build the rest of the computer system, or a new registration system may have to be ready for students. In other projects, cost might be critical. For example, only a specific amount of money is allotted to the project. In this case, a "fence," or limit, may be put on the budget of a project. In some cases, quality may be the most important characteristic, and resources are essentially unlimited. Traditionally, the project management organization focuses on planning and controlling time and cost while assuming that its functional departments will ensure quality through a focus on specifications.

It is now recognized that project management success goes beyond cost, time, and performance. It depends on customer satisfaction and supplier efficiency. In fact, Harold Kerzner, in the fourth edition of *Project Management* (1992), states that

> the definition of project management has been modified to include completion:
>
> - Within the allocated time period
> - Within the budgeted cost
> - At the proper performance or specification level
> - And acceptance by the customer/user
> - With minimum or mutually agreed upon scope changes
> - Without disturbing the main work flow of the organization
> - Without changing the corporate culture

Matrix organization

Project management requires resources. It relies on the specialities in each one of many functions at varying times during the project. This requires a matrix organization to share resources between both functional management and project management. This use of a matrix organizational structure for project management presents a major management challenge. To comprehend this challenge, the differences between the traditional organizational structure and the matrix organization must be known. Figure 2.1 shows a traditional organization. This organizational structure is based on functional organizations. For example, on the chart, engineering, production, marketing, and support are functional organizations. These functional organizations perform all the activities in the organization within their specific areas. This type of organization depends on each separate function performing within its specialty. Little emphasis is placed on cross-functional coordination or on communication with the customer. Each functional organization is responsible for the technical capabilities of its processes and people.

The matrix organization is shown in Fig. 2.2. In a matrix organization, the project managers use resources (people, equipment, materials) from the functional organizations as necessary. This requires using the same resources in both functional management and project

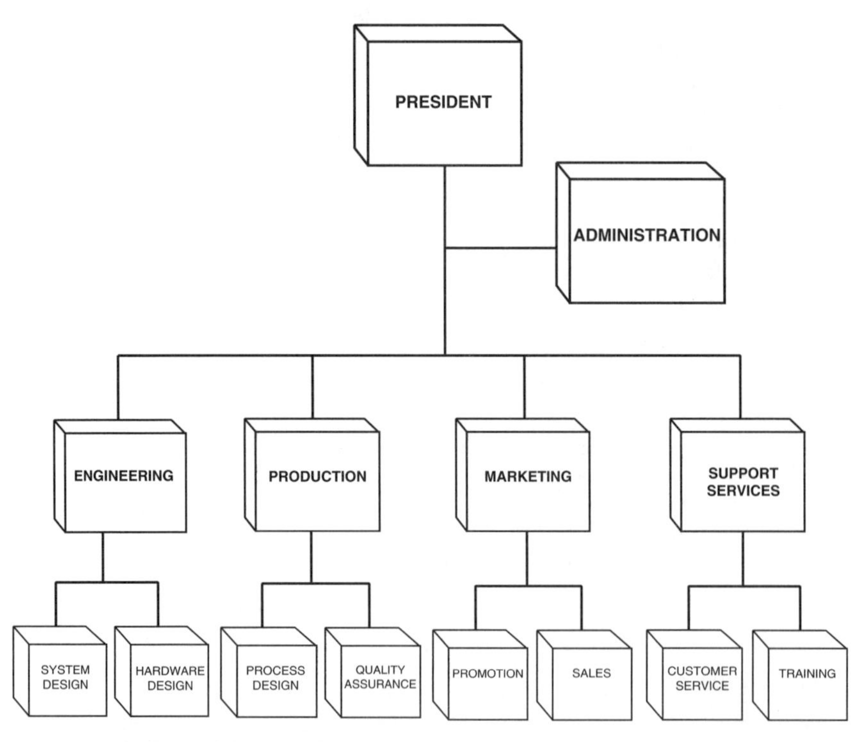

Figure 2.1 Traditional functional organization.

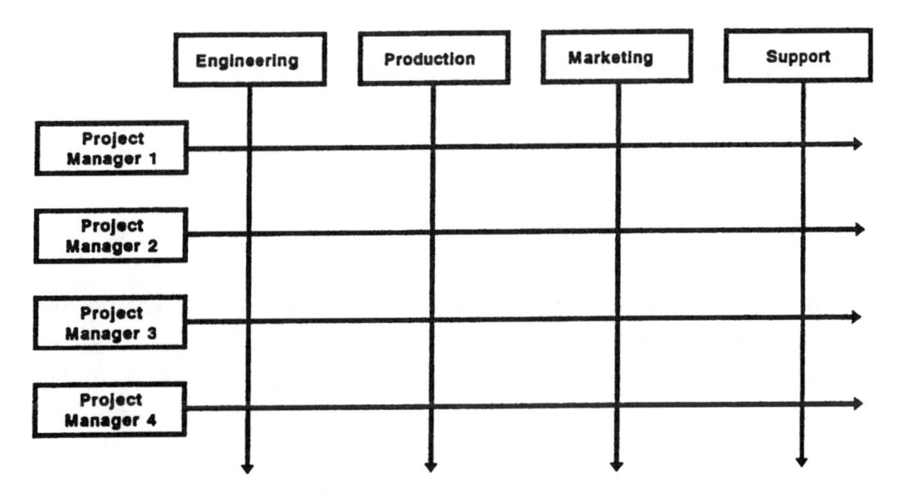

Figure 2.2 Matrix organization.

management. It also means that resources must be distributed among various projects. In a matrix organization, responsibility, authority, and resources flow vertically through the functional organization and horizontally from the project managers. Project managers influence the "what," "when," and "how much." These are the essential elements of the project. Functional managers direct the "how." The "hows" are the processes in the organization. In today's organizations, the functional managers are the overall process owners. They decide on how the process will operate.

Project management requires a full appreciation of the complexity of behavior in organizations. It also recognizes that successful work in an organization is not guaranteed, or even facilitated, by a traditional organization structure. Project work concentrates on pulling diverse activities together into short-term projects. It emphasizes communication and coordination of effort among functional departments (e.g., planning, engineering, production, and marketing). Project management stresses functional departments focusing on distinct short-term outputs and products while performing their traditional continuous long-term operations. Project management facilitates successful negotiation of scarce time and resources, thus strengthening the organization's ability to share responsibility and accountability throughout the organization. Rather than encouraging isolated work in each departmental setting, the project management approach encourages people to aim their functional expertise toward the organizational goal (the project). Thus, the organization can produce products and services on time, within budget and performance standards, and, most important, meeting and satisfying the needs of the customer.

Project management philosophy

The project management philosophy incorporates the following fundamental beliefs:

- The project is the primary focus for organizational activity, with specifications and project tasks driving the work.
- Resources and responsibility can be shared between the functional organization and the project.
- The organization's matrix team completes projects on time and within cost and performance specifications.
- Planning and control are the principal techniques for achieving the project objectives, with tasks completed sequentially through critical-path networks.
- Technology is usually the main method to make improvements.
- Coordination of all project activities is the key to effective use of resources.
- Teams in a matrix are the organizational structure for project management.
- Authority, responsibility, and resources can be spread throughout the functional and project organization.
- Numerous product lines and projects can be managed at the same time.
- An adequate reservoir of functional specialists can be maintained.
- Growth is encouraged through the project management process.

Project management principles

The principles of project management target successful completion of the project. The emphasis is on the project, production, technology, control, responsibility, cost, schedule, and performance parameters, matrix and team organization, and customer satisfaction.

The project management principles are to

- *P*rovide a project focus
- *R*eward production
- *I*nvolve functional organizations
- *N*urture rapid technological change
- *C*ontrol and plan all activities
- *I*nclude authority and resources with responsibility

- *P*rovide time, cost, and quality objectives
- *L*et functional organizations perform processes
- *E*ncourage teamwork and cooperation
- *S*atisfy the customer

Project management cycles

Project management involves a cycle of processes. These cycles for defining, designing, developing, and delivering a deliverable vary by organization. The classic project management approach and Department of Defense cycles provide two examples of commonly used project management cycles. The classic project management cycle has been described by many authors. For instance, David I. Cleland, in *Project Management: Strategic Design and Implementation,* discusses a generic project management life cycle, including conception, definition, production, operation, and divestment. This approach, as shown in Fig. 2.3, is also detailed by Harold Kerzner, in *Project Management.* The five phases generally involve the following functions:

1. *Conception.* This is the phase in which objectives and goals are set and specifications determined. It is in this phase that projects are outlined and modeled to ensure that the project deliverable is understood. Often the assumption is that the customer—the sponsoring agency or firm—has already determined the priority and need for the project deliverable and that the basic role of the project team is to deliver it within schedule and budget. Traditional project management does not make much room for involvement of the project team

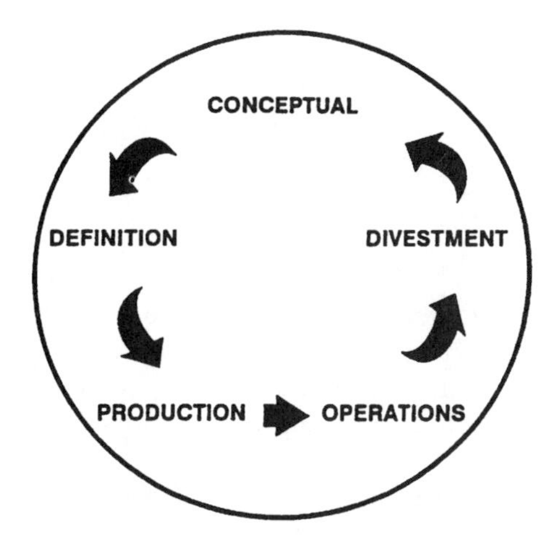

Figure 2.3 Five phases of a project life cycle.

with the customer in selecting the project, much less ensuring that it results from a quality-improvement process performance within the customer's organization and environment.

2. *Definition.* This is the process of defining the project deliverable in terms of a work breakdown structure (WBS), a budget and schedule, and a critical-path network. This is where the WBS provides an organizational and hierarchical look at the project, showing basic interdependencies and interrelationships with the project task structure. A scope of work, budget, and schedule are drawn up in this phase, and the project team is developed around the tasks.

3. *Production.* It is in this phase that the project deliverable is actually produced or "prototyped" so that testing and measuring can proceed. Production involves lining up all the required resources and integrating them according to their interdependencies shown in the WBS.

4. *Operations.* Here the project deliverable is installed, tested, and measured in operation with the customer or user. Operation ensures that the project deliverable, whether a system, product, or new service, conforms to the original specifications.

5. *Divestment.* Divestment involves documenting the project and closing it down. Here the team members are typically selected for other project teams, and the project books are closed.

Within the Department of Defense, the project management cycle is described as the seven-phase acquisition cycle. The seven-phase acquisition cycle, as described in James V. Jones's *Integrate Logistics Support Handbook,* includes

- Preconcept
- Concept
- Demonstration and validation
- Full-scale development
- Production
- Deployment and operations
- Disposal

The *preconcept phase* begins the acquisition cycle. In this phase, the need is identified through an analysis of missions and/or systems. This triggers the identification of operational deficiencies, operational needs, system or equipment development, modifications, and improvements.

The *concept phase* involves developing alternative approaches to satisfying the need identified during the preconcept phase. During the concept phase, all possible alternatives are analyzed to determine the alternative or alternatives best capable of satisfying the need.

During the *demonstration and validation phase,* the alternative or alternatives developed during the concept phase are evaluated to determine their feasibility for actually accomplishing the requirement. The demonstration and validation phase has two purposes: (1) to demonstrate that the concept can actually work and (2) to validate that the alternative can meet the need defined earlier.

Once an alternative passes the demonstration and validation phase, *full-scale development* begins. During full-scale development, the deliverable is designed. The *production phase* involves the actual development and/or manufacturing of the deliverable.

The *deployment and operation phase* begins after the item is delivered to the customer. During this phase, the customer assumes ownership with the support of the supplier.

Eventually, the item may need to be replaced. This is when the *disposal phase* begins for the old item. The disposal phase involves removing the item from inventory.

Whichever traditional project management cycle is used, the thrust is the same—producing a deliverable. The deliverable is essentially already identified by the sponsor or customer. This project management approach is based on traditional quality control. Its innovation was the matrix team that integrated internal functions to complete a project.

Foundation of Total Quality Management

Total quality management (TQM) has it foundation in the quality movement. The quality movement began with the application of statistics process/quality control by Dr. Walter A. Shewhart after World War I. During World War II, the federal government required industries to use statistical controls to ensure the quality of weapons systems. This led to many quality and productivity improvements.

At the end of the war, the United States was the leading producer in the world, and its industrial leaders did not feel the need to continue the quality push. However, Japan demanded an economic rebirth. Japan viewed quality as an essential catalyst to its economy. Japan sought the assistance of many U.S. quality experts.

This next stage in the quality movement was stimulated by Japan with the assistance of U.S. quality experts. One of these experts, W. Edwards Deming, helped the Japanese focus on their quality obsession. The primary motivation for the quality vision in Japan was the creation of jobs. The Japanese determined that to recover from the war, they had to transform their industries to enable them to produce "quality" commercial products. Deming showed the Japanese how they could improve quality and productivity through statistical techniques to capture more business and create jobs. His 14-point approach to quality was originally detailed in his book *Out of the Crisis*. His updated 14 points are as follows:

1. Create and publish to all employees a statement of the aims and purposes of the company or other organization. The management must constantly demonstrate their commitment to this statement.
2. Learn the new philosophy—top management and everybody.
3. Understand the purpose of inspection—for improvement of process and reduction of cost.
4. End the practice of awarding business on the basis of price tag alone.
5. Improve constantly and forever the system of production and service.
6. Institute training.
7. Teach and institute leadership.
8. Drive out fear. Create trust. Create a climate for innovation.
9. Optimize toward the aims and purposes of the company the efforts of teams, groups, staff areas.
10. Eliminate exhortations for the work force.
11. (a) Eliminate numerical quotas for production. Instead, learn and institute methods for improvement. (b) Eliminate management by objectives (MBO). Instead, learn the capabilities of processes, and how to improve them.
12. Remove barriers that rob people of pride in their work.
13. Encourage education and self-improvement for everyone.
14. Take action to accomplish the transformation.

There were many others who also assisted the Japanese in pursuing their "quality" vision during the succeeding decades after World War II. The most notable were Joseph M. Juran, Armand V. Feigenbaum, Kaoru Ishikawa, and Genichi Taguchi.

Joseph M. Juran, a leading quality planning advocate, was another American, like Deming, who was instrumental in Japan's early success. He taught the Japanese his concepts of quality planning. Both Juran and Deming stressed traditional management as the "root" cause of quality and productivity issues. Juran focuses on a disciplined planning approach to quality improvement.

Armand V. Feigenbaum, also an American, was the first to use the term *total quality.* His book, *Total Quality Control,* is one of the best early works on total quality improvement. His quality-improvement approach involves a systematic, integrated, organizationwide perspective. He also originated the concept of the cost of quality, which monitored cost of failures, quality appraisal, and prevention costs. This aimed managers toward quality improvement through quality cost reductions.

Kaoru Ishikawa, Japan's leading expert, geared the quality vision to the masses. He stressed the seven basic tools of quality used for problem solving. He believed that almost all quality problems could be solved using the seven basic tools. These tools include Pareto charts, cause-and-effect diagrams, stratification, checksheets, the histogram, scatter diagrams, and control charts.

Genichi Taguchi, another one of Japan's top quality experts, was the first to stress proper design strategy. He redefined the concepts of design specifications. Simply being within specifications is not good enough. He introduced a methodology that focuses on optimizing the design. According to Taguchi, any variation of performance from best target values is a loss, and loss is the enemy of quality. The goal is to minimize loss by focusing on the best target value.

In general, the Japanese adapted, developed, and continuously improved the quality approaches of these early pioneers. They formulated the concept of continuous improvement. As a result of the continuous-improvement philosophy, they advanced, through many small innovations, many of the products originally developed in the United States. The list is endless. It includes such items as televisions, automobiles, cameras, small electronics, air conditioners, video- and audiotape recorders, telephones, semiconductors, and so on.

The next stage of the quality movement started in the United States in the late 1970s. Until this time, the United States did not feel the need to embrace the quality vision. The United States was the number one economic power in the world, and the world bought whatever the United States produced. During the late 1970s, the threat of competition from many other countries became apparent to many U.S. industries. These industries started to investigate ways to become more competitive. This resulted in a renewed interest in the quality management techniques being used by the competition—mainly Japan.

During this stage, the U.S. quality movement was reborn. Many of the early quality experts' teachings were updated. In addition, many others joined the march to quality and provided additional insights into a transformation process for the United States. This new movement in the United States goes beyond the quest for quality. It calls for a transformation of U.S. management. This new management philosophy evolved into total quality management (TQM). American TQM strives to develop an integrated system that takes advantage of America's strengths. In particular, TQM seeks to maximize our people resources. America's strength lies in its people's diversity, individuality, innovation, and creativity. This targets maximizing all our resources, especially our people resources, on quality. This means total customer satisfaction from every internal customer in an organization to the ultimate external customer. The quality obsession in the United States targets customer satisfaction. To focus the entire organization on total customer satisfaction requires management to create a TQM organizational environment. This TQM work environment, where everyone can contribute with pride, is developed through leadership. As it exists today, this U.S. style of TQM stresses a totally integrated, systematic, organizationwide perspective. It requires a transformation of many of the ways the United States traditionally does business.

This transformation process has been evolving since the late 1970s through the efforts of many U.S. organizations. Some of the early leaders were organizations that realized that their survival in the new global economy required some fundamental changes. These were organizations such as Xerox, Motorola, IBM, and Hewlett-Packard. By necessity, they quickly learned the lessons of continuous improvement. In the government area, the Department of Defense is one of the leaders in total quality management. In the middle 1980s, the Department of Defense was facing an ever-declining budget. It also sought assistance from the quality experts to help it determine how it could protect the United States at a lower cost. In addition to the many organizations, many people also contributed to refining the new U.S. management movement. Besides the quality masters like Deming, Juran, Feigenbaum, Ishikawa, and Taguchi, TQM embodies the ideas of many others from quality management, organizational development, training, engineering, and other disciplines. The ever-evolving list of contributors is too numerous to mention. Some of the better-known early contributors include Philip B. Crosby, Tom Peters, Robert H. Waterman, Jr., H. James Harrington, and A. Richard Shores.

An early proponent in the late 1970s, Philip B. Crosby outlined the "zero defects" system in his book *Quality Is Free.* His program was based on many years with Martin Marietta and ITT. It was embraced by many U.S. companies and the U.S. government. The program was instrumental in uncovering many defects in the U.S. industrial process. The Crosby approach is based on four points: (1) quality is conformance to requirements, (2) prevention is the key to quality, (3) zero defects is the standard, and (4) measurement is the price of nonconformance.

In the early 1980s, more attention was directed toward quality in the media and popular literature. Tom Peters and Robert H. Waterman, Jr., provided U.S. business with more ideas on what contributed to the success of the top companies in the United States. Their book, *In Search of Excellence,* presented an initial inside look at what made these companies so competitive. They determined the following eight attributes that distinguish excellent, innovative companies. First, these organizations were geared for action. They preferred to do something rather than going through endless analysis and committee reports. Second, they strove continuously to meet the needs and expectations of their customers. Third, innovative organizations were structured with smaller organizations within, allowing internal autonomy and entrepreneurship. Fourth, successful organizations fostered the ability to increase productivity through people. Fifth, they were value-driven through management setting the example with the application of hands-on attention to the organization's central purpose. Sixth, they built on the organizational strength by sticking to what they did best. Seventh, they had few layers of management and few people in each layer. Eighth, they created an atmosphere of dedication to the primary values of the company and a tolerance for all employees who accepted those values. These eight attributes still characterize many capable organizations today.

In 1987, Tom Peters' book, *Thriving on Chaos,* proclaimed no excellent companies. In it he said, "There are no excellent companies." He attributed this new statement to the constant changes in the new environment. A company in the new environment must continuously improve or other companies will replace it. Therefore, a company cannot ever achieve excellence. Peters called for a management revolution in the United States—a revolution that is constantly adapting to the many challenges of the economic environment of today. Tom Peters details a total management transformation process with prescriptions for proactively dealing with today's environment.

In the late 1980s and early 1990s, many other Americans updated the teachings of the earlier quality experts to the environment of today. Again, they are too numerous to mention. For proven approaches, H. James Harrington from IBM, A. Richard Shores from Hewlett-Packard, and James H. Saylor from GenCorp Aerojet provide excellent, proven road maps based on practical experience for total quality management in U.S. industry today. Harrington's approach is described in his books, *The Improvement Process* and *Business Process Improvement.* Shores' book, *Survival of the Fittest,* outlines his teachings. Saylor's approach is given in the *TQM Field Manual.* The Harrington approach focuses on the entire improvement process, whereas Shores stresses total quality control. The *TQM Field Manual* defines the VICTORY-customer approach for an organizationwide, consistent effort. This VICTORY-C approach is described in Chap. 3. These quality practitioners provide many specific methods geared to U.S. industry today.

In 1987, the need for quality improvement was formally recognized by many industry leaders and the U.S. government with the creation of the Malcolm Baldrige National Quality Award. As part of a national quality-improvement campaign, the Malcolm Baldrige National Quality Award was created by public law to foster quality-improvement efforts in the United States. The annual award recognizes U.S. companies in the categories of manufacturing, service, and small business that excel in quality achievement and quality management. This award's criteria for leadership, information and analysis, strategic quality planning, human resource utilization, quality assurance of products and services, quality results, and customer satisfaction have been improved continuously since its inception. The award is recognized as the standard for all organizations in the United States trying to pursue excellence.

In the 1980s, the federal government began its quest for quality and productivity improvement. In the early 1980s, some defense logistics organizations began exploring methods to enhance their performance. This resulted in the eventual spread of systematic improvement efforts to other Department of Defense organizations throughout the decade. In 1988, the Department of Defense (DoD) adopted the total quality management approach. TQM was to be the vehicle for attaining continuous quality improvement within DoD and its many contractors.

Also in the 1980s, other federal agencies initiated productivity-and/or quality-improvement ventures. These included the Internal Revenue Service, NASA, the General Services Administration, and the Departments of Agriculture, Commerce, Energy, Interior, etc. By mid-1988, total quality management had evolved into a government-wide effort. This was formalized by the establishment of the Federal Quality Institute as a primary source of information, training, and TQM services. In the late 1980s, a President's Award for Quality and Productivity Improvement was created to recognize "the agency or major component of an agency that has implemented Total Quality Management (TQM) in an exemplary manner, and is providing high quality service to its customers." This award, like the Malcolm Baldrige National Quality Award, formally established the commitment to quality of the federal government.

In the 1990s, total quality management is being adapted in many other government agencies, communities, and private industries in the United States under many names, such as total quality leadership (TQL), total quality improvement (TQI), continuous quality improvement (CQI), total customer satisfaction, and so on. In particular, educational institutions and state, county, and city government agencies are discovering that change is necessary. For instance, many higher-education institutions are implementing total quality management. Small businesses are discovering TQM. In recent years, health care organizations have started on the road to continuous quality improvement. As they are finally understanding, the United States can procrastinate no longer. The "sense" of this movement is that all of America must be transformed to restore its position as a leader in the world economy.

What total quality management is

Total quality management is a recent management concept evolving from a wide range of earlier management practices, productivity enhancements, and improvement efforts. There are almost as many definitions of TQM as there are organizations using it. That is so because each organization's transformation must be personalized to establish ownership for creating the commitment needed for success.

Although there are many applied definitions of total quality management, the basic essence of TQM involves the elements of continuous improvement, a people orientation, quantitative methods, and a focus on customer satisfaction. TQM is a management philosophy and set of guiding principles that stress continuous improvement through people involvement and measurements focusing on total customer satisfaction.

TQM is both a philosophy and a set of guiding principles that represent the foundation of a continuously improving organization. TQM is the application of quantitative methods and human resources to improve the material services supplied to an organization, all the processes within the organization, and the degree to which the needs of the customer are met—now and in the future. TQM integrates fundamental management techniques, existing improvement efforts, and technical tools under a disciplined approach focused on continuous improvement. This definition is offered in the Draft Department of Defense Total Quality Management Guide. The Federal Quality Institute's definition is as follows: "TQM is a strategic, integrated management system for achieving customer satisfaction which involves all managers and employees and uses quantitative methods to continuously improve an organization's processes." These are only a few of the many applied definitions of TQM. These examples show that the definitions of total quality management include all the essential elements of TQM.

The definition of total quality management we like is: TQM is a leadership philosophy and set of guiding principles that stress continuous improvement through people involvement, and a disciplined, structured methodology, emphasizing process measurement, and focusing everything on total customer satisfaction.

A further understanding of TQM comes from the words total quality management:

Total in this context means the involvement of everyone and everything in the organization in a continuous improvement effort. Everyone is committed to "one" common organizational purpose, as expressed in the vision and mission. They are also empowered to act to make that vision a reality. Besides people, everything in the organization, including systems, processes, activities, tasks, equipment, and information, must be aligned toward the same purpose.

Quality is total customer satisfaction. Total customer satisfaction is the center or focus of TQM. The customer is everyone affected by the product and/or service and is defined in two ways. The customer can be the ultimate user of the product and/or service; this is known as an *external customer*. The customer also can be the next process in the organization; this is known as an *internal customer*. TQM focuses on satisfying all customers' expectations, both internal and external.

Management means creating and maintaining the TQM environment. This involves the leadership of an organization. In fact, many organizations use the term *total quality leadership* to emphasize the need for leadership throughout the whole organization to guide the transformation. Further, management manages quality through the invention and improvement of processes.

Total quality management philosophy

The total quality management philosophy provides the overall general concepts for a continuously improving organization. The TQM philosophy stresses a systematic, integrated, consistent, organization-wide perspective involving everyone and everything. It focuses primarily on total customer satisfaction (both the internal and external customers) within a management environment that fosters continuous improvement of all systems and processes. The philosophy values empowering people. Teams, predominately with a multifunctional emphasis, are an important primary method used to lead improvement from within the organization. The TQM philosophy stresses optimal life-cycle cost. It uses measurement within the disciplined methodology to target improvements. Prevention of defects and quality in design are key elements of the philosophy. Elimination of losses and reduction of variability are its aims. Further, it advocates developing relationships—internal, supplier, and customer. Finally, the philosophy is based on an intense desire to succeed.

TQM guiding principles

The TQM guiding principles are the essential, fundamental rules required for total quality management. The first principle is to provide a TQM environment, requiring a committed, continuous pursuit by everyone in the organization. The TQM environment is characterized by a foundation of ethics, integrity, and trust. Through this basic foundation, the organization builds a TQM environment of open communication, people involvement, ownership, and pride of accomplishment. The TQM environment provides a place where people truly want to contribute. When people contribute, they must be rewarded or recognized appropriately. Everyone and everything must be involved, requiring an integrated organizationwide approach to optimize people and technology toward one focus. Supplier partnerships must be nurtured for the long-term benefit of both suppliers and customers. Customer relationships are the foremost consideration in all situations internal and external to the organization. Keeping and maintaining customers must be everyone's primary purpose. A continuous-improvement system must be instituted throughout the organization. Continuous improvement through a disciplined approach that is used everywhere in the organization is essential for total quality management. The continuous-improvement system focuses on optimizing internal processes in the organization. An essential principle of TQM requires including quality as an element of design. This involves designing products with robustness. Further, customer satisfaction must be designed into all products and services. This requires a relentless devotion to identifying the customers and then creating the capability to convert customer needs and expectations into appropriate products and services. To develop this internal capability to rapidly respond to the customer, training and education must be provided to maximize the human potential of the organization. A critical principle of TQM is an organizational focus on long-term improvement efforts geared toward prevention of problems through systems thinking. TQM gears an organization toward focusing on fixing its systems though a reduction of variation or improvement in the processes. This is accomplished through cooperation and teamwork. Teams, especially multifunctional teams, are the principal organization structure. The most important principle of TQM is satisfying both internal and external customers. Customers are the primary focus of the organization.

The TQM guiding principles involve continuously performing the following actions:

- *P*rovide a TQM environment.
- *R*eward and recognize appropriate actions.

- *I*nvolve everyone and everything.
- *N*urture supplier partnerships and customer relationships.
- *C*reate and maintain a continuous-improvement system.
- *I*nclude quality as an element of design.
- *P*rovide training and education constantly.
- *L*ead long-term improvement efforts geared toward prevention.
- *E*ncourage cooperation and teamwork.
- *S*atisfy customers (both internal and external).

The TQM umbrella

The TQM umbrella includes the integration of all the fundamental management techniques, existing improvement efforts, and technical tools under a disciplined approach focused on continuous improvement. All existing improvement efforts fall under the TQM umbrella. One notable exception is project management. Figure 2.4 shows some of the current improvement efforts, including concurrent engineering, robust design, statistical process control, just-in-time, cost of quality, total production maintenance, manufacturing resource planning, computer-aided design, computer-aided engineering, computer-aided manufacturing, computer-integrated manufacturing, information systems, total integrated logistics, and total customer service.

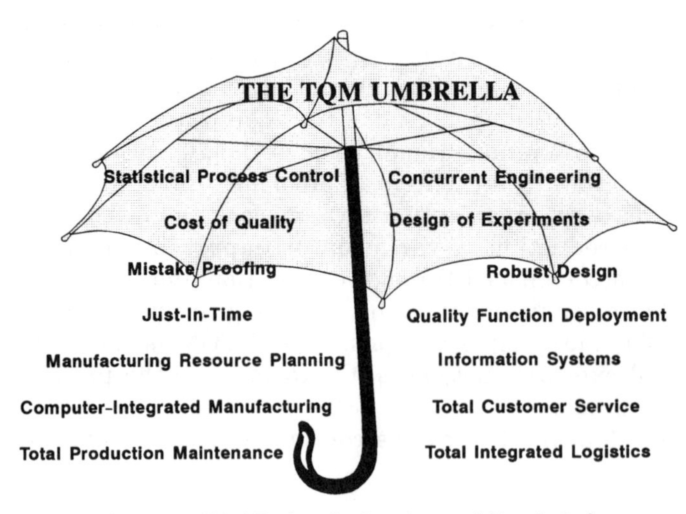

Figure 2.4 TQM umbrella.

The improvement efforts are geared toward improvement of one or more of the aspects of an organization. For instance, robust design, statistical process control, just-in-time, cost of quality, total production maintenance, manufacturing resource planning, computer-aided design, computer-aided engineering, computer-aided manufacturing, and computer-integrated manufacturing are oriented toward engineering and manufacturing. These existing improvement efforts can show some visible results by themselves. However, TQM integrates all these improvement efforts to enhance the overall effectiveness of the entire organization, focusing on customer satisfaction.

TQM, a unique management approach

TQM is a unique management approach. As explained under philosophy and guiding principles, TQM represents a change from the traditional U.S. management mind-set. To get a clearer understanding of what TQM is, Table 2.1 provides some comparisons between traditional management and total quality management.

TQM is a people-oriented, measurement-driven, customer-focused, long-term, strategically oriented management philosophy using a structured, disciplined, continuous-improvement operating methodology. It is not a "quick fix" using firefighting techniques. TQM uses

TABLE 2.1 Comparison of Traditional Management and Total Quality Management

Traditional management	Total quality management
Looks for a "quick" fix	Adopts a long-term, strategically oriented philosophy
Firefights without an analytic component	Uses a disciplined methodology of continuous improvement
Operates the same old way with a commitment to stability	Advocates breakthrough, innovation, and creative thinking
Adopts improvement randomly	Systematically selects improvement
Inspects for defects and errors	Focuses on prevention
Decides by using opinions	Decides by using facts
Throws money and technology at tasks	Maximizes people resources
Controls resources by function	Optimizes resources across the whole organization
Controls people	Empowers people
Targets individual performance to meet job description requirements	Focuses on team performance to meet customer expectations
Is primarily motivated by profit	Strives for total customer satisfaction
Relies on programs	Is a never-ending process

many small continuous improvements targeting breakthroughs, innovations, and creativity rather than simply operating the same old way with business as usual.

With TQM, management must systematically select the long-term continuous improvement efforts. In the past, many improvement efforts were randomly adopted in the organization for a short period.

TQM focuses on "doing the right thing right the first time." This emphasizes prevention of errors and quality of the design. Traditionally, inspection was the method used to find and eliminate errors.

Further, TQM bases decisions on facts instead of opinions, as traditional management often does. TQM's use of people's capabilities as a primary means to add value to a product or service is also a major variation from the traditional approach. Normally, management tends to increase resources or technology to add value to its product or service.

With TQM, the objective is to optimize resources across the whole organization. Traditional management suboptimizes resources by function. TQM fosters the empowerment of people to perform and improve their processes rather than controlling people.

Team performance focused on total customer satisfaction is valued, rewarded, and recognized in TQM. Traditionally, individual performance targeted to meeting specific job descriptions was the organization's goal.

Above all, the TQM philosophy focuses on customer satisfaction. It is not solely motivated by profit. Finally, TQM is not simply a new management program but a never-ending way of life for the future.

The total quality management process

Total quality management focuses on the continuous improvement of all systems and processes in an organization. In fact, TQM is a process itself. TQM is a process within the overall system of the organization. The entire organization is a system made up of many processes to accomplish the functions of the organization.

But what is a process? A *process* is a series of activities that takes an input, modifies the input, and produces an output. The TQM process transforms all the inputs in the organization into a product and/or service that satisfies the customer. In Fig. 2.5, the overall TQM process consists of the inputs received from a supplier, the process itself, and the outputs supplied to the customer. The most important inputs include the wants, desires, needs, expectations, and requirements of the customer. These inputs are combined with many other inputs to the process, including people, materials, supplies, methods, machines, technology, and the external environment, to cre-

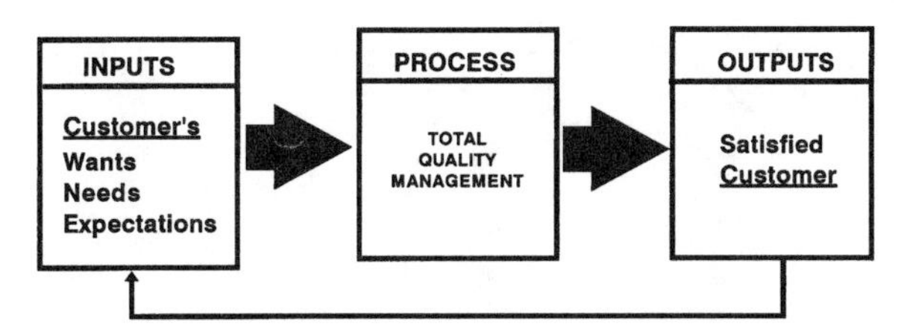

Figure 2.5 TQM process.

ate an internal TQM organizational culture and a deliverable. The output of the process is increased financial performance, improved operating procedures, better employee relations, and greater customer satisfaction. The TQM process starts with the customer, and its output focuses on the customer.

TQM methodology

The TQM methodology involves a disciplined continuous improvement approach. This methodology provides an organizationwide approach to process improvement and problem solving. The specific approach varies among organizations. The approach usually is a derivative of the Shewhart/Deming cycle. Figure 2.6 shows the Shewhart/Deming cycle and some other common derivatives. They are:

1. Figure 2.6*a* is the Shewhart/Deming cycle. This consists of plan, do, check, and act (PDCA):
   a. *Plan* involves developing an approach.
   b. *Do* the approach on a trial basis.
   c. *Check* to see if the approach works.
   e. *Act* to implement the approach if it works
2. Figure 2.6*b* is plan, do, study, and act (PDSA). This cycle is the same as PDCA except that instead of checking the approach, the approach is studied. This is to see if it works and to determine what is learned from doing the approach. This cycle is found in many organizations of the Department of Defense.
3. Figure 2.6*c* consists of check, act, plan, and do (CAPD). This variation of the Shewhart/Deming cycle is usually used to improve established processes.
   a. *Check* to make sure the right parts of the target process or problem are being measured and monitored to detect the root causes.
   b. *Act* to determine the root causes.

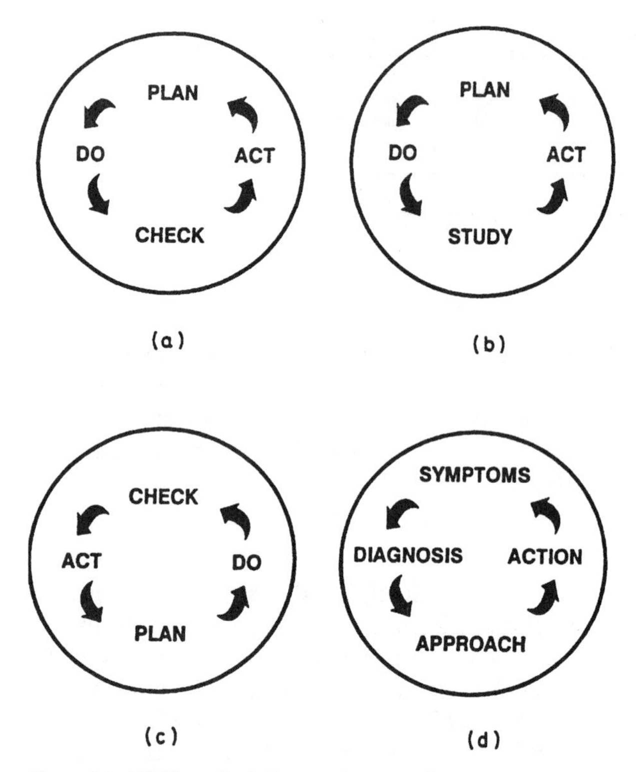

Figure 2.6 TQM methodology cycle examples.

   c. *Plan* an approach to deal with the root causes by identifying preventive and corrective actions.
   d. *Do* the action to solve the problem.
4. Figure 2.6*d* is another common approach. This approach consists of symptoms, diagnosis, approach, and action.
   a. List the symptoms of the problem.
   b. Do an analysis of the problem and ask why.
   c. Look at and define the approach to deal with the root cause.
   d. Take the appropriate action.

In many organizations, the basic PDCA cycle is expanded into detailed activities. Figure 2.7 shows an example from the Navy Personnel and Development Center. It provides a complete flow diagram of the entire PDCA process. In this model, *planning* involves

1. Stating the goal for the process improvement.
2. Describing the process flow.
3. Describing desired changes in outcomes.

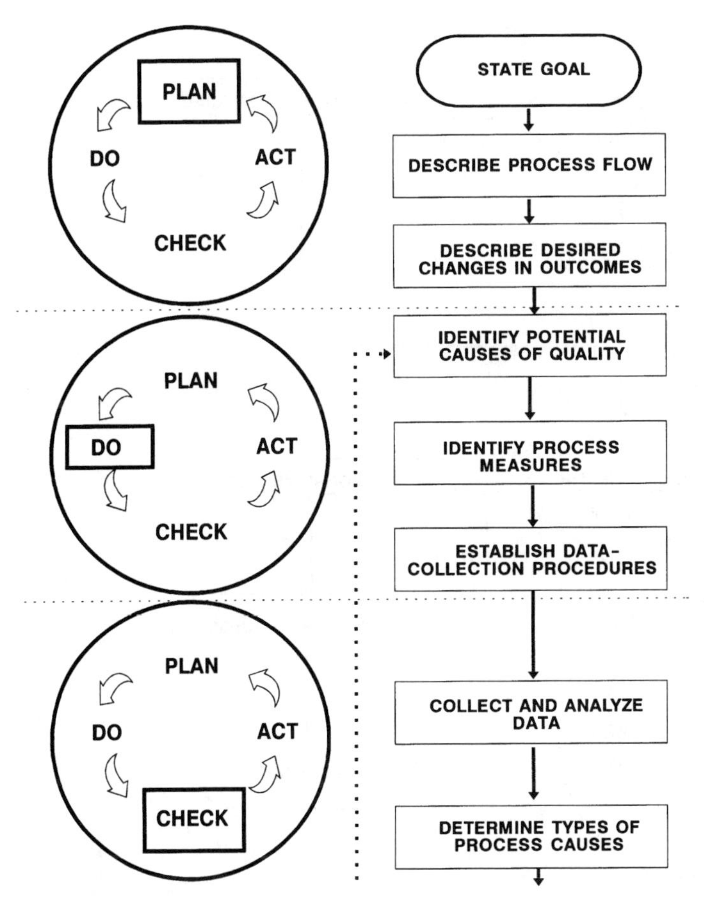

Figure 2.7 TQM methodology example from Navy Personnel Research and Development Center.

Doing includes

1. Identifying potential causes of quality.
2. Identifying process measures.
3. Establishing data-collection procedures.

The steps for *checking* are

1. Collect and analyze data.
2. Determine types of process causes.

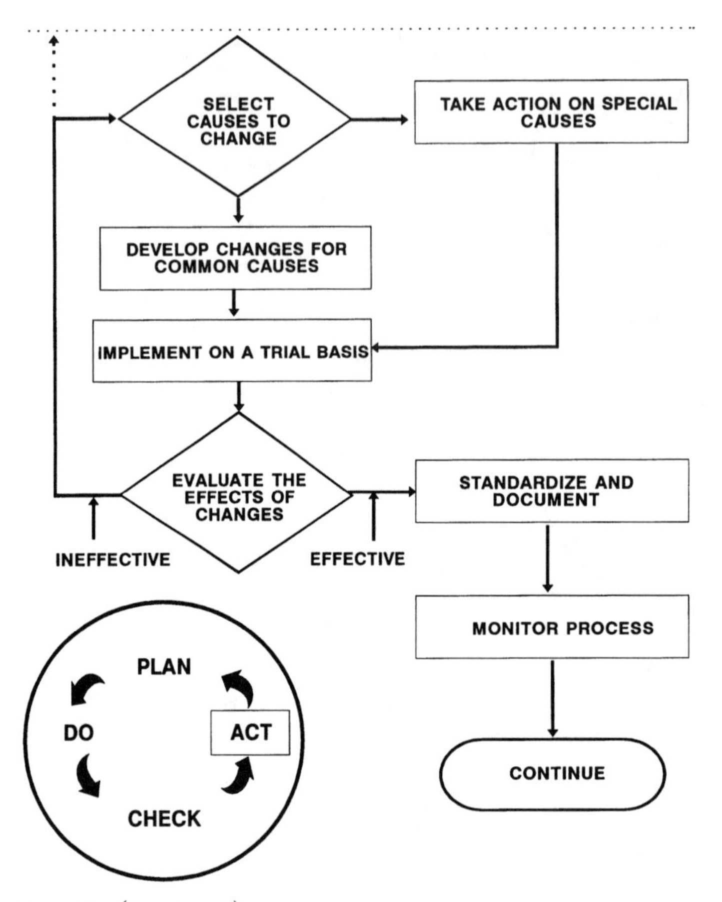

Figure 2.7 *(Continued.)*

The *action* steps involve

1. Selecting a special or common cause to change.
2. If a special cause is selected, taking action on the special cause.
3. If a common cause is selected, developing changes for the common cause.
4. Implementing the change on a trial basis.
5. Evaluating the effects of the change.
6. If ineffective, going back to identify the potential causes of quality.
7. If effective, standardizing and documenting.
8. Monitoring the process.
9. Continuing with continuous improvement.

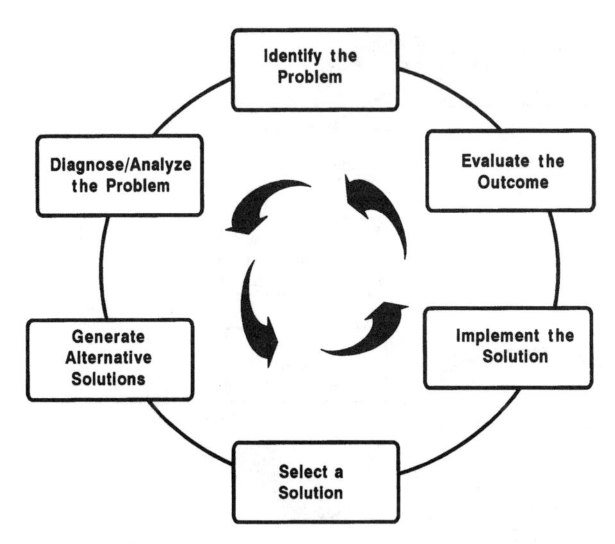

Figure 2.8 Typical problem-solving methodology.

Another common approach is the basic problem-solving model shown in Fig. 2.8. Basic problem solving involves

1. Identifying the problem.
2. Diagnosing/analyzing the problem.
3. Generating alternative solutions.
4. Selecting a solution.
5. Implementing the solution.
6. Evaluating the outcome.

Figure 2.9 shows another example of a typical TQM basic improvement methodology. This example, from the *TQM Field Manual,* combines the PDCA cycle with the problem-solving model. This is a basic eight-step approach similar to the eight-step customer-driven project management methodology described in Chap. 1. Steps 1 through 8 are the basic continuous process improvement/problem-solving method. Steps 5*a* to *d* form the plan, do, check, and act (PDCA) cycle; they are optional and are used to test or pilot an alternative. For simple or minor-impacting alternatives, steps 5*a* to *d* can be skipped. Use the PDCA cycle to determine whether the alternative will provide the desired outcome. Once the alternative is verified to accomplish the goal, the remaining steps are accomplished to institutionalize the improvement.

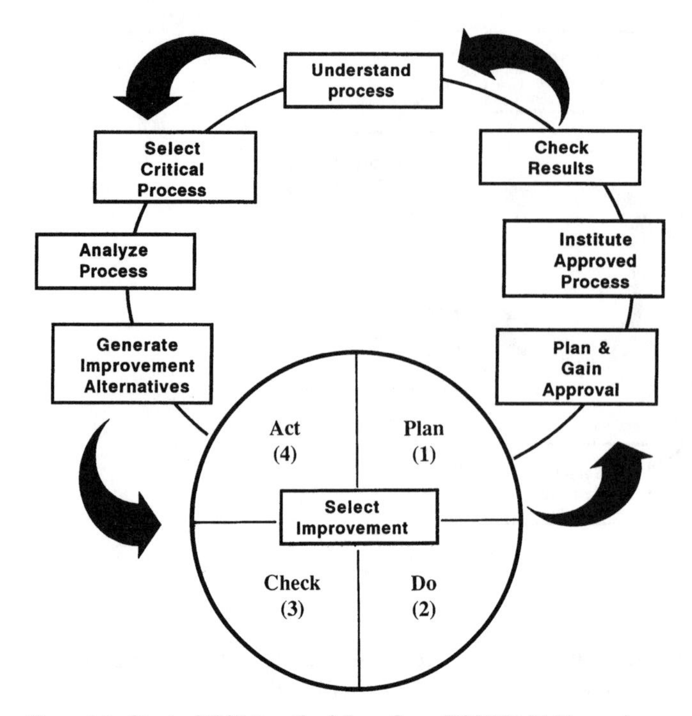

Figure 2.9 Typical TQM methodology from *TQM Field Manual.*

Basic improvement methodology steps are

1. Understand/identify the opportunities.
2. Select an opportunity for improvement.
3. Analyze the selected opportunity.
4. Generate improvement alternatives.
5. Select an improvement alternative.
   a. Plan the improvement on a test or pilot basis.
   b. Do the improvement on a test or pilot basis.
   c. Check results of test or pilot against desired outcome.
   d. Act to make the improvement permanent or repeat, starting with step 5*a,* or go back to step 1.
6. Plan and gain approval for the selected improvement.
7. Institute the selected improvement.
8. Check the results for the desired outcome.

As shown above, there are many TQM improvement methodologies. The specific methodology is not as important as the discipline to use the same improvement methodology throughout the organization. Usually, an organization selects an improvement methodology that can transfer from its major organizational focus to other organizations in the company. For instance, an organization with a major organizational focus on engineering would select a TQM improvement methodology consistent with the engineering discipline, such as the basic PDCA methodology. The PDCA TQM improvement methodology would then be used in all improvement activities, whether in finance, human resources, or cross-functional teams. Each of the TQM improvement methodologies is effective when applied with persistence toward accomplishing a specific goal.

Why integrate project management and total quality management?

Today's environment of rapid change, rising complexity, and rabid competition has caused organizations to examine approaches that enable them to survive. Project management was adopted in the 1950s to adapt to the environment of managing large-scale, complex projects. Today, it has become the preferred approach of producing and delivering products and services. Total quality management was adopted in the 1980s to meet the challenge of increased competition by focusing on the quality issue of customer satisfaction. Project management and total quality management together provide an approach to adapt to the global economic environment of the 1990s and beyond.

Project management provides the management techniques for delivering a project, program, process, task, or activity. Total quality management furnishes the environment for selection and continuous improvement of the right project, program, process, task, or activity. In combination, they provide a systematic, disciplined, flexible, adaptable approach for producing deliverables, improving organizational performance, and moving toward continuous improvement focused on customer satisfaction.

The paradigm that evolved in most of today's organizations stressed internal needs rather than external customer needs or expectations. The driving forces were internal objectives, often driven by a management-by-objectives system that ignored customers. The decisions were normally made by management without input from the people who knew the processes. In fact, people were controlled by a closed system that generated fear and alienation. In this system, nobody was identifying what would satisfy the customer, and no one asked employees at the operating levels to participate in any problem solving or decision making despite the fact they were the ones who knew the customer requirements and ways to make their processes work the best.

Further, most managers believe in the fundamental concepts of total quality, but they have extreme difficulty applying the concepts in their day-to-day operations because organizations, both public and private, all have well-developed paradigms. These paradigms have driven managers to manage ineffectively and inefficiently. The philosophy in this paradigm has been that management's chief function is to control. The theory goes that the most successful and productive organizations are the ones with the best control systems. This might be true in a mass-production society, provided they are focusing on and controlling the right things. Unfortunately, in these kinds of organizations, it is difficult to determine if the right thing is being controlled or that the right thing is being done right the first time.

In addition, many organizations have not been able to find meaningful work for middle managers, once thought to be the backbone of such organizations. As Warren Bennis stated in *The Unconscious Conspiracy:*

> The bigger the bureaucracy, the greater is the danger that it may yield to a kind of incestuous inward-dwellingness, with middle management spending all of its time writing self-justifying memos to each other and, as far as the outside world is concerned, scarcely knowing whether it is raining or is Thursday.

New paradigms of management thought are periodically presented as radical shifts in management approaches. Today, the need to shift to fundamental concepts that focus on the customer should be obvious to most managers. Since customers ultimately influence the survival of an organization, customer satisfaction should be the focus of all organizational efforts. Continuous improvement with measurable indicators of customer satisfaction is the only way to fulfill changing customer needs and expectations. Empowered people are the best strategic route to rapid response to the customer. Multifunctional teams are the structure of choice to solve problems and improve processes to ensure that the right things are done right the first time. Complex undertakings require a disciplined, structured, project-oriented approach to optimize the resources of the project and the organization.

The refreshing new paradigm for success today is encompassed in the combination of project management and total quality management into customer-driven project management. It emphasizes management's purpose to lead and empower highly qualified people, usually through a team structure, in the continuous improvement of processes and products focused on total customer satisfaction. It stresses management supporting its people in performing, measuring, and improving their processes. It advocates teams, with leaders at all levels in the organization, functioning through commitment and support to solve problems and make decisions. It gives project management and matrix teams a new set of internal values driven by the customer. Management truly works to balance the cost, schedule, and performance of the deliverable. The cost is optimized by doing the right thing right the first time. Schedule is met by doing it on time all the time. And performance/quality is accomplished by always striving for improvement and always satisfying the customers' expectations.

Main Points

Customer-driven project management is a management approach that integrates and expands both total quality management and project management.

Project management involves the management of defined, nonroutine activities aimed at distinct time, cost, and performance goals.

Project management is the management of an activity that has a defined completion.

Project management is also called *program management, product management,* and *construction management.*

The unique aspects of project management are

- Defined end.
- Borrowed resources.

A successful project involves

- Completion of the project within cost.
- Completion of the project on or before schedule.
- Completion of the project within performance criteria.

The matrix organizational structure is used in project management. This organizational structure requires shared resources and responsibilities.

The project management principles are

- *P*rovide a project focus.
- *R*eward production.
- *I*nvolve functional organizations.
- *N*urture rapid technological change.
- *C*oordinate and control all activities.
- *I*nclude authority and resources with responsibility.
- *P*rovide time, cost, and quality objectives.
- *L*et functional organizations perform processes.
- *E*ncourage teamwork and cooperation.
- *S*atisfy the customer.

The classic project management cycle includes conception, definition, production, operation, and divestment.

Total quality management, with its roots in the quality movement, modernizes many of the traditional philosophies, principles, methodologies, tools, and techniques.

U.S. TQM seeks to optimize through a disciplined organization-wide approach the strengths of its people's diversity, individuality, innovation, and creativity.

TQM is a management philosophy and set of guiding principles stressing continuous improvement through people involvement and measurements focusing on total customer satisfaction.

Total means involving everyone and everything.

Quality is total customer satisfaction.

Management refers to creating the environment through leadership and empowerment throughout the organization. Further, management manages quality through the invention and improvement of processes.

The *customer* is everyone affected by the product and/or service and is defined in two ways. The customer can be the ultimate user of the product and/or service; this is known as an *external customer.* The customer can be the next process in the organization; this is known as an *internal customer.* TQM focuses on satisfying all customers, both internal and external.

The TQM principles are

- *P*rovide a TQM environment.
- *R*eward and recognize appropriate actions.
- *I*nvolve everyone and everything.
- *N*urture supplier partnerships and customer relationships.
- *C*reate and maintain a continuous-improvement system.
- *I*nclude quality as an element of design.
- *P*rovide training and education constantly.
- *L*ead long-term improvement efforts geared toward prevention.
- *E*ncourage cooperation and teamwork.
- *S*atisfy the customer (both internal and external).

The TQM umbrella includes the integration of all the fundamental management techniques, existing improvement efforts, and technical tools under a disciplined approach focused on continuous improvement. All existing improvement efforts fall under the TQM umbrella.

TQM is a unique management approach. As explained under philosophy and guiding principles, it is a change from the traditional U.S. management mind-set.

The TQM process starts with the customer's wants, needs, and expectations, and the outputs of the never-ending TQM process are satisfied customers.

The TQM methodology involves the application of a disciplined continuous approach throughout the entire organization. The methodology is used for structured process improvement and problem solving.

The combination of project management and total quality management provides the approach for meeting the challenges of the global competition environment. The combination provides the approach for both performance and continuous improvement, focusing on results leading to organizational success and total customer satisfaction.

Chapter

3

Customer-Driven Project Management: An Integrated Approach

Focus: A detailed description of the integration of a total quality management environment, project management system, and customer-driven project management structure to form customer-driven project management is contained in this chapter.

Introduction

CDPM is a sensible extension of the total quality and project management paradigm that has the customer as the driver. CDPM is a natural progression for many organizations building long-term relationships with their key customers.

To put customer-driven project management in action, the following basic concepts are necessary:

1. Customer-driven project management requires leadership and top-management support as the way to do business. This is necessary because CDPM requires sharing resources and responsibilities. It also requires unique relationships wholly different from those found in the classical organization. Top management must provide visible and continuous leadership to create and sustain success.

2. Customer-driven project management assumes that virtually all an organization's work can be structured into team projects. Thus, the organization must be committed to empowering and training everyone in the organization to apply CDPM techniques.

3. Customer-driven project management means that accountability for the outcomes of projects is shared with the customer. The cus-

tomer and the organization define the measures of total customer satisfaction, the criteria and standards for processes, and the controls for schedule and cost. The customer project leader guides the customer-driven project teams. The project team facilitator creates and maintains teamwork. The project manager manages the internal aspects of the project. The process owners, functional managers, and employees ensure technical soundness through the support, training, and education of their functional department people.

4. Customer-driven project management requires a matrix organizational structure. This provides for shared responsibility among the project customer leader, the project team facilitator, the program manager, and the functional managers/process owners, and other team members.

5. Customer-driven project management makes the customer drive the project through all the phases of project management. Of particular importance is the customer's active involvement in the conceptual phase and continuous improvement during all phases. Customer-driven project management is truly driven by customer expectations and satisfaction.

6. Customer-driven project management requires the development of empowered project teams. These teams must have the responsibility, authority, and resources to perform and improve the project and/or their processes.

7. Customer-driven project management presumes that project management phases can be performed simultaneously. This produces a truly iterative process unencumbered by traditional linear constraints.

8. Customer-driven project management assumes optimum use of information and telecommunications systems. This enables the customer to drive the project and share all necessary information.

CDPM requires an environment that stimulates growth for everyone. The growth protential is magnified in CDPM through understanding customer needs and expectations, organizational processes, and individual competencies. The structure allows a greater interaction along a wider scope of analysis, definition, production, operations, and improvement with the customer in the driver's seat.

CDPM implies a major shift in thinking about how an organization conducts business. Currently in many organizations, project management teams produce deliverables and quality improvement teams improve processes. Quality improvement teams are seen as front-end priority setters, finding the right job to do right to improve customer satisfaction. Quality improvement teams seldom are included in implementation, especially if the improvement involves a new or reengineered process. Implementation is usually reserved for the functional or project teams. CDPM requires the customer-driven teams to perform and improve processes that produce results and deliverables focusing on total customer satisfaction.

CDPM requires the foundation of total quality and project management to produce deliverables, improve processes, and perform work. Therefore, the implementation of customer-driven project management requires the integration of a total quality management environment, project management system, and a customer-driven management team structure.

CDPM Requires a Total Quality Management Environment

Customer-driven project management first requires the creation and maintenance of a total quality management (TQM) environment that focuses on customer satisfaction. The TQM environment is an internal organizational culture of openness, honesty, trust, communication, involvement, ownership, pride of workmanship, accomplishment, individuality, innovation, creativity, strategic thinking, and personal commitment to be the "best."

A customer-driven project management environment must be established and maintained over the long term. This environment requires a systematic, integrated, consistent, organizationwide perspective. It will not just happen. The total quality management environment must include the entire organization and be shared by everyone in the organization. This requires the creation of a VICTORY environment focused on total customer satisfaction. The creation of a VICTORY environment requires the following elements:

*V*ision and the leadership to make it happen

*I*nvolvement of everyone and everything

*C*ontinuous improvement system

*T*raining and education

*O*wnership

*R*ewards and recognition

*Y*ears of support and commitment

This is the VICTORY-C model from the *TQM Field Manual.* All the elements of VICTORY focus on total customer satisfaction, as shown in Fig. 3.1, and are absolutely essential for survival today and success in the future. This TQM environment starts with top leadership and

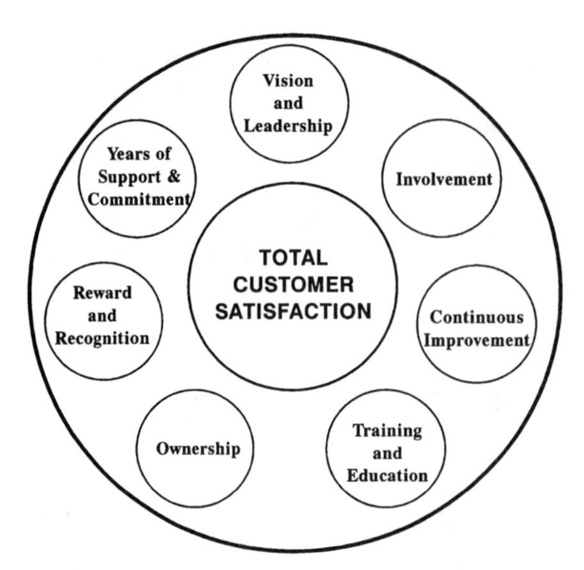

Figure 3.1 The VICTORY-C total quality management model.

key people throughout the organization. The initial thrust must be spread from a cadre team of dedicated champions with the top manager in the organization as the leader.

TQM must be manifested in every aspect of the organization. First, a vision must be developed and stated by top management. The vision is the purpose of the organization and desired future state, and it must be shared by everyone in the organization. In addition to a vision, the environment must be developed and maintained by leadership throughout the organization. Equally essential is the involvement of everyone and everything and the continuous improvement of all systems and processes. A fourth necessity is training and education, which must be provided constantly. Fifth, "ownership" must be established and fostered for all systems and processes. Sixth, rewards and recognitions must be systemized to reinforce desired behavior. And seventh, years of personal management commitment and support must be provided to ensure VICTORY. Finally, all the essential elements must be focused on total customer satisfaction.

Vision

The *vision* provides a future state for the organization to strive to reach. As shown in Fig. 3.2, the vision, mission, and values provide a common sense of focus. A *vision* is the future "image" of the organization, the scenario of where the organization leadership wants to go.

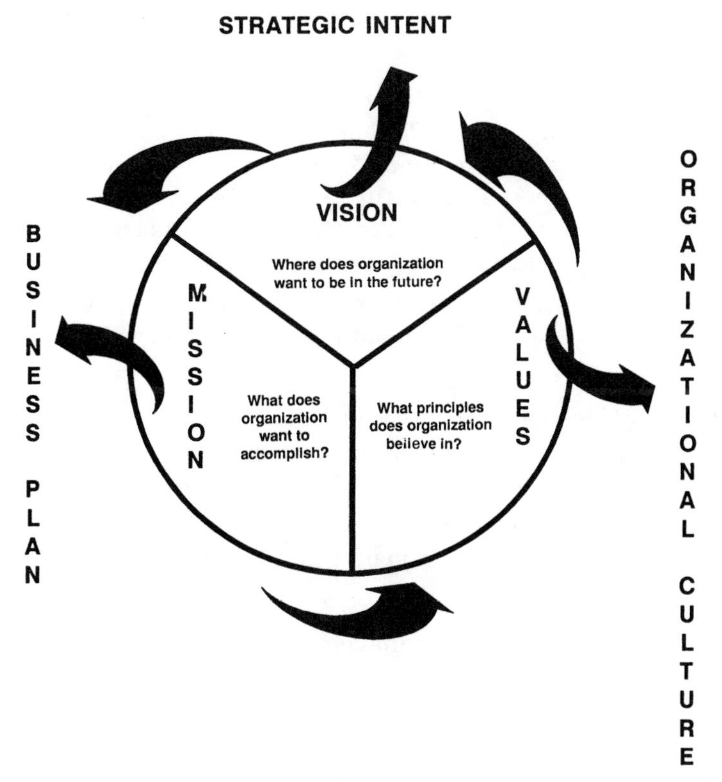

Figure 3.2 Vision, mission, and values.

Vision reflects the potential of the organization to establish long-term relationships with its customers and to participate in and share their continuous improvement. Vision is articulated first by top management, and then it is worked through the organization by means of "focus" teams. Vision is the outlook of the organization to be prospering long into the future.

Steps to the vision. The most effective way to develop vision is for top management to develop a sense of the organization's long-term future look through top-management and middle-management teams. This view of the organization's outlook is then opened to dialogue with a representative sample all employees to obtain feedback. From this understanding of the organization, top management defines the vision. The vision becomes the guiding focus of all efforts in the organization. It aims toward *excellence.* It gives a statement of what the future looks like. The vision provides the basis for continuous improvement as well, since in the visioning process, the leadership begins to outline the path along the way to fulfilling the mission. Here are the steps for creating a vision:

1. Leadership visualizes the future of the organization. This future must be defined in years. Depending on the dynamics of the organization, this can be anywhere from 5 years to an unlimited time in the future. This view of the future should be developed only after a thorough understanding of the organization, its customers, and its environment opened through strategic planning.
2. The initial vision is evaluated to ensure that it meets the organization's criteria for the future. It should answer the following questions:
   - Is it an attainable view of the future?
   - Does it clearly state to all people in the organization a common purpose of the organization?
   - Does it convey where the organization wants to go?
   - Does it provide an understanding of what to do?
   - Is it oriented toward customers?
3. The vision is established by top management/leadership.
4. The vision is instituted throughout the organization. This is accomplished through constant communication at every opportunity. In addition, the vision must be set and sustained by the example of everyone in the organization, especially by the leadership who created the vision.
5. The vision leads to a strategic intent statement that places content on the vision. For instance, a company might state in its vision statement that its future outlook involves becoming a globally recognized software engineering organization.

Besides the vision, the mission and values of the organization combine to provide guidance for the overall focus of the organization. Top management is responsible for formulating and fostering the vision, mission, and values of the organization.

Mission

The *mission* describes the basic corporate view of the role and function of the organization in satisfying customers' expectations today and in the future. For a public agency, mission derives from constitutional or legislative mandates and executive requirements and is further elaborated by written mission statements that provide more detail on how the organization is going about fulfilling its public mandate. For a business concern or a public enterprise such as a transportation or power authority, mission is more a function of the ownership's view of what lies at the core of the firm. Mission is "what we are in business for," and for a private firm, this statement results from a complex articulation of markets, customers, internal capability, boundaries, and performance. Mission should be stated clearly in documents that every employee can use as the "anchor" for his or her performance. This is especially true if the organization is intending to drive a new view of mission, such as is suggested in the concept of customer-driven project management.

Steps to the mission. The development of a mission statement starts at the top. The corporate mission is the expression of the will and approach to serve customers. For a public agency, mission states the public-policy goals and objectives. Once developed by top management, a mission statement should be reviewed by every employee and feedback incorporated in a continually updated version. Here are the steps to mission:

1. All current mission documents are assembled and synthesized.
2. Top management brainstorms about critical elements.
3. Measures of customer satisfaction and corporate performance are identified.
4. The mission statement is written. The mission statement should answer these questions:
   - What is the purpose of the organization?
   - What are the customer values driving the mission?
   - What is the common direction for the organization?
   - What are the expected results of the organization?
5. The mission statement is confirmed with all employees through a participative process.
6. The mission leads to a business plan.

Values

Values are important to guide the conduct of the organization. *Values* include the principles the organization believes and follows. Values derive from the ethics of the organization. Values are the collective concept of what is important and what is "right" about the organization. Typically, values bring to the surface issues of honesty, trust, and integrity. They describe ways of communicating within the organization. They guide relationships with the competition, suppliers, and customers. Values generally establish ground rules for producing on the promise of the organization. Values have to do with the rights and privileges of management and employees and set the tone for policy and procedure. For instance, if the organization values an internal communication system that designs and develops projects through "concurrent and parallel work," it states this value in its core value document.

Steps to the value statement. An organization sets its values through top-management initiative, backed up by actual behavior and business practice. A statement of values speaks to right and wrong, to honesty and integrity, to communication and trust, to competence and the work ethic, and to ways to resolve conflicts and problems. Here are the steps:

1. Top management identifies basic values through brainstorming.
2. Priorities are set.
3. Each value is written into a statement, and all values are shared with employee focus groups to achieve consensus.
4. Everyone in the organization is encouraged to live the values.
5. Values lead to a clear organization culture.

Leadership

Leadership is essential to creating and maintaining the TQM environment. Leaders are responsible for all the elements of VICTORY. They make the vision, mission, and values a reality. The leaders set the example for actions throughout the organization. Leaders create the reason to work productively to satisfy customers. They inspire the imaginations of people in the organization. They bring meaning to the workplace by allowing each person to focus on his or her contribution to the greater societal good—the needs of customers. Specifically, leaders do the following:

1. *Leaders create synergy.* This synergy is developed by the creation of meaning and value beyond the self and into the team setting of the workplace. Leaders call up the inherent needs of people to work cooperatively to create values, services, and products that go beyond the sum of their parts. Leaders inspire coordination, collaboration, and cooperation because they are successful in establishing a shared vision of mission and values focusing on progress and growth. Leaders create the sense of a common purpose.

2. *Leaders create vision.* This vision provides an image of change in which the quality of life is enhanced. Visions are pictures of how the world could be and allow people to connect their own *personal* contribution to the realization of the vision. Leadership establishes

its vision of how the organization will serve customers and what business practices and technologies it will use to accomplish the vision. Here is where a customer-driven project organization would indicate the significance of penetrating the customer quality improvement process, placing customers in charge of project teams, and nesting product and project development in customer planning. The vision promotes the view of excellence in the business by perfecting products and services so that a supplier meets customer requirements in a productive and profitable way. Suppliers create value added in customer performance and value added creates wealth. The vision becomes meaningful to all members of the organization because they align themselves with it.

3. *Leaders give structure.* Leaders give purpose and definition to the vision by ensuring that there is meaning and relevance to the vision. For example, a leader establishes customer-driven teams to evidence the key importance of understanding and meeting customer needs as the principal activity of the supplying firm or agency. The way the organization works is a function of the leadership decision. Leadership answers the following questions:

"How will the internal and external customer be treated?"

"How will we treat each other?"

"How will the organization empower employees and teams to carry out its vision of meeting customer needs and expectations by placing the customer in the key role in programs and projects?"

4. *Leaders set the example.* Leaders live their vision by incorporating it into their daily behaviors. Leaders are inherently directed people who evidence their vision in the choices they make in the use of their own time. This becomes visible to employees and to customers, thus reinforcing the value of the vision itself. Leaders ensure applicable processes are used through the whole organization. They pay constant attention, so that continuous improvement is a way of life.

5. *Leaders grow other leaders.* Other leaders are grown in the organization by mentoring, coaching, and monitoring by the current leaders. The key long-term contribution of leaders to the organization comes in growing new leaders. This involves training and educating people at all levels to exercise their leadership potential. One way leaders are able to do this is in challenging people to work with customers to establish the customers' visions of quality. This requires education and training, since the supplier organization must understand the customer organization almost better than it does itself.

6. *Leaders establish and maintain organizational systems.* These systems must support the CDPM environment. There must be systems for all elements of VICTORY. The organizational systems must

promote maximization of people resources and technology. This means the leader must know the work force. Such leaders must continually survey the work force to stay in touch with changing values. The organizational systems for performing work, improving processes, communication, involvement, leadership, management, ownership, accountability, reward, recognition, training, and education must keep pace with organizational and individual needs.

The leadership issue: Quality, performance, cost, and schedule

The most challenging forces against CDPM and its putting the customer in the driver's seat are from within the organization. Often this occurs because internal requirements for meeting schedule and cost goals become the focus of the organization. Frequently, time and budget constraints are established with no communications with the ultimate customer, or managers consult with customer representatives who do not know the total implications of their decisions.

Leadership sets the tone in an organization for how quality, performance, cost, and schedule are "traded off" in the daily work of the organization. In CDPM, leadership must allow work groups to decide on quality and performance under the direct guidance of the customer. Everyone must work together to develop schedules, budgets, technical specifications, and other issues of importance to both the organization and the customer. It is important to state clearly that total customer satisfaction is always the most important consideration.

Leading a total quality effort and developing customer-driven project teams will require leaders to establish a new value system in an organization. This value system places top priority on total customer satisfaction. The commitment must be to customer satisfaction over all other constraints, including schedule and cost. That is, while the customer is always interested in timely and cost-effective delivery of the deliverable, the customer will rarely be satisfied with a poor product or service that does not meet his or her needs, regardless of how efficiently and quickly it was produced.

Involvement of everyone and everything

The TQM environment requires the total involvement of everyone and everything in the organization. Everyone includes the entire organization, including management/leadership, all the people in the organization, suppliers, customers, and teams dedicated to the ultimate goal of customer satisfaction. Everything comprises systems,

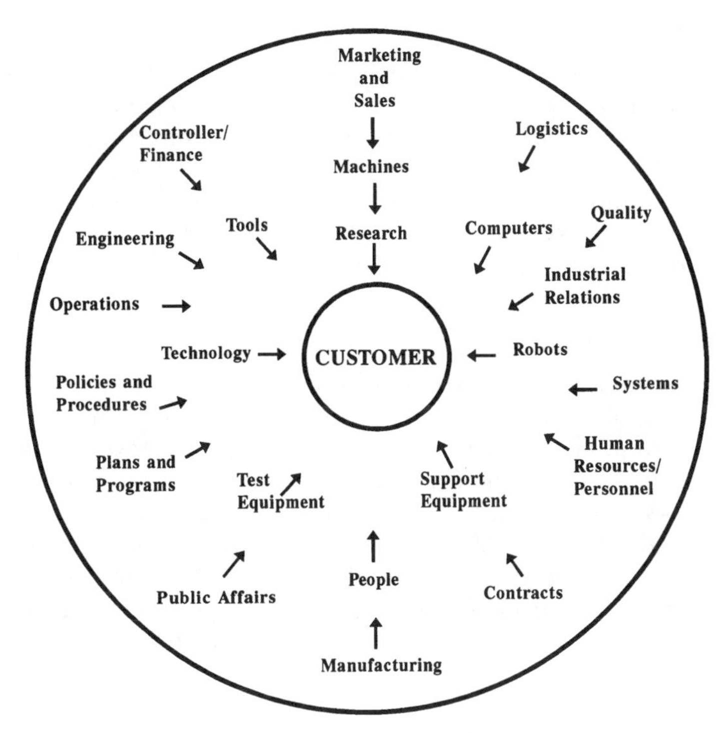

Figure 3.3 Involvement of everyone and everything.

equipment, and information. Figure 3.3 shows everyone and everything included in the TQM environment.

Management

Management or those who control the responsibility, authority, and resources of the organization ensure the total organization is geared to total customer satisfaction. They provide both management and leadership to the organization. They provide the focus of the organization. Management designs the processes that are used to perform the work. Management must provide an environment where people can perform to the best of their capabilities. This involves providing everyone with the means to run their specific process. In addition, management fosters the development of a sense of pride and "ownership" of processes. Management empowers the work force. Management invests in education and training, guides cooperation and teamwork, and motivates actions through rewards and recognition.

People in the organization

In a TQM environment, people are the most important resource. Therefore, people must be encouraged to be creative and innovative within all areas of their work. All the people in the organization must be empowered to perform their work with excellence. They must be allowed to make whatever changes are necessary to within regulatory guidelines and boundaries negotiated with management—to perform the work and improve the process providing customer value.

Suppliers and customers

Suppliers and customers are also important players in the TQM environment. Both suppliers and customers must be integrated into the TQM process. Suppliers must understand the requirements of the organization. This is often done through "certification," as a supplier firm. The ISO 9000 standards are one form of certification for meeting international quality standards. Also, the organization must weave the customers' needs and expectations into all its processes. Further, the organization must develop a continuing relationship with its suppliers and customers to ensure long-term customer satisfaction.

Teams

The involvement of teams is critical to the success of a total quality management environment. Teams should be the primary organizational structure to accomplish critical missions. Teams involve the internal organizational groups and include functional and especially multifunctional teams. In addition, suppliers and customers also should be participating in teams within the organization.

Include everything

The TQM management environment must include not only everyone but everything in the organization. Everything includes all the systems, processes, activities, jobs, tasks, capital, equipment, machines, vehicles, support equipment, facilities, tools, and computers. This management environment requires providing the proper means to perform the job. A proper balance of technology and people is essential.

The TQM management environment must include all systems, processes, activities, jobs, and tasks within the organization. A TQM system integrates all elements of the organizational environment and all functions of the organization. This is absolutely essential. Many past improvement efforts focused only on one area of the organization, such as manufacturing, design, or marketing. In addition, many policies and procedures did not allow the required improvements. Today, a total integrated improvement effort is the only way to success.

The organizational environment includes such items as communications, policies and procedures, training, rewards and recognition, benefits, accountability, evaluation, and marketing. The functions of the organization encompass engineering, marketing, human resources, manufacturing, finance, information systems, and logistics. In addition, all improvement efforts become elements of the TQM environment. These improvement efforts include such items as those shown under the TQM umbrella in Fig. 1.2.

All the equipment must contribute to the attainment of the purpose of the organization. The equipment must assist people in performing processes while allowing them to add value to the product or service through their own ideas.

Information must be shared by everyone who has use for it. The information must reflect the current status of the organization as well as projections for the future. It must provide an accurate and comprehensive picture of all supplier requirements, internal process performance, and satisfaction of customer needs and expectations.

Information sharing is critical to showing management's commitment. This requires management to open up all information channels. It sometimes helps for management to translate traditional management information into a form that makes it easier for everyone to understand what they need to do. For instance, if inventory information is presented, it should include an indication of appropriate inventory levels with the amount in inventory. There should also be clear steps regarding what is needed in order to ensure the inventory supports requirements.

Critical performance information must be available to all people who need it. This type of information should be on charts that can be easily read and updated. When possible, the performance feedback should be constant and immediate. This must include all critical performance information in the organization. Performance feedback is essential to the TQM environment.

Continuous improvement

Continuous improvement of product, processes, and people in an organization is essential (see Fig. 3.4). Customer expectations drive a product deliverable. This drives the people on customer-driven teams. Customer-driven teams drive the improvement of processes. Improved processes drive better product deliverables that exceed customer expectations. This leads to the need for continuous improve-

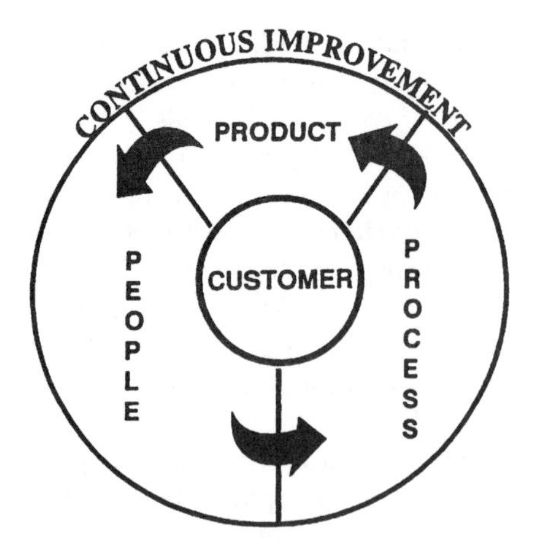

Figure 3.4 Continuous improvement of processes, people, and product.

ment of product, people, and processes. And this requires the establishment and maintenance of a disciplined continuous improvement system such as the customer-driven project management improvement methodology.

The continuous improvement system applies all the fundamental aspects of the TQM definition. First, people are not the problem. People are the solution. Almost all "root" causes of problems in an organization or variation in a process can be traced to the system or process itself. Therefore, the continuous improvement system uses people to focus on the system, process, issue, and problem; it does not look for fault in the people. Second, quantitative methods are the principal means to make decisions. Measurement is basic to all TQM activities in the entire organization. Third, the continuous improvement system with an appropriate improvement methodology is used to improve all material services supplied to an organization, all the processes within the organization, and the degree to which the needs of the customer are met—now and in the future.

Training and education

A training and education system must be instituted. Training and education comprise a never-ending process for everyone in the organization. This is an investment that must be made. Training and education provide skills and knowledge—the ability to make it happen.

Training is geared toward developing and improving specific knowledge and skills. The TQM environment requires everyone to gain additional capabilities to improve the process and perform the work. This requires TQM and job skills training. Training in TQM philosophy, guiding principles, and tools and techniques is never-ending.

Interpersonal and team interaction skills must be refined continually. Specific job skills training must be provided and constantly updated to reflect the improved processes. All training must be geared to specific, clearly defined objectives. The training must be performed as close as possible to the time it is required. It needs to be used immediately by the trainee. Finally, all training requires reinforcement to ensure the results needed to achieve victory.

The educational system must support the goals of the organization and the individual. Each organization must provide opportunities for individual growth through education. In addition, each individual in the organization should be encouraged to pursue a life long educational process to foster future success for the organization and the individual.

To meet these needs, leaders must establish programs of education and training in the normal and routine structures of the workplace. In other words, education and training become one of the core functions of the organization. Workers seek training and education to improve their performance on their teams and their understanding of customer issues, and leaders encourage such growth and development by empowering the teams to perform and improve their work.

Specifically, the key skills that must be developed for a TQM environment include communication, especially listening; teamwork; conflict management; problem solving; consensus decision making; critical and systems thinking; understanding customer needs; and process improvement. In addition, leaders must cultivate skills to motivate and coach people and to facilitate meetings and continuous improvement.

Ownership

Ownership comprises the ability to perform and improve work. It involves encouraging and empowering people to think and to make decisions. Ownership is important to ensure pride of accomplishment. Everyone must have ownership of his or her work. Ownership implies responsibility, authority, and resources. Responsibilities, authority, and resources encompass the boundaries of empowerment. People must assume responsibility for work performance. In addition, they must have the authority to take the necessary actions. Also, they must have the required resources. This leads to empowerment to do whatever is necessary to do the job and improve the system within the defined responsibility, authority, and resources. Individual ownership must include everyone in the organization, both top management and all the workers. Besides individual ownership, team ownership is important in a TQM environment. With empowerment and ownership, the entire organization can work with pride and commitment toward satisfying customers.

Empowerment

Empowerment is an important element of the TQM environment. *Empowerment* means a person can do whatever is necessary—within its responsibility and authority and with available resources—to perform or improve the process or deliverable to satisfy the customer. This is an optimistic and positive approach to leading people. The assumption is theory *Y,* that people want to contribute and work hard and that management's job is to provide the environment where they can do so. The challenge is to find the balance between empowerment and control. *Empowerment* is the process of enabling employees at all levels to exercise wide discretion in meeting customer needs, both within and outside the organization. *Control* is the process of setting boundaries on that discretion, through guidelines, so that employees are clear on the extent of empowerment.

The leader's role in empowerment should be a supportive and enabling one. Leaders develop their people's capacity to perform and assume more responsibility to serve customers, whether inside or outside the organization. Leaders remove the barriers to empowerment and draw the boundaries of discretion with each employee and team.

Reward and recognition

Reward and recognition must be instituted to support the TQM environment. Although rewards and recognition are elements of any organization, the TQM environment mandates changes to the usual reward and recognition systems. Rewards and recognition are shown in Fig. 3.5. A reward is given for performance of some specified action. Rewards can be extrinsic, such as compensation, promotion, and benefits, or they can be intrinsic, such as feeling of accomplishment, improved self-esteem, personal growth, or a sense of belonging. Recognition is given for special or additional efforts. Recognition takes the form of praise or a celebration. Praise should be the normal method to reinforce the right behavior. Celebrations can be individual or group oriented. The reward and recognition systems of the organization must foster the TQM philosophy and guiding principles. They must constantly and immediately reinforce leadership, teamwork, individual contributions, continuous improvement, and customer satisfaction behavior.

The TQM environment requires people to add new responsibilities. The reward systems must recognize this with new rewards. Any new

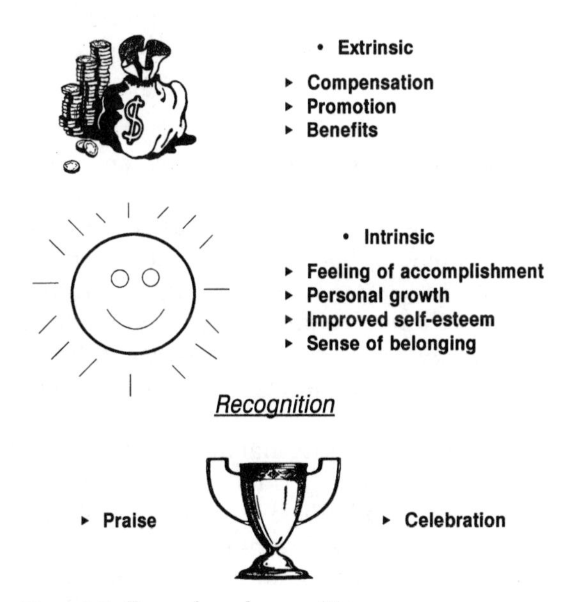

Figure 3.5 Rewards and recognition.

reward system must be equitable and just. Further, it should include an appropriate combination of extrinsic and intrinsic rewards.

Years of commitment and support

Leaders must commit to long-term support. They must be willing to make the investment in their personal time and the organization's resources. They must understand that although some results will be realized quickly, permanent changes will take many years. This involves setting the example by displaying the expected behaviors day after day. It also includes providing resources and a support system. The support system must provide direction, guidance, and support to the overall TQM effort. Leaders must be active, highly visible participants in all aspects of the TQM process.

Leaders also must make a commitment to total customer satisfaction as the primary focus of the organization. Total customer satisfaction must take precedence over all other influences, including costs and schedules. Leaders must have the discipline to make this long-term commitment for the future of the organization. They must first thoroughly understand the TQM philosophy and guiding principles.

Then, the TQM philosophy and guiding principles must be constantly and consistently applied throughout the organization. Leaders must devote personal attention to the implementation of the TQM philosophy and guiding principles throughout the entire organization.

Focus on the customer

All the elements of success focus on total customer satisfaction. Total customer satisfaction is the focus of the entire CDPM process. Total customer satisfaction is quality. Quality includes all elements required to satisfy the target customers, both internal and external, and can include such items as product quality, service quality, performance, availability, durability, aesthetics, reliability, maintainability, logistics, supportability, customer service, training, delivery, billing, shipping, receiving, repairing, marketing, warranty, and life-cycle cost.

CDPM focuses on the satisfaction of both internal and external customers. Figure 3.6 shows the relationships of internal and external customers. Each process is the "customer" of the next process. These processes are the internal customers. If each internal customer satisfies the next internal customer while focusing on external customer satisfaction, the ultimate customer—the external customer—will be satisfied. CDPM teams ensure that each process is linked to customer-driven projects.

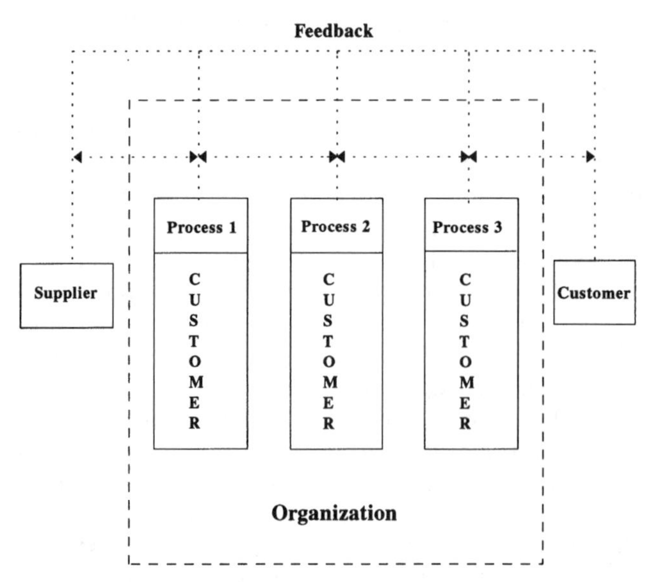

Figure 3.6 Customer relationships.

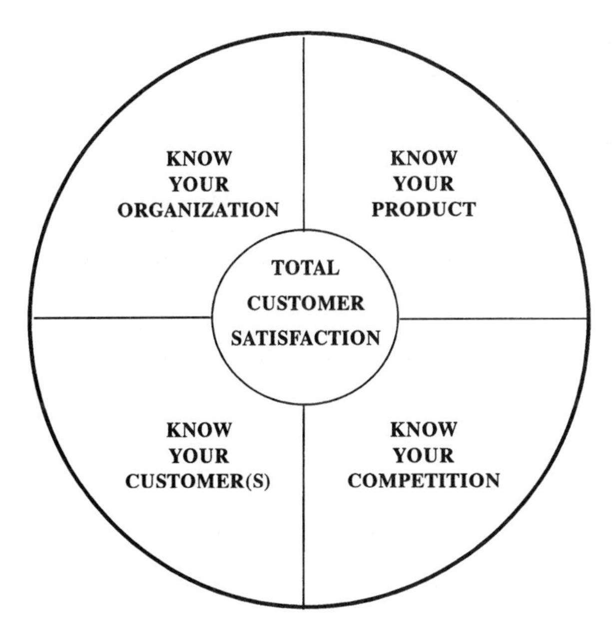

Figure 3.7 Elements that must be observed for total customer satisfaction.

Besides an obsession with quality from the inside out, total customer satisfaction requires the organization to know itself, its product, its competition, and its customers. Figure 3.7 shows the elements that must be observed in order to achieve total customer satisfaction. First, knowing your own organization is an important element of customer satisfaction. In knowing itself, the organization understands what it can do to satisfy its customers. Second, knowing the product allows the organization to position the deliverable so as to maximize total customer satisfaction. Third, knowing the competition provides targets for service and quality. Fourth, knowing the customer focuses the organization on the development of relationships to keep and gain customers. It also provides an advantage when identifying customer needs and expectations. Further, it allows anticipation of changing customer requirements.

Project Management System

Once a TQM environment is established, CDPM requires the development of a project management system. Specifically, customer-driven project management requires an organization to develop a "world class" project management system to complete projects within cost, schedule, and performance parameters while satisfying customer expectations. Therefore, the organization must establish a system to effectively perform all the tasks in each project management stage and transition from one stage to the next in the project. The project management system is the overall internal management system used for analyzing, planning, implementing, and evaluating a project to ensure that the deliverable satisfies the customer. The project management system provides the methods, tools, and techniques to manage a project in the particular organization through all the stages of the project.

The focus of a project management system is customer satisfaction. As shown in Fig. 3.8, a project management system involves the following major processes:

1. Analysis
2. Planning
3. Implementation
4. Evaluation

Analysis

The analysis process in a project management system provides specific methods to assist the project team in defining particular project requirements and improvement opportunities. The analysis process provides the approach to assess the customer, itself, the deliverable, and the competition. It must provide a process to

- Identify the target customers
- Determine customer wants, needs, and expectations
- Define how the organization must adapt to changing customer requirements

Figure 3.8 Project management system.

- Evaluate customer and supplier relationships
- Determine the processes in the organization that are needed to meet customer expectations
- Assess management support and commitment
- Assess the performance of critical processes
- Benchmark processes
- Judge if process performance is adequate
- Establish process improvement goals
- Identify the particular deliverable(s) required for customer satisfaction
- Recognize risk
- Determine competitive advantage
- Develop metrics
- Perform tradeoffs

Planning

The planning process in a project management system provides tools to help the project team in the design and development of the project and its resulting deliverable. The planning process must provide guidance for

- Relating with the customer
- Preparing proposals
- Planning strategy
- Documenting project information
- Developing mission objectives and goals
- Setting priorities
- Establishing an organizational structure
- Utilizing resources, including people, technology, facilities, tools, equipment, supplies, and funds
- Selecting and training people
- Setting up the project management information system
- Managing the project
- Identifying roles and responsibilities
- Empowering teams and people
- Developing supplier relationships

- Funding the project
- Measuring and reviewing progress
- Designing and developing the deliverable
- Investigating risk
- Solving problems
- Improving processes
- Maintaining accurate configuration information
- Providing and communicating necessary information
- Supporting the deliverable
- Scheduling the work
- Building teamwork
- Closing the project

Implementation

The implementation process in a project management system equips the project team management with approaches to ensure that the project is successful. The implementation process must provide approaches that allow the team to

- Set performance measures
- Direct use of resources
- Handle project changes
- Provide negotiation methods
- Manage risk
- Control costs
- Manage conflict
- Motivate team players
- Take corrective action
- Deliver the outcome to the customer
- Support deliverable

Evaluation

The evaluation process in a project management system provides a methodology to assist the project team in assessing progress and performance. The evaluation process must provide techniques to

- Measure customer satisfaction
- Document and report the status of such parameters as cost, schedule, and performance
- Conduct project progress reviews
- Keep track of risk
- Test the deliverable
- Gather lessons learned
- Determine the impact on the business
- Continuously improve

CDPM Team Structure

The customer-driven project management team structure involves establishing an infrastructure within which the customer is the primary focus of all projects, processes, tasks, and activities. This requires establishing an organizational structure using project teams that links all essential players from the external customer to all teams in the supplier's organization.

The first requirement of a customer-driven structure involves using the existing top-level framework to create and maintain the customer-driven project management approach. This top-level leadership team provides the direction for all other activities.

The second requirement of a customer-driven structure involves a process for the customer and suppliers to work together to identify the project using the customer-driven project management improvement methodology. A structure for jointly analyzing customer wants, needs, and expectations should be accomplished by a joint customer and supplier project steering team. The outcome of this process is the agreement to proceed on a specified project.

The third part of a customer-driven structure—recognizing the customer as the leader of the team—uses the customer or customer's voice to drive the project. First a customer-driven project lead team, led by the customer or customer's voice, is created. Preferably, the customer-driven project lead team leader is also a member of the joint customer and supplier projects steering team. In addition, the customer-driven project lead team should consist of at least the project supplier's program manager and process owners. (The program manager could also be a member of the joint customer and supplier projects steering team.) The process owners at the lead team level are the project supplier's major functional/process managers. In addition, other team members can be added as appropriate.

The fourth part of the customer-driven structure requires an established customer-driven project team organization for projects, processes, and quality improvement as appropriate. Customer-driven project teams primarily perform project work. Their major emphasis is on a successful project. Examples of a customer-driven project team include project management teams, product teams, and concurrent engineering teams. Customer-driven process teams are work teams. They work continuously on a specified process. Customer-driven work teams are the natural work team, self-managed work teams, commodity teams, and functional teams. They are the teams that own the processes to get the work done. Customer-driven quality-improvement teams focus on working specific quality, process, or productivity issues. The specific organizational structure depends on the magnitude and complexity of the project. For some projects, one team can accomplish all requirements. For other projects, many teams are needed to complete the project.

The customer-driven project management framework in its most complex form, as shown in Fig. 3.9, includes the following essential teams:

- Customer and supplier strategy teams
- Customer and supplier project steering teams
- Customer-driven project lead teams
- Customer-driven teams

Customer and supplier strategy teams are top-level teams in each organization. Each strategy team consists of the leadership in the organization. It is led by the top executive in the organization. It provides the overall direction, guidance, and support for its organization. The strategy team does not include the customer as a permanent member. However, this team maintains an ongoing relationship with its customers. Key objectives of this team are

- To develop vision, mission, and values for the organization
- To create and maintain a total quality management environment, project management system, and customer-driven project management structure
- To provide strategic direction for the organization
- To involve everyone and everything in common purpose
- To institute continuous education and training programs
- To ensure appropriate reward and recognition systems
- To act to give support
- To foster customer relationships

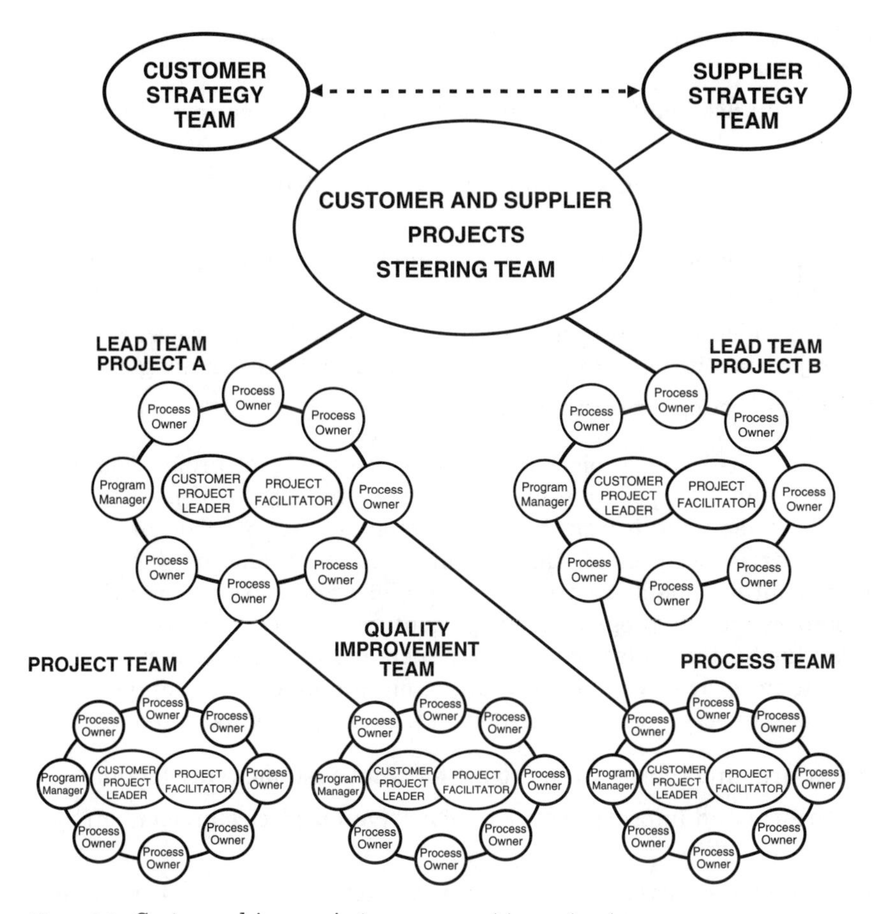

Figure 3.9 Customer-driven project management team structure.

Customer and supplier project steering teams are top-level project management teams. Each team consists of project management leadership from both customer and supplier organizations. Each team is jointly led by customer and supplier. The leaders should be members of respective customer/supplier strategy teams. The customer and supplier steering teams provide the overall direction, guidance, and support for each project. They include both customer and supplier team members. Key objectives for such teams are

- To develop customer relationships
- To analyze customers, the marketplace, the organization, and the project
- To define customer expectations and satisfaction issues with development of a conceptual model of the project
- To select projects of mutual benefit to customers and supplier

- To ensure effective and efficient customer-driven project management
- To enhance communication, cooperation, and teamwork between customers and supplier
- To develop systems to support customer-driven project management
- To provide leadership, direction, guidance, and support to project lead teams
- To develop education and training, especially for project leaders, project facilitators, program managers, and process managers
- To foster supplier partnerships

Customer-driven project lead teams are top-level project management teams for specific projects. Such teams consist of project management leadership from both customer and supplier organizations for a particular project. Each team is led by a customer. And each team contains members from the supplier organization. The principal members include program management and process owners. A project team facilitator is also essential for such a team. The customer-driven lead team provides the overall direction, guidance, and support for a specific project. The customer-driven lead team is empowered to

- Perform and improve the project for total customer satisfaction
- Perform and improve the customer-driven project management system
- Establish a customer-driven project management structure appropriate to the specific project
- Develop teams with adequate responsibility, authority, and resources
- Determine appropriate rewards and recognition
- Analyze customer needs and expectations and suppliers' processes
- Plan the requirements to meet total customer satisfaction, which includes definition, design, and development of the project deliverable
- Implement the plan, focusing cost, schedule, and performance on delivery for customer satisfaction
- Evaluate the project for continuous improvement

Customer-driven teams are the working teams throughout the organization. These teams are led by the customer of each team's process. Each team consists of process owners and process suppliers.

Each team also could have program management representation. In addition, a project team facilitator advises the team. There are three types of customer-driven teams. They are the customer-driven project team, the customer-driven process team, and the customer-driven quality-improvement team. The customer-driven project team is empowered to product deliverables and improve the project. Also, the customer-driven process team is empowered to perform and improve a specific process. The customer-driven quality-improvement team is empowered to make quality improvement their stated mission.

Customer as leader

The customer-driven project management approach advocates using the customer as leader. This makes the customer a driver for the entire life cycle of the project. The customer-as-leader concept makes customer satisfaction the primary aim of the project deliverable. With the customer as leader, the major problems in current project management processes that tend to cause customer disengagement are overcome. For instance, this process allows the customer to pull the deliverable through the project phases. In the past, project management processes frequently pushed the deliverable on the customer. In addition, the customer as leader stresses continuous communication among all project team members. Today this is facilitated by modern telecommunications systems. In the past, customers and suppliers were separated by relationships and geography. The customer-as-leader concept greatly increases the probability of providing a deliverable that satisfies the customer.

Integrating the customer into the project

In CDPM, the customer is the ultimate beneficiary of the deliverable. The ultimate beneficiary of the project is the external customer. The receiver of the output of a process is the internal customer. The customer can be inside or outside the organization. A full, vertical linkage is established among all the intermediate organizations, firms, or agencies involved in the project with downstream consumption or use of the product or service by the customer. This means, for instance, that if a CDPM project is being performed for a construction firm, the customer is the ultimate user of the building, the tenant. The builder is a critical link in the process, but not the customer. For a management information system project, the customer is the user of information provided to make decisions; the intermediate provider of the system product is the provider. For an army research, development, and life-cycle engineering center, project management is designed to maintain an effective interface with the user community to meet the requirements of the soldier in the field. Customer-driven, then, will often mean that consumers or customers of products and services actually drive projects at critical stages during the life cycle.

The customer is integrated into the project in CDPM through an agreement up front between the customer and the project management supplier(s). The agreement is based on a commitment to use the customer-driven project management approach. The approach will select projects and identify customers as project leaders. This agreement between a sponsoring firm or public agency (the customer) and a project firm (the supplier) establishes the groundwork for a continuing relationship to meet customer needs. The relationship is triggered by the customer's decision to employ customer-driven project management to continuously improve the deliverable.

The agreement incorporates the customer's vision for customer satisfaction, and it commits the two organizations—the customer and the supplier—to a customer-driven project management system. Such a system is described in terms of a process beginning with total quality management, a project management system, and a customer-driven project management structure. It starts with analysis and transitioning into project design and development. It involves the use of customer-driven project teams composed of technical and administrative staff from the customer and project supplier organizations. These teams are headed by customer project team leaders, with a team facilitator provided usually by the project supplier. The team is focused early in the process on a structure that addresses the customer's interests, needs, and requirements on a continuous basis throughout the project life cycle. The team's day-to-day basic responsibility is to provide a deliverable that achieves total customer satisfaction.

Main Points

The implementation of customer-driven project management requires a total quality management environment, project management system, and a customer-driven management structure.

The TQM environment is an internal organizational environment of openness, honesty, trust, communication, involvement, ownership, pride of accomplishment, and personal commitment to be the "best."

The TQM environment requires all the elements of VICTORY focused on customer satisfaction. These are

- *V*ision and leadership
- *I*nvolvement of everyone and everything
- *C*ontinuous improvement system
- *T*raining and education

- "*O*wnership"
- *R*eward and recognition
- *Y*ears of commitment and support

The vision is the view of the organizations's future.

The mission describes the basic corporate view of the role and function of the organization in satisfying customers today and in the future.

Values are the collective concept of what is important and what is "right" about the organization.

Leadership guides the organization to accomplishment of the vision.

The TQM environment requires the total involvement of everyone and everything in the organization. Everyone includes management/leadership, all the people in the organization, suppliers, customers, and teams. Everything includes all the systems, processes, activities, jobs, tasks, capital, equipment, machines, vehicles, support equipment, facilities, tools, and computers.

Continuous improvement of product, processes, and people in an organization is essential. This requires the establishment and maintenance of a disciplined continuous improvement system.

The continuous improvement system includes

- Defining the vision/mission
- Determining improvement opportunities
- Selecting improvement opportunities
- Improving, using an improvement methodology
- Evaluating the results
- Doing it again and again and again

Training and education are a never-ending process for everyone in the organization to gain skills and knowledge—the ability to make it happen.

Ownership comprises the ability to perform and improve the work so as to provide pride of workmanship.

Empowerment means a person can do whatever is necessary—within his or her responsibility and authority and with available resources—to perform or improve process to satisfy customers.

Rewards and recognition are essential to stimulate and maintain expected performance.

Years of commitment and support are required to create and maintain the TQM environment.

Total customer satisfaction is the focus of the entire organization.

Total customer satisfaction requires an organization to know itself, its product, its competition, and its customers.

Customer-driven project management requires an organization to develop a "world class" project management system to complete projects within cost, schedule, and performance parameters while satisfying the customer.

The project management system is the overall internal management system used for analyzing, planning, implementing, and evaluating a project to ensure that the deliverable satisfies the customer.

The analysis process in a project management system furnishes specific analysis methods to assist the project team in defining particular project requirements and improvement opportunities.

The planning process in a project management system provides tools to help the project team in the design and development of the project and its resulting deliverable.

The implementation process in a project management system equips the project team management with approaches to ensure that the project is successful.

The evaluation process in a project management system provides a methodology to assist the project team in assessing project progress and performance.

The customer-driven project management structure involves establishing an infrastructure within which the customer is the primary focus of all projects, processes, tasks, and activities.

The customer-driven project management framework includes the following types of teams:

- Customer and supplier strategy teams
- Customer and supplier project steering teams
- Customer-driven project lead teams
- Customer-driven teams

Customer and supplier strategy teams are top-level teams in each organization that provide the overall direction, guidance, and support for the organization.

Customer and supplier project steering teams are top-level project management teams that provide the overall direction, guidance, and support for each project.

Customer-driven project lead teams are top-level project management teams for specific projects that provide the direction, guidance, and support for that specific project.

Customer-driven teams are the working teams throughout the organization that perform and improve the projects and processes and drive quality improvements.

The customer-driven project management approach advocates using the customer as leader. This makes the customer a driver for the entire life cycle of the project.

Customer-driven project management links the ultimate beneficiary of the deliverable, the external customer, with every process within the supplier organization, the internal customers.

Chapter

4

Customer-Driven Project Management in Action

Focus: This chapter provides a description of the application of customer-driven project management.

Introduction

Integration of the total quality management environment, project management system, and customer-driven project management structure establishes the foundation for the application of the customer-driven project management approach. Next, customer-driven project management is applied to perform and improve any project—large or small. Putting customer-driven project management in action involves focusing on a specific project, developing teamwork in customer-driven teams, and using the CDPM improvement methodology. The customer and supplier must target a common focus for the project. Further, customer-driven teams must be empowered to perform and improve the project and its processes. In addition, perpetual improvement of quality and productivity must be pursued by both the customer and the supplier.

Before putting customer-driven project management into action, the concepts of continuous improvement in quality and productivity in customer-driven project management must be understood by all players. The concept of continuous improvement in quality goes beyond the traditional definition of quality as conformance to requirements—to *total customer satisfaction.* This requires an enhanced view of everyone's quality responsibilities. In the past, quality responsibilities for a project were simple: Monitor production output and nonconformance by making sure that the product stays inside its tolerance limits as originally set in the concept phase.

Now, with quality defined as total customer satisfaction, traditional quality responsibilities expand to include a continuous effort to provide the best value to the customer. In a customer-driven project environment, the target is always optimal output, as defined by the customer, at minimum cost. This means that everyone is always going to be looking for continuous improvements in performance levels of systems, processes, people, and product deliverables.

Total customer satisfaction requires the unrelenting pursuit of productivity and quality growth. This, again, requires enlarging traditional strategies for productivity improvement past an emphasis on acquisition of new systems or equipment. The traditional strategies for productivity improvement often involved the premature acquisition of new equipment or systems. Internal management considerations usually drove these decisions. Typically, there was no analysis of optimal capacity. In addition, there was not any assessment of how the people running the equipment felt it could be better used without expensive new acquisitions. To management, new technology meant higher speeds, better reliability, and better productivity. It also meant stricter enforcement of employee activity, with strict output standards and supervision. This tendency to rely on capital expenditures has created a major market for product, component, and system project suppliers. They have developed designs and produced deliverables against specifications set by these customers, leading to an ongoing cycle. Repeatedly, manufacturing engineers design new systems, outside supplier companies build them, industrial engineers help install them, and support engineers train the people to use them. In this process, the acquisition of new systems is usually the major cause of productivity increases—leading to an improvement pattern now called *step-function change.*

Besides this step-function change, normal project award activity also does not foster continuous improvement. It consists of the following steps:

- Customers request procurement bids from suppliers.
- Project and product specifications are developed and incorporated into bid invitations.
- The project begins upon award of a contract.

The winning project firm and the customer never discuss the options of the procurement bid. There is no conversation on the specifications or their origins. There is never an indication of any customer quality-improvement goals. These were typically not shared with a project supplier firm. It was the project supplier's job to produce to specification—period.

Customer-driven project management seeks a disciplined, structured, people-powered approach to project success. This means stressing an optimal mix of technology and people based on a thorough, ongoing analysis of the product, customers, organization, and competition. This approach involves unleashing the human capital of the organization through leadership and empowerment, which represents a significant shift in strategy. People within the customer and supplier firms perform as a customer-driven project team to improve the project. They have the "ownership" to find the best solutions. Such a team continuously explores new opportunities for improvement of the project and deliverable. Frequently, the solutions do not require expensive systems or equipment. When the requirement is for new equipment or system enhancements, the acquisition results from a thorough analysis using a disciplined, structured approach.

Customer-driven project management advocates giving all people involved in a project the power and the skills to analyze customer satisfaction issues, discover opportunities for improvement, and institute improvements. Once all people are involved in the improvement process through customer-driven teams, creative and innovative ideas emerge at a steady rate. Every day new methods are found for eliminating waste. New techniques are developed for getting even more performance out of existing equipment. A new improvement pattern is created. Continuous improvement methods lead to continuous improvement results.

The preceding paragraphs provided some key concepts about quality and productivity improvement for total customer satisfaction in customer-driven project management. Besides understanding these key concepts, putting customer-driven project management into action requires a focus, teamwork, and an improvement methodology. These three elements form the basic pillars of the application of customer-driven project management. They provide the purpose, structure, and system for customer-driven project management. The focus is the specific mission of the project. This focus should be customer-driven, but it requires consensus among the customer-driven project team. All customer-driven teams involved in the project require teamwork. The improvement methodology provides the process for performance and improvement of the project.

Focus

Focus is the first step in any customer-driven project management endeavor. There must be a constancy of purpose. This focus provides a common reason for joint action. The focus of the customer and supplier organizations starts with their respective visions. The visions provide a view of the future for each organization. Although the visions normally will be geared to each individual organization, the visions provide the initial understanding of each organization's views of the future. The visions of the customer and supplier organizations involved in customer-driven project management must be at least aligned to some common long-range outlook. From this understanding of a collective affiliation, the customer and supplier can build a mutually beneficial relationship using customer-driven project management. This initial relationship building provides the foundation for joint customer-driven project management efforts. Customer-driven project management starts with an agreement between a customer and a supplier for a specific project. The mission statement for the project documents the project agreement. The mission becomes the focus of all project efforts. The customer and supplier steering team usually develops the draft mission.

As stated above, before the customer and supplier can agree on a specific project, a customer and supplier relationship must be established to ensure that both customer and supplier understand each other enough to target projects with maximal mutual benefits for both parties. Developing this relationship requires both the customer and the supplier do the following:

1. Completely understand themselves and their environment
2. Thoroughly define their key processes

Once each organization knows the preceding information about its own organization, customers and suppliers can decide appropriate matches for developing long-term customer-driven project management efforts. The initial undertakings start with the formation of a customer and supplier steering team. This team's purpose is to find and select projects that offer win/win ventures for both the customer and the supplier. These projects should provide value for both customer and supplier over time. In addition, the customer and the supplier must be committed to providing the up-front investment required for total customer satisfaction.

The setting of a focus for projects is the result of continuous activity between customers and suppliers. It involves matching visions, missions, and values. It depends on the competitive market environment and the capabilities of the organizations. Total customer satisfaction is built on both organizations' potential. The continuous activity of the customer's organization to reach out to the supplier and the supplier reaching back to the customer to develop a mutual relationship producing valuable deliverables set the focus of customer-driven project management.

The customer's and the supplier's top-management teams guide the project focus by determining the vision and draft mission, approving recommendations, and generally monitoring results. They decide on the key projects and processes of the business, which sets priorities for the customer-driven project management organization. In the initial efforts or startup of customer-driven project management, this is particularly important. The implementation of the customer-driven management structure initiates the formation of customer-driven project management teams throughout the organization.

In customer-driven project management, supplying project management organizations do not simply respond to all demands for doing business in their market area. The successful organizations develop long-standing relationships with key customers and stay with them. CDPM with a common focus is the method for doing so.

The customer and supplier project steering team initiates the project mission statement. This initial draft mission statement must

- Be customer-driven.
- Include the purpose for establishing a customer-driven project lead team.
- Set a common direction.
- Establish expected results.
- Involve both the customer and the supplier.
- Provide a long-term orientation.

Focus setting is described in detail in Chap. 7.

Teamwork

Once a focus is established through a mission statement, the next step in customer-driven project management is the development of teamwork. The first team to be formed to perform the project is the customer-driven project management lead team. Next, other required customer-driven project management teams are established depending on the size, complexity, and life cycle of the project. For some projects, only one team may take the project from cradle to grave. In other projects, many teams will be developed and disbanded over the life cycle. No matter if there is one or many teams focusing on a common mission, the development of teamwork is critical in customer-driven project management.

Developing teamwork initially involves

- Developing a code of conduct.
- Reaching consensus on the mission.
- Defining roles and responsibilities.

The first step in developing teamwork is the establishment of a code of conduct. The code of conduct provides an opportunity for the team to form using a non-mission-related task. The code of conduct provides some general practices to be followed for all team activities. The code of conduct must be accomplished by consensus. It should consider the following:

- *Commitment.* The amount of commitment from both the customer and the supplier must be determined. This involves use of resources. Of particular concern should be people, money, and time.
- *"Owners" of team roles.* This code of conduct rule establishes the team leader, team facilitator, program manager, recorder, and team members.
- *Negotiation and decision-making process.* An agreement must be reached on how negotiations and decision making will be performed by the team.
- *Unity.* Specific conditions of cooperation must be defined. This is particularly important in the early stages of customer-driven project management until trust can be built between the customer and supplier.
- *Communications.* A communication process must be established.

After establishing a code of conduct, the next step in building teamwork is reaching consensus on the mission statement. Since the mission statement provides the focus for all customer-driven project management lead team activities, it must

- Have the same meaning for all team members.
- Be perceived as achievable by the team.

The customer-driven project management lead team clarifies and expands the original mission statement from the customer and supplier project steering team to include the following:

- Magnitude of the project
- Statement of boundaries of empowerment with specific responsibility, authority, and resources

After the team agrees on the mission statement and gets confirmation from the customer and supplier project steering team, the next step for the customer-driven project management lead team is to define roles and responsibilities. As part of the definition of roles and responsibilities, each team member must know the following about himself or herself and all other team members:

- What they can contribute to the team
- Expected outcomes
- Amount of empowerment
- Measures of performance

The first step, identifying each team member's potential contribution, should involve a team session in which each team member specifies what he or she can contribute to help the team accomplish its mission. This can be done by simply having each team member, one at a time, list one item on a chart. Everyone repeats the process until all team members list all their contributions. After listing all team members' contributions, the team discusses the contributions to clarify and understand each other's roles and responsibilities. The roles and responsibilities list is updated as necessary throughout the life of the project. Customer-driven project teams are described in Chap. 6, and teamwork is detailed in Chap. 7.

Customer-Driven Project Management Improvement Methodology

With a purpose and structure established, the third critical pillar in using customer-driven project management is the customer-driven project management improvement methodology. Figure 4.1 shows the complete customer-driven project management improvement methodology with the eight phases for improvement, the five phases of project management, and the expected outcomes of the eight major phases. The customer-driven project management improvement methodology blends the continuous pursuit of total customer satisfaction with the control of a disciplined, structured management system for successful project completion. This allows for progressive improvements while still managing the project. This new view for both continuous improvement and project management is facilitated by the use of a disciplined, structured customer-driven project management methodology, as described below.

The eight phases of the customer-driven project management improvement methodology are the general guidelines for the customer-driven team to proceed through all the processes of customer-

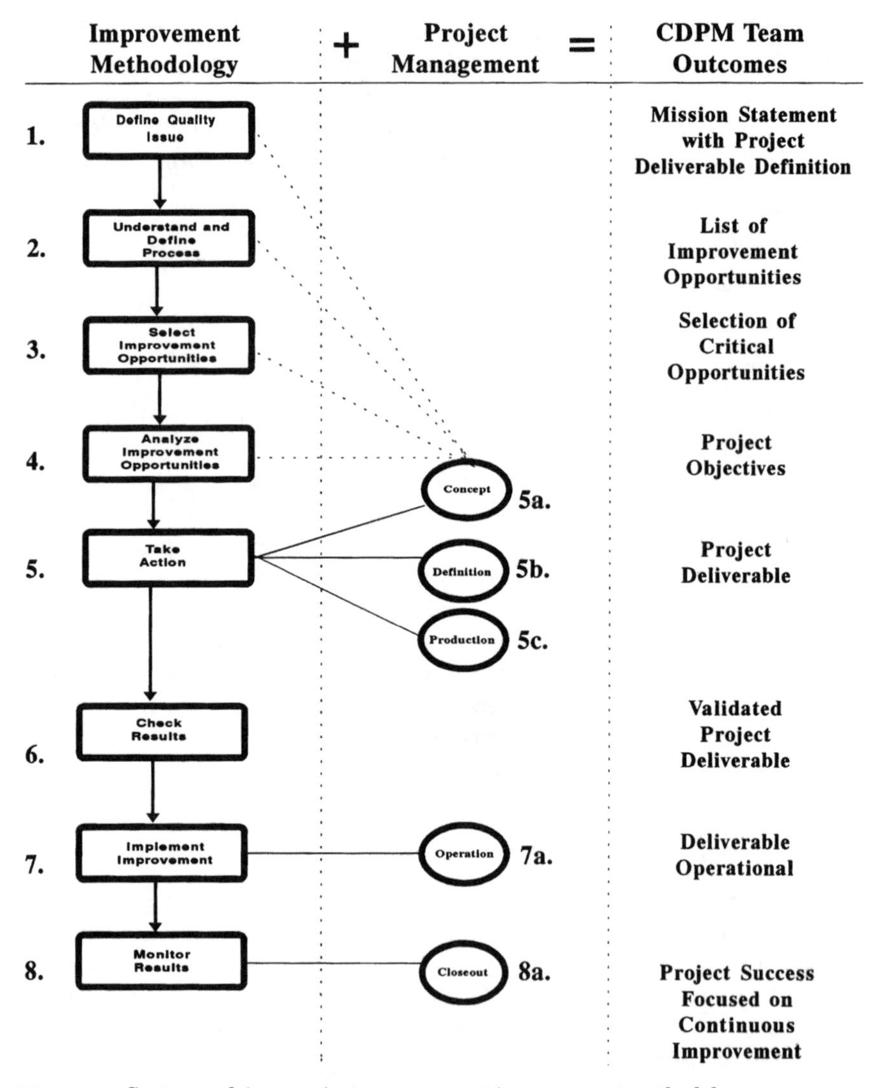

Figure 4.1 Customer-driven project management improvement methodology process.

driven project management. The eight phases go from defining the issues to monitoring the quality of outcomes. The customer-driven project management improvement methodology finds, selects, analyzes, and implements project and process improvement opportunities concurrently and continuously.

The eight phases of the customer-driven project management improvement methodology are

Phase 1: *Define the quality issue.* During this phase, the customer's needs and expectations are developed. This phase defines the "what" of the project. The focus and priorities of the project are determined in this phase. The mission of the project is clarified, and specific customer requirements are detailed enough to establish priorities. In addition, critical processes are identified that contribute to the success of the project.

Phase 2: Understand and define the process. This phase thoroughly examines all the specific processes required to totally satisfy the customer. During this phase, the "how" of the project is determined. This phase establishes performance outputs and opportunities for improvement. It is important to understand all the processes required for the project. This includes current and required process performance. This phase involves determining input requirements with specific suppliers and output expectations with all customers and includes all internal and external suppliers and customers. It also may include establishing performance expectations based on benchmarking information. In addition, any potential problems, shortfalls, and improvement opportunities are identified at this time.

Phase 3: Select improvement opportunities. The third phase involves the listing of all improvement opportunities and the selection of high-priority opportunities. It is critical to focus on the processes with significant impact on total customer satisfaction.

Phase 4: Analyze the improvement opportunities. This phase uses disciplined analytical tools and techniques to target specific improvements in the selected processes. This phase involves knowing the exact performance of all processes, the variations in the processes, and the underlying or "root" causes of problems. This phase provides the objectives for the project and project processes.

Phase 5: Take action. This phase relates most closely to the traditional project management phases of concept, definition, and production. During this phase, actions are taken to explore alternatives, state the project concept, define the project deliverable, demonstrate and validate the project, and develop and produce the project deliverable. During this phase, the team actually prepares plans and organizes, staffs, controls, and coordinates the project deliverable. This includes preparing the scope of work, project work breakdown structure, schedule, and budget. The customer-driven project lead team ensures that the necessary human resources and financial, contracting, and other support systems are in place. Ultimately, this is the phase in which the necessary deliverable is produced and deployed to the customer.

Phase 6: Check results. During this phase, the improved processes are measured after the improvement action is taken. The project and project process outcomes based on indicators of customer satisfaction are judged by the customer-driven management project lead team.

Phase 7: Implement the improvement. After validating the project and project process, the team ensures that the new process steps or project deliverables become a permanent part of the appropriate systems. This usually involves supporting the deliverable. This process includes all or some of the following: customer response, training, documentation of administrative procedures, maintenance and operations, people and/or organizational development, maintenance, supply, transportation, facilities, and computer services.

Phase 8: Monitor results. The customer-driven team stays in business over as many cycles as needed to monitor performance, solve problems, and continuously improve the deliverable. This is a long-term function. It requires the use of customer-driven project management improvement methodologies for understanding, analyzing, and accomplishing continuous improvement. It involves benchmarking and metrics. It ensures that the customer and supplier do not have to "reinvent the wheel" every time an improvement is needed. This phase assumes a longer-term relationship between customers and project suppliers. Eventually, there may be a time for project closeout.

The eight phases of the customer-driven project management improvement methodology correspond with the "plan, do, check, and act" (PDCA), or Deming, cycle. This is shown in Fig. 4.2. The "plan" has the following phases: (1) define the quality issue, (2) understand and define the process, (3) select improvement opportunities, and (4) analyze improvement opportunities. The "do" is the take-action phase. "Check" involves checking results. The "act" involves implementing the improved process and monitoring results. The planning stage ensures that the right project is selected and analyzed. The do stage is where the project is conceived, defined, produced, and delivered. The check stage ensures that the deliverable meets the customer's needs and expectations. The act stage institutionalizes and continuously measures for the desired outcome.

Phase 1: Define the quality issue

The process of defining the quality issue is shown in Fig. 4.3. The input is a draft mission statement from the customer and supplier project steering team. This input is used in the process to (1) establish the project mission, (2) form the customer-driven project lead team, and (3) define the project deliverable. The output of this

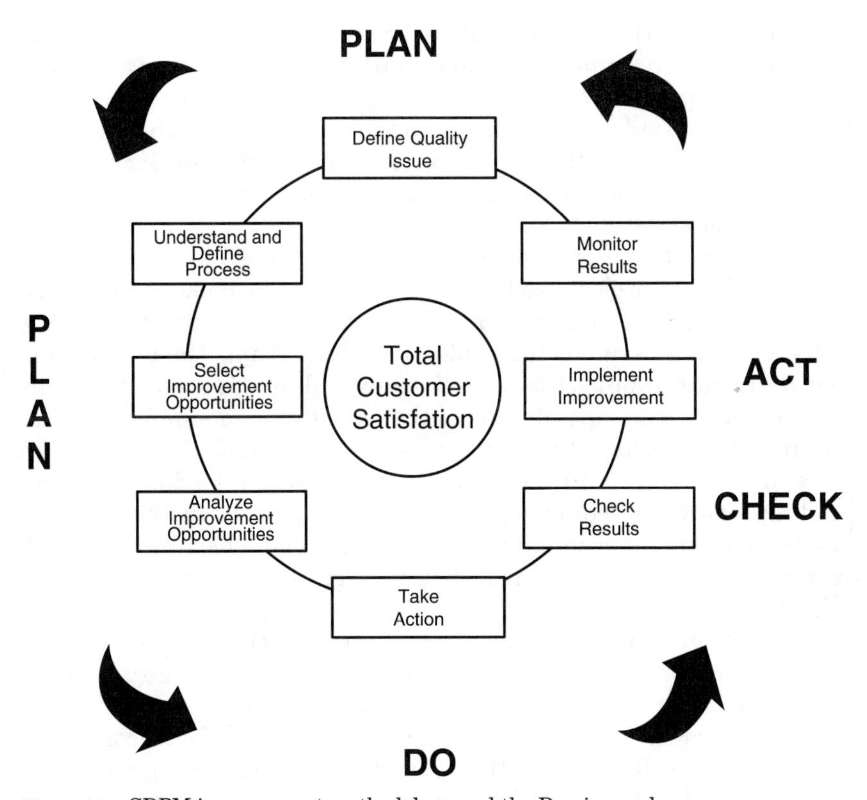

Figure 4.2 CDPM improvement methodology and the Deming cycle.

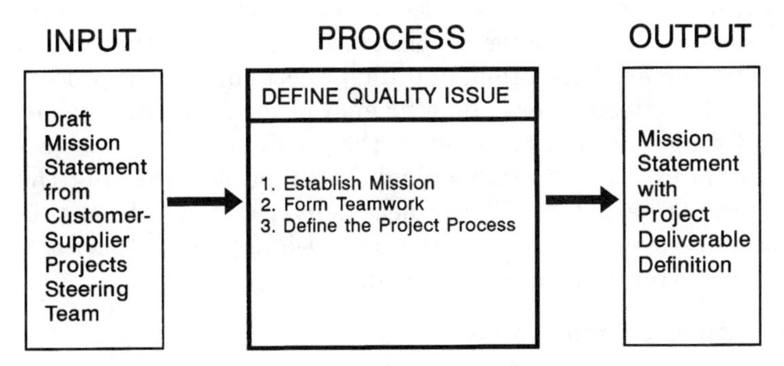

Figure 4.3 CDPM improvement methodology phase 1 process.

process is a project mission statement with a specific project deliverable definition.

The focus of phase 1 is a statement of the quality issue or a problem statement from the customer-driven project lead team, which forms the basis of the mission of the project. This statement may be changed later in the process, but it serves as the focus for the all the customer-driven project management activities. The statement describes the project process in narrative terms and indicates why the project process under review is important from the standpoint of the customer. It should define the desired state of the project process, that is, a view of how it is expected to perform when all customer needs are being met.

In addition to the mission statement, building teamwork in the customer-driven project lead team is essential in phase 1. The customer-driven project leader from the customer's organization or the customer's voice assumes the leadership of the team. The project supplier designates a project team facilitator and program manager. The team members are appointed from the customer and supplier organizations as necessary. In addition, building teamwork involves an agreement on roles and responsibilities, as stated earlier. Further, building teamwork requires training. It is important at this stage to ensure that all team members start to develop the interpersonal, team dynamic, and customer-driven management skills necessary for success.

In this initial phase, the customer-driven project leader coordinates communication among all other team members, including the project team facilitator and program manager. This new arrangement between customer and supplier is facilitated by today's telecommunications systems. In essence, the customer and supplier agree to "co-produce" the deliverable, the deliverable being the output or outputs of the project process focusing on total customer satisfaction. The deliverable can be a product, a service, or any combination of products and services. The customer as project leader drives the project toward total customer satisfaction, and the project supplier as process owner ensures the effectiveness and efficiency of the final deliverable.

Part of defining the quality issue in the customer-driven project management improvement methodology encompasses creating the project deliverable definition. The project deliverable definition provides the baseline specifications for the project and process to accomplish the project. This involves the following:

- Define the top-level project process.
- Determine the boundaries of the project process.
- Specify the output(s) of the project process (deliverable).
- Identify a customer or customers other than the customer-driver.
- List other customer needs and expectations.
- Identify requirements for input into the project process.
- Determine suppliers of inputs.

Table 4.1

Top-level process	Action	Descriptor
Serve food	Serve	Food
Build car	Build	Car
Train people	Train	People
Design hybrid	Design	Hybrid
Construct building	Construct	Building
Program computer	Program	Computer
Balance accounts	Balance	Accounts
Sell real estate	Sell	Real estate

- Determine the customer's measure of project process performance.
- Establish "ownership" of the project process by the customer-driven project management lead team.

1. The team must define the top-level project process. This is an overall description of the completed project. The definition of the top-level project process contains a statement of the process action and process descriptor. Table 4.1 shows examples of top-level project processes.

This top-level process needs to be further understood in relation to the satisfaction of customer needs and expectations. For instance, the process of serving food for an organization feeding the homeless centers on satisfying the basic hunger needs of the customers. For an upscale restaurant, the process of serving food involves meeting customer expectations for a dining experience. Although the top-level process is the same, the actual performance of the process is different.

2. The boundaries of the project process must be determined. It is important to specify the boundaries of the process. This involves deciding the start and finish of process diagrams for the project process. For instance, suppose a customer-driven team in a financial institution is assigned a project to improve the loan process. For the team process diagrams, the team must decide where the process starts and finishes. Does the process start with the customer inquiry, or does it start at the receipt of the loan papers? Does the process end with loan approval, or does it end when the loan is paid off? Depending on the decision of the team, the process diagram will be different. In addition, the team may inevitably find itself in other related processes, such as interest charge determinations, marketing, or promotion. The team must decide if it will look into the other related process or processes. If the team goes too far with this, it could find itself with too large a process. However, if the team does not do a thorough, detailed review, it may miss parts of the process with the most to offer in continuous improvement.

3. The team specifies the output of the project process. This is the project deliverable. Today, most project deliverables contain both

products and services. At this time, the major project deliverables should be identified.

4. All customers must be identified. In customer-driven project management, the major customer is usually leading the project. However, the team may have some difficulty defining the other customer(s) of the process. For instance, in our bank example, the loan process involves making both mortgage and commercial loans. Who is the customer? Typically, the end user of the service or product is the ultimate customer. In this case, the customer is the receiver of the loan. In addition to the end user, there may be many internal and external customers between the project supplier and the end user. Everyone affected by the process must be identified as potential customers. In customer-driven project management, frequently, the customer-driver of the project may not be the ultimate end user of the deliverable. The customer-driver may be a representative of the ultimate end user. For example, in the defense industry, the customer-driver of the project may be the acquisition organization, and the end user may be a soldier, sailor, airman, or marine. In the construction industry, the customer-driver may be the developer, and the end user may be the building owner or buyer.

5. A list of all customer needs and expectations is compiled. Customer expectations may be solicited through surveys, interviews, current data, focus groups, brainstorming, quality function deployment, and other sources. Customer expectations are the specific requirements to satisfy the customer. This is what the customer wants, desires, needs, and expects from the deliverable. This is where the customer and supplier further develop the relationship that started as part of a customer and supplier project steering team. During this phase, a list of customer satisfaction issues is developed by the customer and supplier. From these early communication activities, the initial customer expectations list was developed. This embryonic list sets the stage for continuous improvement by the customer-driven project lead team. Ultimately, the output of this phase is a list of the customer's expectations for the deliverable. This initial list forms the basis of defining the deliverable.

6. The team identifies input requirements for the project process. This list includes the inputs plus all the specific requirements for each input.

7. The suppliers of the inputs are determined. From this list, the customer-driven project lead team identifies possible supplier partnerships for further refinement of requirements.

8. The team determines the customer satisfaction measures. These measures are stated in the form of a metric. These meaningful measures form the basis for project performance evaluation.

9. "Ownership" of the project process is established by the customer-driven project management lead team. This involves commitment of resources. In addition, the customer-driven lead team sets boundaries of empowerment during this stage. The boundaries of empowerment include specific responsibilities, authority, and resources.

The result of phase 1 is a clear definition of the quality issue. This definition of the quality issue must be documented. The documentation contains the following:

- Mission statement
- Customer-driven project lead team membership
- Roles and responsibilities matrix
- Project process top-level diagram
- Start and finish of project process
- The project process deliverable(s)
- Customer(s)
- List of customer needs and expectations
- List of inputs for the project process
- List of suppliers of the inputs
- Measures of customer satisfaction

Phase 2: Understand and define the process

Figure 4.4 shows the steps in understanding and defining the processes. The input is the top-level project process. The process in phase 2 involves (1) deciding critical processes to perform the top-level project process, (2) forming or identifying additional customer-driven teams to support the customer-driven project lead team, (3) understanding these critical processes, and (4) defining the performance of these critical processes. The output of this process is a list of improvement opportunities.

The first step in phase 2 involves deciding which process or processes are critical. A critical process is one of the top six to eight processes whose outputs have the most impact on total customer satisfaction. The customer-driven lead team determines that special attention to a process would make major improvements for the customer, supplier, or deliverable(s). In addition to the critical processes, the customer-driven lead team initiates action to determine all the processes needed to perform the project.

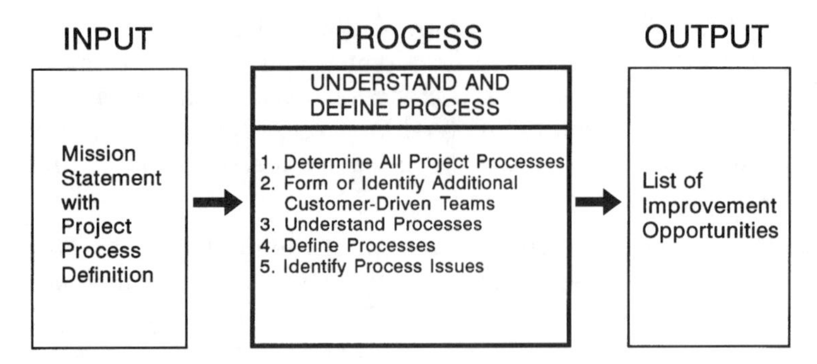

Figure 4.4 CDPM improvement methodology phase 2 process.

In the second step, the customer-driven project lead team forms or identifies additional customer-driven teams. Each of the critical processes requires an additional customer-driven team. The customer-driven project leader for these teams must be a member of the customer-driven lead team. This is the link to the critical-process level. In addition, these customer-driven teams form or identify other customer-driven teams as necessary. This hierarchy of customer-driven teams is shown in Fig. 4.5. In this hierarchy, the customer-driven project leaders are the links between and across each level. A customer-driven project leader is the leader of one team on one level and a member of the team on the next-highest process level.

The third step is gaining a thorough understanding of the specific impact of each process on total customer satisfaction. This involves a complete review of all processes required to perform the project

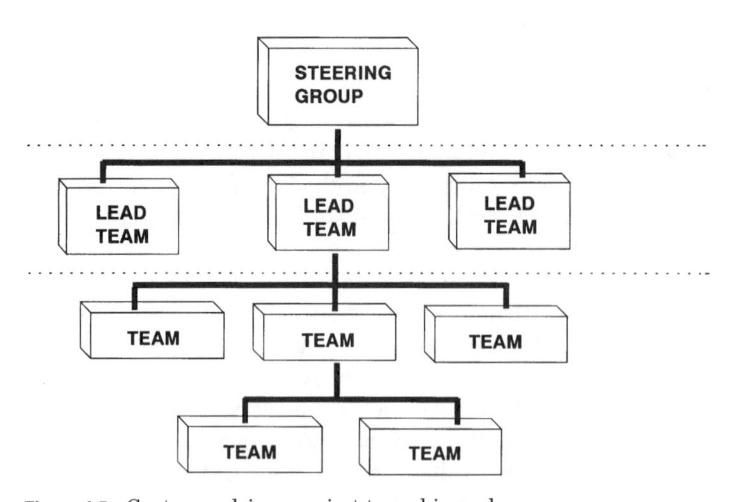

Figure 4.5 Customer-driven project team hierarchy.

process. The customer-driven lead team focuses on the critical processes, and other customer-driven teams are identified and formed for other processes as necessary. Understanding the process involves ensuring that the right processes have the right output with the right supplier requirements satisfying the right customer. This involves the following for each process:

- Benchmarking
- Diagraming the processes at the top and top-down levels
- Specifying the customer or customers
- Listing all customer needs and expectations
- Determining whether the process is meeting customer expectations
- Discovering who "owns" and influences each process
- Determining all the inputs and outputs of each process
- Understanding the relationship between inputs and outputs
- Listing the suppliers of the inputs
- Determining whether the suppliers are meeting the requirements
- Determining how to measure the process
- Measuring the process to determine how it is performing
- Understanding the value of the process to the deliverable
- Determining whether the process can be eliminated
- Listing the problems, issues, and opportunities

Benchmarking involves measuring the process against the recognized best performers of the process. The purpose is to provide a target for ideal performance.

Diagraming the processes at the top and top-down levels provides a broad picture of the process content. The top and top-down processes provide an indication of activities requiring further review. Such an approach targets issues that are the most significant to total customer satisfaction. It provides a full view of the processes before starting a detailed, process-by-process analysis.

Next, in specifying the customer or customers of each of the processes, the focus is on total customer satisfaction of all the customers of each process. It is also important to identify the customers as customer-driven project leaders to form teams to work out the details of all required processes. These customer-driven project leaders provide the guidance to formulate a list of all customer needs and expectations. From this list, the customer-driven team determines whether the process is meeting customer expectations.

Another important aspect of understanding the process is discovering who "owns" and influences each process. These people become the team members. They know the specific details of their process. As part of a customer-driven project team, they help determine all the inputs and outputs of each process. From this list of inputs and outputs, the team examines the relationships between inputs and outputs. The impact of each input on specific outputs is reviewed and discussed until the team has a complete understanding. This is particularly important when performing the process. Unless the correct inputs are received for the process, the process cannot be performed successfully. For instance, a real estate transaction requires an accurate appraisal as an input. Without an accurate appraisal, the real estate agent cannot satisfy the customer. To ensure that the inputs are correct, the team needs a list of suppliers of the particular inputs along with the specific requirements for the inputs. This information is used to determine whether the suppliers are meeting the requirements. It also can be used to develop supplier partnerships.

From the inputs, the team focuses its attention on the outputs of the process. The team needs to determine if the outputs are meeting customer expectations. To do so, the process must be measured against specific customer satisfiers. This requires the team to specify performance measures. The team determines how to measure the process, which requires a systematic plan for collecting, sorting, and presenting information about the process. The measures may be as simple as the difference between the input and output of the process, since every process has an input and output. Performance measures provide a baseline for process improvement. These performance measures provide the team with a process performance level before making any changes. Later such measures will be used as the basis for comparison after changes are instituted.

Performance measures of administrative processes take the form of error rates, turn-around times, accuracy, delivery, cost, etc. They can represent the performance of the entire process from input to output, or they can be made at different points within the process. There are a lot of interesting ways to look at process performance, but the customer's viewpoint should always be given top priority.

All process performance must be supported by measurement of data. Since this data collection consumes resources, it is advisable to target the data collection to measurements having the most impact on total customer satisfaction.

Another key to process understanding is to recognize the value of the process to the deliverable. Each process must have a value to completing the project. If the process has no value or added value, the team determines whether it can be eliminated. If the process is required, the next step is to list the problems, issues, and opportunities for improvement of that process. This is usually a list from all the previous steps plus any additional issues the team brainstorms.

The fourth step defines the process. This involves determining the target performance of the process and specifying metrics. The target performance is based on an understanding of the process and an interpretation of the process requirements for the project. During this step, the team brainstorms the ideal process. From this information, the team specifies metrics for the process. Metrics are the meaningful measures of total customer satisfaction targeting action. The metric must focus on meeting the customer's expectations and the organization's objectives. Metrics go beyond measurement. Metrics define the process and provide motivation to continuously improve the process.

The outcome of phase 2 is a distinct understanding and definition of all the processes, especially the critical processes, involved in the overall project. Again, the team needs to document this phase. The process understanding and definition documentation contains the following:

- Process diagrams
- Input/output analysis
- Supplier/customer analysis
- Process performance measure(s)
- Problems, issues, and opportunities
- Process targets
- Process metrics

Phase 3: Select improvement opportunities

The process of selecting improvement opportunities is shown in Fig. 4.6. The input is the list of improvement opportunities. The process in phase 3 involves (1) specifying selection criteria, (2) determining a selection method, and (3) making selections. The output of this process is improvement opportunities for analysis.

During phase 3, the customer-driven project management lead team selects an agenda of high-priority improvement opportunities. This agenda is based on the information collected in phase 2. All the improvement opportunities identified in phase 2 are listed on the agenda. The selection phase involves choosing high-priority processes and/or problems for further analysis.

The first step in the selection phase is specifying the selection criteria, which are determined by the project mission. Specifically, two kinds of opportunities are normally addressed: (1) actions that will

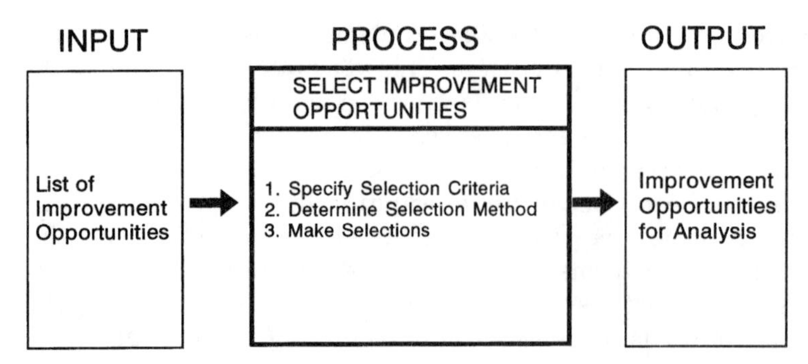

Figure 4.6 CDPM improvement methodology phase 3 process.

increase customer satisfaction through internal project supplier process improvements and (2) project deliverable opportunities which are needed to satisfy the customer.

In this step, the team decides to focus on a narrowly defined point in the process. The team decides where major improvement opportunities are likely to occur. It is in this step that the team decides on the improvement opportunities for processes and project deliverables, especially if phase 2 has revealed a new ideal process that can take advantage of new approaches. Remember, the focus of quality improvement is prevention and design in quality, rather than detection and correction. Once this is instituted, opportunities for fundamentally increasing process capacity are examined. These opportunities could include new information systems, new facility development, new training programs, or new complex communication systems. Many improvement opportunities involve design and production of complex systems and integration of components, and thus the application of project management is important for quality improvement and project success.

The criteria for selection depend on the project, but they should consider cost, resources, importance, time, effect, risk, integration with the organization's objectives, and authority. Once the criteria are determined, the team needs to focus on consensus using appropriate selection methods. The team decides the selection methods so as to target the choices. The selection method the team uses could be one or more of the following: voting, selection matrix, or selection grid.

After the team limits the choices to a manageable number, it seeks consensus on improvement opportunities for further analysis. Consensus decision making involves all team members having the opportunity to discuss and understand each of the improvement opportunities. From this understanding, each team member reaches a decision that he or she can support without compromise. This does not mean that everyone on the team totally prefers the improvement opportunities, but it does mean that everyone on the team agrees to support the team's decision.

The outcome of phase 3 is a list of improvement opportunities for further analysis. The documentation for the selection phase contains the following:

- List of improvement opportunities
- Selection criteria
- Selection methods
- List of improvement opportunities for further analysis

Phase 4: Analyze the improvement opportunities

The process of analyzing improvement opportunities is shown in Fig. 4.7. The input is the selected improvement opportunities. These selected improvement opportunities are further analyzed through the following: (1) process analysis, (2) cause-and-effect analysis, and (3) data statistical analysis. The output of this process is project(s) objectives.

This diagnostic phase of the customer-driven project management improvement methodology requires a thorough use of analytical tools to focus on the root causes of problems, process variations, and customer dissatisfaction. The team asks "why" and then asks "why" again and again until it is satisfied that the underlying causes are identified. This phase accomplishes its diagnostics through process analysis, cause-and-effect analysis, and data statistical analysis. Process analysis assesses the content of the process. Cause-and-effect analysis determines the underlying or root causes of problems. Data

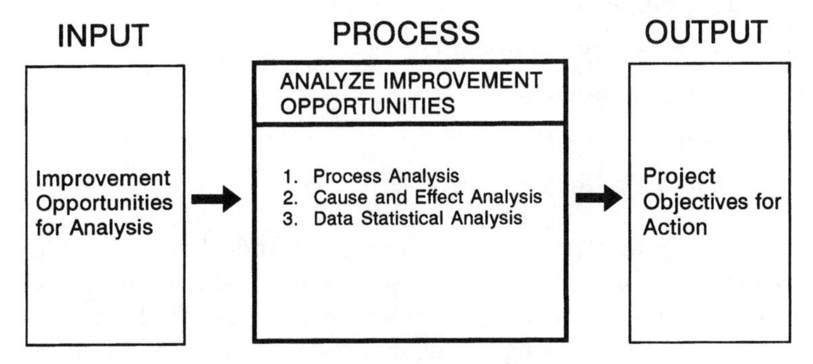

Figure 4.7 CDPM improvement methodology phase 4 process.

statistical analysis evaluates process performance. During this phase, the customer-driven project team discovers the issues that prevent the project processes from performing in an optimal way.

The analysis phase of the customer-driven project management improvement methodology is one of the essential phases of the process because it ensures that the team does not go from symptoms to solutions without looking at underlying causes. It is underlying causes that represent the agenda for process improvement and project development.

The first step in analyzing the improvement opportunities is process analysis. Process analysis involves a thorough review of the processes from a detailed process diagram that includes cost and time elements. Specifically, the analysis targets high-time and high-cost processes, activities, and tasks. Process analysis looks for immediate ways to improve the process by focusing on eliminating non-value-added processes and simplifying necessary processes at little or no cost. The initial focus is on eliminating processes, tasks, and activities that do not contribute worth to the project. Some examples of non-value-added activities are layers of signature approvals, unnecessary paperwork, some inspection, inefficient status reporting, and ineffective meetings. Next, the process is examined with an eye to removing complexity. Some examples of complexity include unnecessary loops, duplicate efforts, too many steps, a lot of rework, and many people involved in an activity. The steps to process analysis are as follows:

1. Detail diagram the selected process(es).
2. Look for ways to eliminate non-value-added steps.
3. Eliminate or reduce high-time and high-cost steps.
4. Look for ways to simplify.
5. Remove any unnecessary loops.
6. Decrease any complexity.
7. Get rid of unnecessary paperwork.
8. Analyze frequency changes.
9. Purge or lessen waste.
10. Look for better ways to do the process.

The second step in analysis is determining the underlying causes of any problem areas. Using cause-and-effect analysis, this step gets to the bottom of problems and issues uncovered in phase 3. This root-cause analysis involves choosing a particular symptom or problem for detailed analysis. The team asks, "Why does that symptom occur in the process, and what are its underlying causes?" Causes are categorized in a cause-and-effect diagram. Within the specified categories, the team members brainstorm a wide variety of ideas about the causes of the effect (symptom or problem). Because many internal and external customers and suppliers have insights about causes and effects, the team is encouraged to bring in outside people at this step. After the root causes are identified, priorities are set for each change to the process.

The third step in analysis is data statistical analysis. During data statistical analysis, the target is process performance. This step analyzes specific data to verify underlying causes discovered during the second step. It also may look into other areas affecting process performance. The foundation of all the data statistical analysis of the process is the performance metrics and measurements determined in phase 2 of the customer-driven project management improvement methodology. Data statistical analysis involves the following:

- Determining what data to find or collect
- Completing the data gathering
- Organizing the data
- Defining the expected outcomes or goals
- Analyzing the data
- Specifying the specific issue(s) for action
- Establishing priorities for action

To start a data statistical analysis, the team determines what data to find or collect. In many cases, data are readily available for the team to analyze. In instances where the team must collect data for analysis, the team answers the following questions before starting the data gathering:

- What is the purpose of the data collection?
- What data are needed?
- Where should the data be collected?
- What sampling scheme should be used?
- How much data must be collected?
- When should the data be collected?
- For how long should the data be collected?
- How will the data be collected and documented?
- Who will collect the data?

Once a data-collection plan is established, the data can be collected using checksheets, observation, questionnaires, consultation, documentation reviews, interviews, group discussion, tests, and/or work samples. When collecting data, it is important to keep in mind the principle of simplicity—the more complex the data collection, the greater the potential for inaccurate data.

After collecting the pertinent data, the next step is to organize the information in a meaningful format. This is called *stratification,* arranging the data into classes. Organizing data allows the observation of such things as the highest and lowest values, trends, central tendencies, patterns, most common values, special causes, common cause, and so forth.

Next, after organizing the data, the team presents the data in an expressive manner on charts. Depending on the data, the team displays the data on a bar, line, pie, or pareto chart, histogram, scatter diagram, control chart, or process capability index.

Analyzing the data involves determining specific issues for action. This is accomplished by asking probing questions about the displayed data:

- What is the expected goal?
- Is the goal being met?
- What are the vital few issues?
- Is there a trend?
- Did it pass or fail?
- What is the cause of variation?
- Is it a special or common cause?
- Is the process capable?

Finally, the data analysis provides a list of items for action. From this list, the team establishes the priorities for action. These action items become part of the project objectives along with the customer's needs and expectations.

The result of phase 4 is a list of project process objectives. The analysis process documentation includes the following:

- Detailed process diagrams
- A process analysis report
- Cause-and-effect analysis diagrams
- A data statistical analysis report
- List of project objectives for action

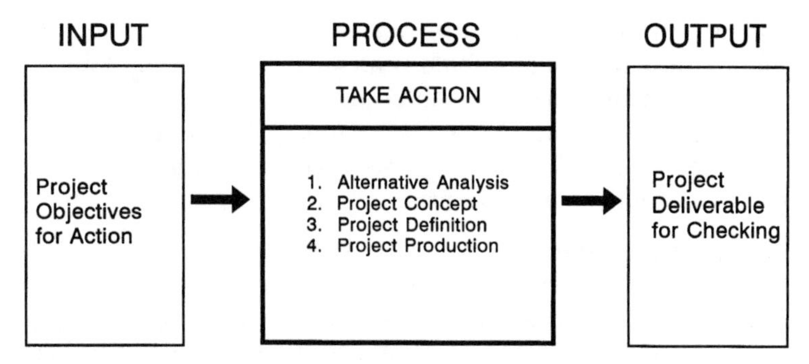

Figure 4.8 CDPM improvement methodology phase 5 process.

Phase 5: Take action

Figure 4.8 shows the "take action" process. The inputs for taking action are the project objectives. The project objectives start the following activities: (1) alternative analysis, (2) project concept, (3) project definition, and (4) project production. The output of this process is a project deliverable.

Phase 5 consists of the following:

- Alternative analysis
- The first three stages of project management:
  - Project concept
  - Project definition
  - Project production

This phase produces a project deliverable. The deliverable can be a product, service, system, program, report, component, analysis, procedure, improved process, corrective action, etc. The deliverable may be something as complex as a new computer system or as simple as installing a new procedure. Complex or simple, it involves project management consisting of concept, definition, and production to provide the deliverable. Although all these practices are essential for taking any action, the tools and techniques to accomplish the processes will vary depending on the project.

This phase transitions from quality management to project management. This is where the project deliverable takes shape. Prior to this phase, the customer-driven project management teams focused on customer requirements and suppliers processes. This is where the project deliverable is designed, produced, tested, and delivered.

Alternative analysis. The first step in the "take action" phase is alternative analysis. Alternative analysis consists of the generation, evaluation, and selection of a project alternative. Alternative generation involves the production of as many ideas as possible to accomplish the project objective(s). These ideas are then assessed against the team's criteria to determine the best solution. Finally, the team reaches consensus on one project alternative. This becomes the project deliverable. Alternative analysis involves the following steps:

1. *Generate possible alternatives.* This step involves producing a wide variety of project alternatives through brainstorming or force-field analysis. First, use creativity, innovation, and imagination to generate as many alternatives as possible to accomplish the project objective(s). In force-field analysis, accomplishment of the mission is the goal. The project objective is the restraining force. Brainstorm as many driving forces as possible to eliminate or reduce the restraining force. Second, process and clarify ideas. During this step, avoid criticism while eliminating duplicate ideas and combining similar ones. Ensure that all ideas are clearly related to the project mission. The driving forces in the force-field analysis are the possible alternatives.
2. *Evaluate possible alternatives.* Evaluation of project alternatives involves developing criteria and assessing each alternative against the criteria. This provides a priority listing of each alternative.
3. *Decide on a project alternative.* Once all the alternatives are evaluated, selection of the project is the next step. This should be accomplished by consensus of the customer-driven lead team. The selected project alternative can be further evaluated using force-field analysis to determine restraining forces for implementing the alternative. In this case, implementing the possible alternative becomes the goal. Then brainstorm and evaluate restraining and driving forces to implement the goal. Once the decision is made, the customer-driven lead team should seek support from the customer/supplier steering team. Upon approval, the selected alternative becomes the project deliverable.

Project management. Once a project deliverable is selected, the project must be produced. Again, the production of both a service and a product deliverable requires a structured, disciplined project management approach. Therefore, this step relates most closely to the traditional project management phases of concept, definition, and production. During this phase, besides producing the project deliverable, project processes are performed and improved for completing the project. In addition, problems are solved during this phase. Further, customer-driven project teams actually perform project management activities, including preparing the scope of work, project work breakdown structure, task list, budget, critical-path network, and gantt charts as necessary. Also, the customer-driven project lead team ensures that the essential human resource, financial, contracting, and support systems are in place. Ultimately, the output of this phase is the necessary deliverable accepted by the customer.

This step represents the transition from analysis to implementation, the point in the quality management process where the role of the customer-driven project lead team transforms from an assessment role to management responsibilities. During this phase, the team conceives, designs, and produces a distinct deliverable. The customer-driven project lead team guides the production of the project deliverable through the assistance of other customer-driven project teams as necessary. Customer-driven project teams are the keys to achieving total customer satisfaction with the project deliverable because they are in touch with the whole process of understanding and meeting the real needs of the customer.

The project management steps in the "take action" phase consist of the following:

1. Project concept
2. Project definition
3. Project production

It is critical that the project proceed through all steps, even if some of the activities seem forced because of the nature of the project. Figure 4.9 shows the major subprocesses for each of the processes of project concept, definition, and production. The importance of understanding and using all the project steps, regardless of the size of the project, is that they provide for the following:

1. *Reporting and evaluation.* Completion of each step represents a key milestone in the project itself, thus providing a good time in the project for reporting and assessing progress.
2. *Evaluation of project team members.* Team members are assessed at the end of each step, and necessary rearrangements and transfers of people to or from various tasks are made.
3. *Clear delineation of the work accomplished and the work to be started in the next step.* The end of each step is a good time to reacquaint prospective team members with what is expected from the project team during the balance of the project.
4. *Consideration of options.* Usually some key choices and commitments are made at the end of each step.
5. *Pricing and cost estimating.* The further into the project the team moves, the clearer issues of cost become. Accordingly, at the end of

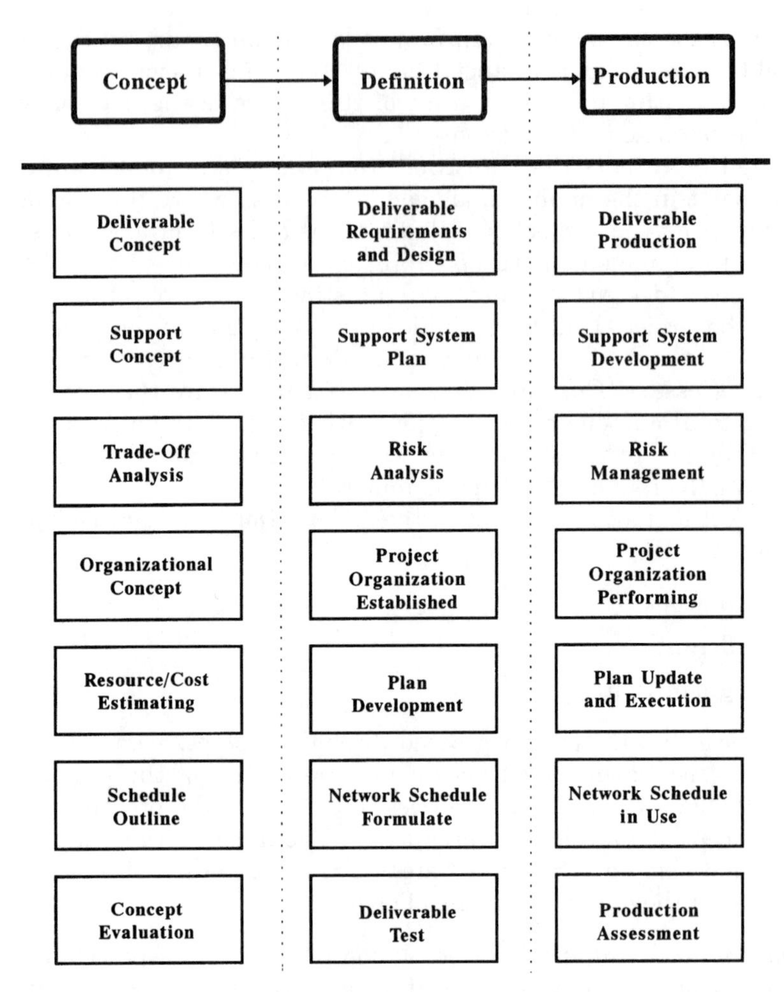

Figure 4.9 CDPM improvement methodology phase 5 project management processes.

each step it pays to update original estimates based on actual unit cost information from the preceding phase.

The concept step. The concept step determines the specific approach to accomplish the project. During the concept step, all possible methods for producing the deliverable and its support are identified and evaluated against benefits, costs, and risks. The concept step defines the project deliverable capable of fulfilling the project mission. During this step, the detailed project management analysis is accomplished, and this readies the project for detailed planning in the definition phase. As shown in Fig. 4.9, the concept step includes the following subprocesses:

1. *Determine the deliverable concept.* This includes deciding on a project deliverable capable of satisfying the customer within known constraints. The concept for the project deliverable depends on the magnitude of the customer's needs and expectations, the potential cost of the project deliverable, the availability of resources, the importance to the customer's and supplier's missions, the ability to provide the deliverable when it is needed, evaluation of the amount of risk, perception of the overall benefits, and the chance of success. Of particular importance in this stage is whether production of the deliverable is achievable. This is equally essential in both a manufactured and a service-oriented deliverable. For a manufactured deliverable, the producibility of the deliverable affects cost and schedule. In many cases, a producibility problem surfaces in later phases, resulting in a redesign that costs many times more than originally estimated which also slips the schedule. For a service deliverable, the producibility of the deliverable frequently affects the implementation. For example, to provide a training service requires producing a training schedule. One concept for producing the training schedule could be use of an automated method. If the automated training scheduling method is not usable because of technological or even social issues, the training service will not totally satisfy the customer. Therefore, the project deliverable concept considers performance and producibility within technology, cost, and resources. In addition, quality objectives are an important element of the deliverable concept, namely, to focus the team on customer satisfaction. The specific deliverable concept is detailed in the project proposal.

2. *Provide a support concept.* The support concept states the basic support considerations. These support considerations influence the project deliverable's design and operation. Some of the specific support considerations include the following:

- How reliable does the customer expect the deliverable to be?
- If the deliverable breaks, does it need to be fixed?
- If the deliverable needs to be fixed, how will it be fixed?
- Who will fix it?
- Will there be a need for any special test equipment to fix the deliverable?
- Are spares required to maintain availability of the deliverable?
- What will be the inventory for the deliverable, supplies, and/or spares?
- Will training be needed to use, operate, or fix the deliverable?

- Does the deliverable require any documentation, i.e., technical manuals, user manuals, job aids, training materials, procedures, policy statements, etc.?
- Are any computer resources required to support the deliverable, i.e., software, hardware, firmware, etc.?
- Does the deliverable require any packaging, handling, storage, and transportation?
- Are any facility construction or modifications necessary?
- How many people will it take to support the deliverable?

3. *Conduct a tradeoff analysis.* Each deliverable alternative with its many options is compared to determine the optimal balance of performance and cost at minimum risk.

4. *Develop a concept for organizational structure.* For customer-driven project management, this involves determining the types and numbers of customer-driven teams required during the life cycle of the project. Some projects involve only the customer-driven project lead team. Other projects include many customer-driven project teams for project management, quality improvement, and process performance. In many cases, the organizational structure changes over the life cycle of the project. The specific customer-driven project management team structure must be outlined in the project proposal.

5. *Perform resource/cost estimating.* During the concept step, the customer-driven project lead team prepares a proposed budget as part of the project proposal. This proposed budget is developed from detailed estimates submitted by customer-driven project teams and process owners. The resource/cost estimate is compiled in the project proposal.

6. *Produce a schedule outline.* Also during the concept phase, the customer-driven project lead team develops the proposed project schedule outline as part of the project proposal. This includes the proposed work breakdown structure and project schedule.

7. *Perform an evaluation of the deliverable concept.* During this subprocess, the customer-driven project lead team demonstrates the feasibility of the project deliverable concept and seeks approval from the customer-supplier project steering team. The deliverable is validated through simulation, pilot projects, prototypes, brass board, and/or bread board. Once proven, the customer-driven project lead team gains approval to develop the deliverable.

The outcome of the concept step is an approved project deliverable concept for full-scale development. This step answers the question, "What is the project deliverable?"

The definition step. This step defines the project in detail. It specifies the "how" and "how to" of the project. It details the deliverable, its producibility, its integration with other systems, and its use. The definition step specifies the performance, cost, and schedule requirements. In addition, the resource requirements of the project are specified in this step. At this step a reliable, producible, and supportable project deliverable is defined and developed to accomplish the mission. An essential element of this step involves the formulation of many plans, such as the quality assurance plan. Also, the detailed project management planning is accomplished, including network schedules and work breakdown structures. In addition, customer-driven project teams are established to perform and improve the project. Figure 4.9 shows the following subprocesses in the definition step.

1. *Perform deliverable design and development.* During this step, the customer-driven project teams formulates the project deliverable. This is the engineering of the project deliverable. It involves determining deliverable specifications, conducting trade studies, evaluating producibility, performing the design process for the deliverable, testing and production, conducting analyses of various design factors, selecting a feasible design, verifying the design and interfaces, establishing and maintaining configuration control, developing the deliverable, and testing the deliverable.

2. *Develop a support plan.* The support plan consists of an integrated plan of all the elements supporting the deliverable. The integrated support plan includes the following:

- Reliability
- Maintainability
- Maintenance
- People resources
- Supply support
- Test equipment
- Training
- Documentation
- Computer resources
- Packaging, handling, storage, and transportability
- Facilities

3. *Conduct a risk analysis.* The project deliverable design must be assessed constantly for risk. This risk analysis includes determining the impact of funding, the selected design, the testing requirements, producibility, facilities, supportability, and project management. Risk analysis involves using the customer-driven project improvement methodology targeting areas of uncertainty.

4. *Establish the project organization.* The customer-driven project management organization is formed during the definition step.

5. *Develop the project master plan.* The project master plan includes plans for project management, organizational structure, the project deliverable, support, test, production, configuration management, risk management, and quality assurance.

6. *Formulate the network schedule.* In this step, the work breakdown structure (WBS) is established. The WBS is an organizational chart for the project, a product-oriented family tree composed of finer and finer levels of definition of the work to be accomplished. The WBS normally establishes at least four levels of the work below the project deliverable: task, subtask, work package, and level of effort.

Definition includes detailed plans for cost, schedule, and quality, and it produces a critical-path plan by placing all elements of the WBS into sequence. A graphic representation of the plan is then made possible, and this representation is used in later stages to show changes and to monitor progress. Administrative details are also developed during definition, including policy and procedure documents, job descriptions, budgets, and tasking papers for team members. The project team is set up, using the WBS as the structure for team members who represent all the functional elements of the organization. The information developed during this subprocess becomes part of the master project plan.

7. *Test the deliverable.* The final subprocess before production is testing the deliverable. This step verifies the performance of the deliverable to actually meet the requirements of the mission.

The outcome of the definition step is a project deliverable certified for production. This step answers the following questions:

- What are the specific processes, tasks, and activities?
- Who is going to perform processes, tasks, and activities?
- How long will the project take to complete?
- When will the project be accomplished?
- How much will the project cost?

The transition from definition to production requires careful attention by the customer-driven project lead team. Typically, the transition from definition to production is not a distinct process, but it is a continuous cycle of activities including the concept, design, test, and production or plan, do, check, and act. To be successful, the customer-driven project lead team instills the discipline of the customer-driven project management improvement methodology.

The production step. The production step involves the actual production of the project deliverable and its support elements. During this step, the customer-driven project teams ready all detailed planning for the operational step by making sure all elements are integrated, interrelated, and interfaced. In addition to the project deliverable, the support elements are produced in this step. This is the process where the customer-driven project teams perform their tasks and report progress to the customer-driven project lead team. Also, the project deliverable is tested. During this step, project management consists of implementation and evaluation. The customer-driven lead team focuses on monitoring performance of the processes and taking corrective action as necessary. Figure 4.9 shows the following subprocesses in the production step.

1. *Produce the deliverable.* During this step, the project deliverable is made. For a product-oriented deliverable, this involves the manufacturing, assembling, and/or production of the product. It also includes developing the product support. For a service-oriented deliverable, this involves putting together all the elements needed to ensure that the service satisfies the customer. For instance, this might include establishing a specific customer's environment conducive to the service. Using the bank loan example, the production of the loan service could include a special modification to the bank loan department, making applying for a loan more inviting to the potential customer.
2. *Develop support.* The supports identified in the support plan are produced in this step. For instance, facilities are built or modified; documentation is prepared, validated, and verified; training is developed; computer resources are procured; and inventory is formulated.
3. *Perform risk management.* Risk management involves developing and implementing a project risk management plan. The focus of the plan is the continuous use of the customer-driven project management improvement methodology aimed at eliminating or reducing project uncertainty. This step targets a disciplined approach for analyzing, understanding, selecting, and taking action on specific project risks.
4. *Perform customer-driven project management.* During this step, customer-driven project teams are continuously performing and improving their processes to produce a deliverable that is totally satisfactory to the customer.
5. *Update and execute project plans.* This step involves implementation of the master project plans formulated in the definition step.
6. *Use network schedule.* During this step, the project is constantly monitored for time and cost using the network schedule. Based on the project progress, the customer-driven lead team takes action to:

- Revise the schedule times.
- Revise the network relationship.
- Revise the resource and/or cost estimate.
- Revise the technical objective.

7. *Assess production and test the deliverable.* The evaluation process in production includes assessment of the production process and testing of the project deliverable before delivery to the customer for operation. In customer-driven project management, customer-driven project teams continuously perform a self-assessment of their processes. In addition, the customer-driven project lead team institutes process metrics to monitor critical project processes. Further, the deliverable is tested prior to any delivery to the customer. This step reinforces this process with a periodic review by the customer/supplier project steering team.

The outcome of the production step is a project deliverable that is acceptable to the customer.

Phase 6: Check results

During the "check results" phase, the customer-driven lead team measures the deliverable and project processes against the customer's needs and expectations. Figure 4.10 shows the process of checking results. The input is the project deliverable. The process in phase 6 involves (1) testing the project deliverable's performance against customer expectations, (2) determining whether process goals are being met, and (3) taking corrective action as necessary to satisfy the customer and ensure process performance. The output of this process is a project deliverable ready for implementation.

As shown in Fig. 4.11, in phase 6 the team starts with a plan with the specific metrics to check the results for the deliverable and the project processes. Next, the team actually assesses the deliverable and the project processes. Are the results as expected? If not, the team uses the customer-driven project improvement methodology to

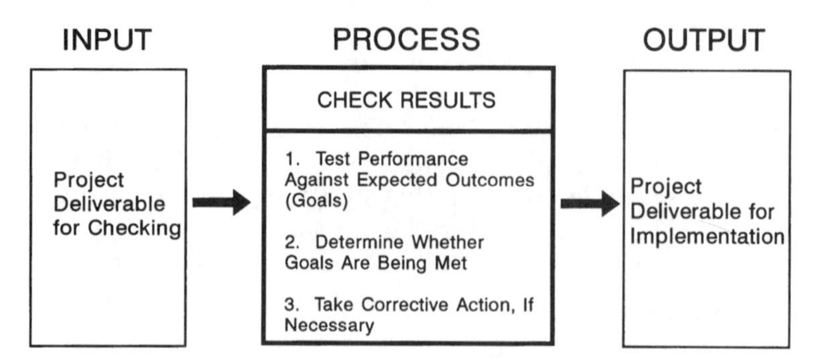

Figure 4.10 CDPM improvement methodology phase 6 process.

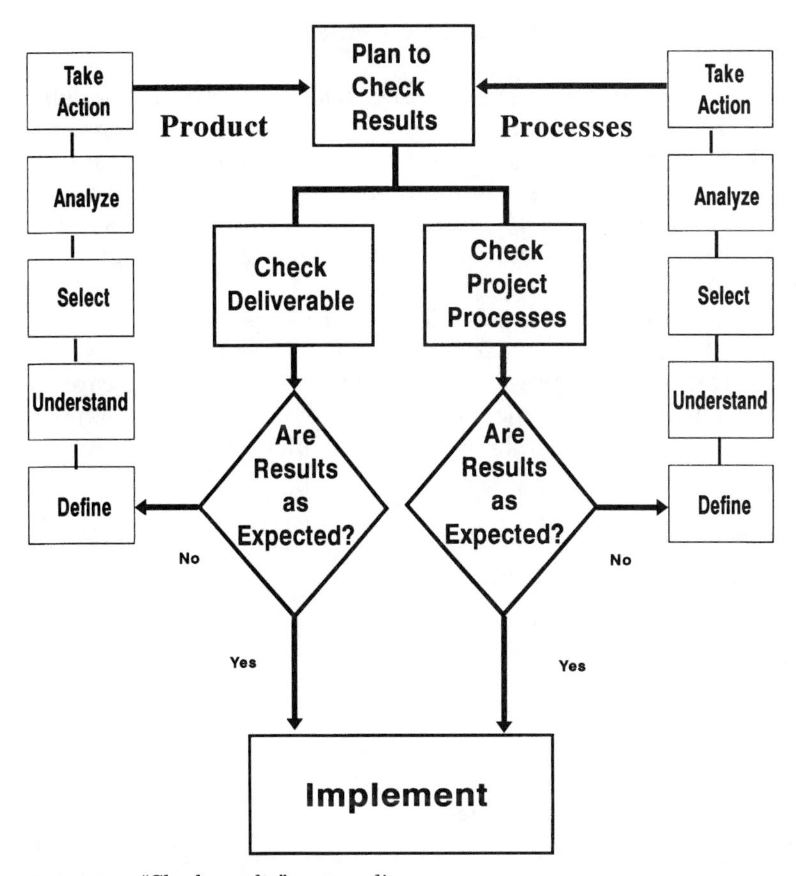

Figure 4.11 "Check results" process diagram.

correct the situation. If the results are as expected, the team continues on to implementation.

The first step in phase 6 of the customer-driven project management improvement methodology is testing of the project deliverable to meet customer expectations. During this step, the team monitors the performance of the deliverable for an appropriate length of time in a customer environment. The deliverable is thoroughly evaluated through a complete cycle of operation, use, and/or service to determine if the deliverable should be implemented as produced or if further improvements are necessary.

The second step is determining whether process goals are being met. The team continues to track the performance of the project processes using the same data-collection scheme developed during the analysis phase. Based on the results of the data statistical analysis, the teams persist with improvement activities to meet process performance expectations.

The third step is taking corrective action as necessary. If the deliverable or any project processes are not meeting expectations, the team makes improvements as necessary. To accomplish appropriate action, the team uses the customer-driven project management improvement methodology. The team reviews information in the take action, analysis, or even as far back as the understanding and defining the process phase of the customer-driven project management improvement methodology to solve the performance problem.

The outcome of phase 6 is a deliverable that is ready for implementation and that satisfies the customer with project processes capable of meeting performance goals. Again, the team needs to document this phase. The "check results" process documentation contains the following:

- Check results plan
- An assessment of the deliverable using metrics
- An evaluation of all project processes
- A list of issues for action
- Documentation of corrective action(s)
- A reevaluation

Phase 7: Implement the improvement

Phase 7 institutes the project deliverable. Figure 4.12 shows the process of implementing the improvement. The input is the project deliverable or improvement. The process in phase 7 involves (1) planning and gaining approval, (2) instituting the project deliverable and/or improvement, and (3) project operation and support. The out-

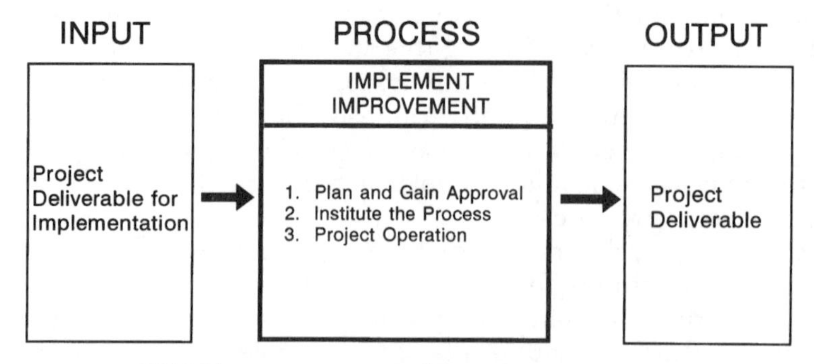

Figure 4.12 CDPM improvement methodology phase 7 process.

put of this process is a project deliverable that continually satisfies the customer.

The first step is planning and gaining approval for the improvement. This involves the following:

- Determining a plan for implementation
- Requesting approval for the plan

Since CDPM focuses on total customer satisfaction, implementation of the project deliverable requires that the improvement be accepted as a customer satisfier over an extended time period. This means that the project deliverable must be valued and internalized. The customer-driven project is not complete until the customer uses it as routine. Depending on the deliverable, this means a varying degree of operation and support. Figure 4.13 shows the operations and support considerations. For instance, in a process-improvement deliverable, the improved process will have to be used as a daily course of action. This may make standard documentation and training a necessity. In the case of a new system deliverable, operation and support are more extensive. The system implementation requires using and maintaining the system, operation and maintenance manuals, training facilities, organizational structure facilities, computer resources, transportation, and support services. At a minimum, implementing an improvement means developing a plan of action. The following should be considered for an implementation plan of action:

- The operational use of the deliverable by the customer
- The documentation needs to standardize and communicate the improvement
- The training required to use or maintain the improvement
- Any organizational structure changes needed to make the improvement work
- Facility additions or modifications to implement the improvement
- Maintenance support to repair, troubleshoot, and prevent discrepancies in the deliverable
- Supply support to keep the deliverable operational
- Transportation service to get the deliverable to the customer
- Computer resources to use and maintain the deliverable
- A continuous feedback system to monitor performance
- A schedule for implementation actions
- A budget to fund the improvement activities

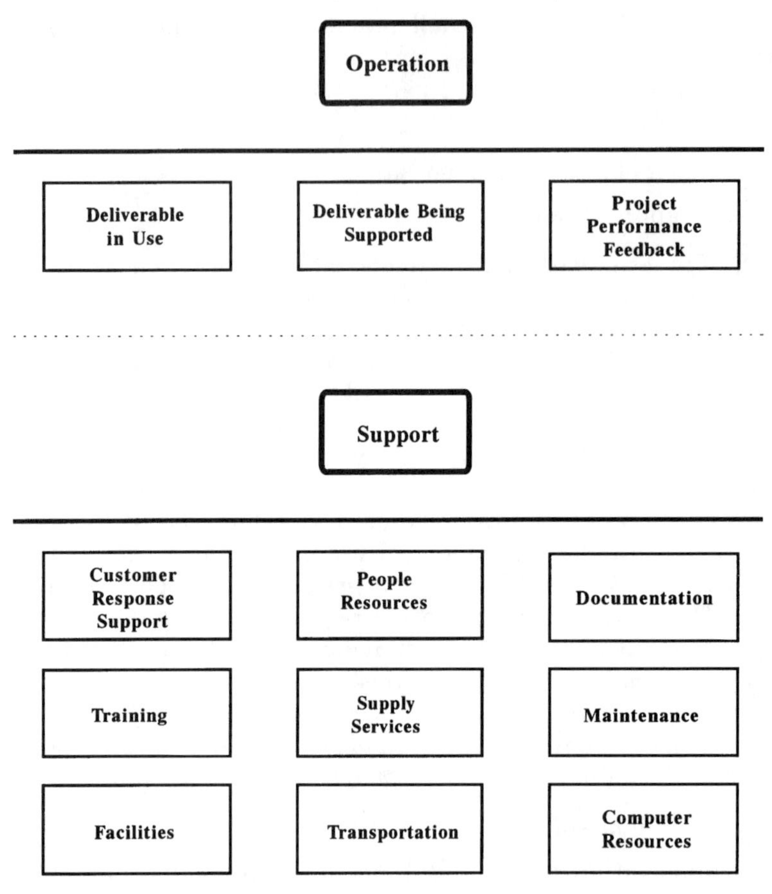

Figure 4.13 CDPM improvement methodology phase 7 project management processes.

Once the team has a plan for implementing the improvement, the team seeks plan approval. This requires gaining support and getting approval. The team gains support for implementation by continuously keeping stakeholders informed about the activities of the project. When the project is ready for implementation, the team coordinates the plan of action with all the stakeholders. The team also ensures that all the stakeholders are present during the approval proceedings. This is normally accomplished by a formal presentation to the approval authority.

Once the plan of action is approved, the team begins instituting the project deliverable or improvement. If the plan of action is complete, this step involves executing the plan to integrate the deliverable into existing organizational and/or customer systems. Of particular importance is the installation of a continuous feedback system of critical customer satisfaction indicators. Again, this should be a continuous improvement of the metrics and measurements developed in the analysis phase, improved in the "take action" and "check results" phases of the customer-driven project management methodology. Although the plan of action implements the improvement, the customer satisfaction indicators are the ultimate proof of implementation.

The third step in implementing an improvement is project operation and support. One major aspect of this step is the continual use of the deliverable by the customer for its intended purpose. Again, the feedback system is essential to ensure that the deliverable's performance continues to meet the customer's needs and expectations. In addition to performance, the other major contributor to total customer satisfaction is support. This means the effective and efficient use of the improvement focuses on the integration of operation and support considerations. The target of operation and support is optimal life-cycle cost. This requires aiming at specific customer needs and expectations. The goal is to provide just the right amount of support for required operations.

The outcome of phase 7 is a deliverable or an improvement. The documentation of this phase is the plan of action. The major topics for a plan of action include

- The specific operation and support actions
- A list of all steps, tasks, and activities
- Assignment of responsibility for each step
- Schedule to start and finish the implementation
- Schedule for start and completion of each step
- Budget

Phase 8: Monitor results for continuous improvement

Figure 4.14 shows the process of monitoring the results for continuous improvement. The input is the operational project deliverable or improvement. The process in phase 8 involves (1) evaluating project performance metrics, (2) assessing the project processes, and (3) seeking continuous improvement of the project deliverable and project processes. The output of this process is a successful project.

Customer-driven project management operates over the long term. Once the deliverable is operational, the customer-driven lead team and other customer-driven teams as appropriate continue operating as long as the mutual relationship provides joint benefits for the customer and the project supplier. The team continues to monitor project and process performance. The team takes action as necessary to

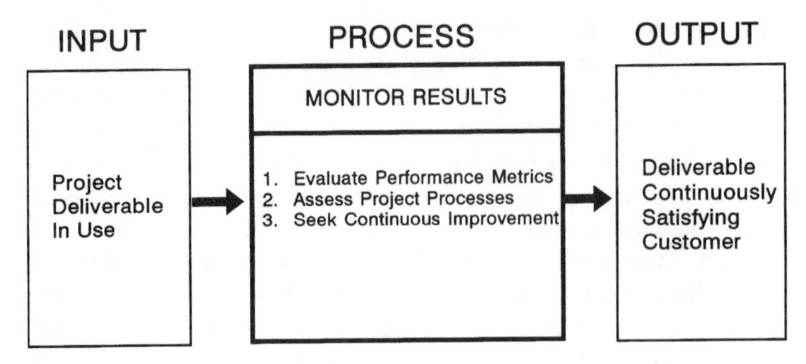

Figure 4.14 CDPM improvement methodology phase 8 process.

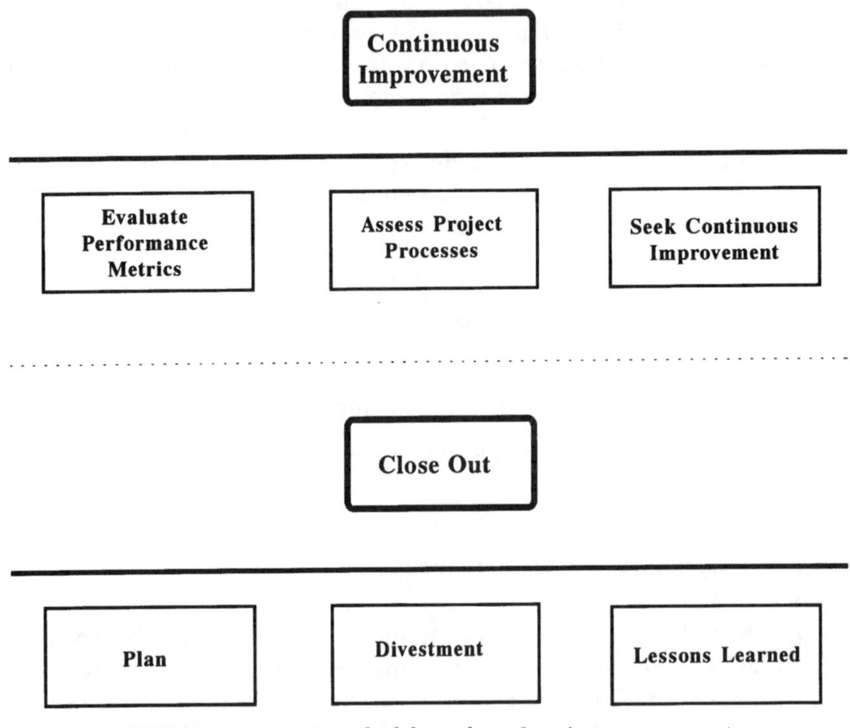

Figure 4.15 CDPM improvement methodology phase 8 project management processes.

maintain total customer satisfaction. The team performs continuous improvement activities, possibly including modification of the existing deliverable or even a completely new deliverable. The project is never finished until there is no deliverable needed by the customer. Figure 4.15 shows the subprocesses of continuous improvement.

The first task in monitoring results is evaluating performance metrics. To evaluate performance, the team asks, "Is the project deliverable meeting the customer's objectives?" If the answer is "yes," the team continues to monitor the changes. If, however, the answer is "no," the team will have to determine what is wrong. This requires the team to start again with phase 1 of the customer-driven project management improvement methodology and continue through all eight phases to make the necessary improvement.

The second task in monitoring results is assessing the project processes, which involves identifying the causes of variations in a process—common or special. Common causes are normal variation in an established process. Special causes are abnormal variation in the established process. Once the cause is classified, the appropriate approach to corrective action can be pursued. The corrective can be aimed at doing nothing, either eliminating or reducing as much as possible the variation in the process, or improving the process. The first option should always be considered when discovering a special variation of a process resulting from chance. This option is often overlooked, resulting in correction of a chance variation in a process, only increasing the variation. Second, eliminate special causes. Third, reduce common causes. Again, once a process variation is detected, the team uses the customer-driven project management improvement methodology to make the correction.

The third task of monitoring results is seeking continuous improvement. This requires cycling through the CDPM improvement methodology as many times as necessary to make perpetual improvements in the deliverable and project processes.

The customer-driven project management improvement methodology becomes a learning process. The organization does not have to reinvent the wheel with each project or process improvement. The organization builds on the information and skills developed during previous projects.

This monitoring phase usually is the time to revisit some important considerations in the customer-driven project management paradigm. The customer-driven project management paradigm must be developed progressively while proving that it does work. It may take several cycles through the customer-driven project management improvement methodology to institutionalize this approach. The customer-driven management organization targets the following key areas that require constant reinforcement:

1. From a belief that "sloppiness" is good enough for "government work" to a belief that excellence through total customer satisfaction should be continuously pursued.

2. From crisis management to prevention management, that is, moving to a style of management that puts the up-front investment into careful planning and training to ensure that the work process meets the customer's requirements, needs, and expectations the first time and every time. This means no longer rewarding managers who leap from crisis to crisis; it means promoting managers who design and redesign better and better systems that prevent crises from happening in the first place.

3. From a focus only on results to stressing the process that produces those results. In other words, this involves moving away from management by objectives (most uncharitably described as managing by the maxim "Don't tell me your problems; all I want is results") to careful attention to the process and doing in-process measuring instead of measuring only the input and the output.

4. From an organization stressing individual accomplishment to an organization that values and rewards teamwork. This type of organization maximizes its people resources for the overall common good. It means constant development of leaders who have a balance of technical and people skills.

This evaluation process continues until the team is satisfied that it has improved the quality of the process in the manner intended and that these changes are permanent. The outcome of phase 8 is continuous improvement of the project deliverable and project processes. Documentation of this step involves:

- Project performance metrics
- Process performance metrics
- Issues for continuous improvement

Phase 8a: Closeout

Figure 4.15 shows the subprocesses of closeout. Closeout consists of (1) the closeout plan, (2) divestment of resources, and (3) report of lessons learned.

The closeout plan outlines the specific tasks, responsibilities, and time phasing for completing the product deliverable's life cycle. The plan should consider reassignment of people resources, disposition of the inventory of product deliverables and spares, and logistics support requirements. The closeout plan should consider a final evaluation of all people participating in the project. This evaluation forms the basis for the next assignment. Beyond people, the closeout plan deals with the disposition of inventory. The inventory disposition must include methods that assess both cost and the environment. Finally, the closeout plan requires a logistics support annex to actually accomplish the closeout. This could include packaging, handling, transporting, disassembly, decomposition, etc.

The second subprocess involves the actual divestment of the project deliverable. This includes the phaseout and/or disposal of the project deliverable and project process capability. Besides the actual inventory mentioned above, this may include the project documentation. Depending on the project, the dismantling of the documentation is a major task by itself.

The third subprocess in closeout is reporting the lessons learned. This is important for future projects. It becomes a database of information for prospective proposals and customer-driven project management endeavors.

Main Points

Putting customer-driven project management in action involves focusing on a specific project mission, developing teamwork in customer-driven teams, and using the improvement methodology.

Any customer-driven project management endeavor requires a common reason for joint action. This is the focus.

The second foundation for putting customer-driven project management in action is teamwork.

The customer-driven project management improvement methodology consists of the following:

Step 1: Define the quality issue
Step 2: Understand and define the process
Step 3: Select improvement opportunities
Step 4: Analyze improvement opportunities
Step 5: Take action
Step 5*a:* Project concept
Step 5*b:* Project definition
Step 5*c:* Project production
Step 6: Check results
Step 7: Implement improvement
Step 7*a:* Project operation
Step 8: Monitor results
Step 8*a:* Project closeout

Phase 1 of the CDPM improvement methodology, defining the quality issue, involves

Input: Draft mission statement
Process: Customer-driven teams focus on the quality issue by analyzing customer requirements and processes to meet the requirements.
Output: Project deliverable definition

Phase 2 of the CDPM improvement methodology, understanding and the process, involves

Input: Project deliverable definition

Process: Customer-driven teams thoroughly examine the critical processes.

Output: List of improvement opportunities

Phase 3 of the CDPM improvement methodology, selection of improvement opportunities, involves

Input: List of improvement opportunities

Process: Customer-driven teams use consensus decision making to select improvement opportunities.

Output: Improvement opportunities for further analysis

Phase 4 of the CDPM improvement methodology, analyzing improvement opportunities, involves

Input: Selected improvement opportunities

Process: Customer-driven teams analyze improvement opportunities for action.

Output: Project objectives for action

Phase 5 of the CDPM improvement methodology, taking action, involves

Input: Project objectives for action

Process: Customer-driven teams use project management techniques to take action to provide a deliverable that achieves total customer satisfaction.

Output: Product deliverable for checking

Phase 6 of the CDPM improvement methodology, checking results, involves

Input: Project deliverable for checking

Process: Customer-driven teams check the project deliverable and processes against external and internal customer's needs and expectations.

Output: Project deliverable for implementation

Phase 7 of the CDPM improvement methodology, implementing the improvement, involves

Input: Project deliverable for implementation statement

Process: Customer-driven teams instituted the improvement.

Output: Project deliverable in operation

Phase 8 of the CDPM improvement methodology, monitoring results for continuous improvement, involves

Input: Project deliverable in operation statement

Process: Customer-driven teams continuously evaluating the deliverable and processes for improvement.

Output: Continuously improving deliverable and processes

Chapter

5

CDPM Tools and Techniques

Focus: This chapter outlines all the customer-driven project management tools and techniques and provides recommendations for use within the customer-driven project management improvement methodology.

Introduction

Customer-driven project management requires an orderly, disciplined, customer-driven improvement methodology to ensure that projects are performed and improved for total customer satisfaction. Specific CDPM tools and techniques are most effective in certain phases of the customer-driven project management methodology. This does not mean that the tools and techniques can be used only as outlined in this book. As your organization gains experience, the tools and techniques can be useful for many different applications. Therefore, each organization and individual is encouraged to use the tools and techniques in any way that seems appropriate to their specific applications. Further, the tools and techniques should be tailored to each specific application. If it works, use it.

When using the CDPM tools and techniques, always remember to analyze the information in terms of the customer-driven project management philosophy and principles. Only when the organization applies the information using the CDPM philosophy and principles will maximal results that are focused on total customer satisfaction be achieved.

There are many tools and techniques for customer-driven project management. The most common are described in this book. For organizational purposes, the tools and techniques are arranged in the following categories:

- Customer-driven teams
- People involvement
- Definition
- Understanding
- Selection
- Analysis
- Project management
- System development

Customer-Driven Teams

Customer-driven teams are the primary technique for performing customer-driven project management. Customer-driven teams are detailed in Chap. 6 and are categorized as follows:

- *Customer-driven project management teams.* These are customer-driven teams whose purpose is to complete a specific project, program, or task.
- *Customer-driven quality-improvement teams.* These customer-driven teams focus on improving a specific process.
- *Customer-driven work teams.* These customer-driven teams constantly perform and improve their particular process.

People Involvement Tools and Techniques

People involvement is one of the key tools of customer-driven project management. People involvement includes both individual and team activities. Since people are the key to success in customer-driven project management, there are many tools and techniques to encourage and enhance people involvement and teamwork. They include individual involvement, teams and teamwork, communication, listening, focus setting, team meetings, brainstorming, and presentation. People involvement is described in Chap. 7, and a brief description of each tool follows:

- *Individual involvement* is the involvement of each person in the organization in the work itself and in improvement of the work.
- *Teams and teamwork.* Teams are a group of people working together toward a common goal. Teamwork is a technique where the individual team members work together to achieve a common goal.
- *Communication* is a technique for exchanging information.

- *Listening* is a communication technique for receiving and understanding information.
- *Focus setting* is a technique to establish a focus on a specific outcome.
- *Team meeting* is a tool for bringing a group together to work for a common goal.
- *Brainstorming* is a tool that encourages the collective thinking power of a group to create ideas.
- *Presentation* is a tool for providing information, gaining approval, or requesting action.

Definition Tools and Techniques

The definition tools and techniques distinguish the quality/customer satisfaction issues. These tools and techniques for definition help the customer-driven project lead team recognize the customer's needs and expectations and critical project processes. The definition tools and techniques are quality function deployment (phase 1), benchmarking, and metrics. These tools are outlined in Chap. 8, and a brief description of each tool follows:

- *Quality function deployment* (phase 1) is a disciplined approach for listening to the voice of the customer to get customer requirements that are converted into deliverable conditions.
- *Benchmarking* is a method of measuring your organization against those of recognized leaders.
- *Metrics* are meaningful measures that target continuous process improvement actions.

Understanding Tools and Techniques

The tools and techniques for understanding help the customer-driven project team recognize the keys to success in focusing resources. The understanding tools and techniques are process diagrams, input/output analysis, and supplier/customer analysis. These tools are outlined in Chap. 9, and a brief description of each tool follows:

- *Process diagrams* are tools for defining the process.
- *Input/output analysis* is a technique for identifying interdependency problems between the input and the output of the process.
- *Supplier/customer analysis* is a technique for obtaining and exchanging information to convey your needs and requirements to

suppliers and to mutually determine the needs and expectations of your customers.

Selection Tools and Techniques

Selection tools and techniques are used at each decision point during customer-driven project management to help clarify assumptions and focus on consensus. Selection tools are used during day-to-day customer-driven project management operations. They are also useful when selecting a project-improvement opportunity. The selection tools and techniques are voting, selection matrix, selection grid, and consensus decision making. They are detailed in Chap. 10, and a brief description of each tool follows:

- *Voting* is a technique to determine majority opinion.
- *Selection matrix* is a tool for rating problems, opportunities, or alternatives based on specific criteria.
- *Selection grid* is a tool for comparing each problem, opportunity, or alternative against all others.
- *Consensus decision making* is a technique for getting a team to accept and support a decision.

Analysis Tools and Techniques

A critical step in customer-driven project management involves thorough analysis of the project and all the processes within the project. The tools and techniques for analysis help improve the project and/or process, determine underlying causes, identify the vital few, measure performance, and describe both sides of an issue. The analysis tools and techniques are process analysis, work-flow analysis, cause-and-effect analysis, data statistical analysis, and force-field analysis. These tools are outlined in Chap. 11, and a brief description of each tool follows:

- *Process analysis* is a tool to improve the process and reduce process cycle time by eliminating non-value-added activities and/or simplifying the process.
- *Work-flow analysis* is a tool that shows a picture of how work actually flows through an organization or facility.
- *Cause-and-effect analysis* is a technique for helping a group examine underlying causes.
- *Data statistical analysis* is actually several tools for collecting, sorting, charting, and analyzing data.

- *Force-field analysis* is a technique that describes the forces at work in a given situation.

Project Management Tools and Techniques

Customer-driven project management retains the traditional project management tools and techniques for planning, organizing, and controlling projects. The project management tools and technique include contract, work breakdown structure, task list, network schedule, and risk management. These tools and techniques are detailed in Chap. 12, and a brief description of each tool follows:

- *Contract.* A contract defines the deliverable specification, statement of work, and information requirements.
- *Work breakdown structure.* The work breakdown structure defines the organization and coding of the deliverable.
- *Task list* includes the development of the task involved in the project.
- *Network schedule* is a technique for planning, scheduling, and controlling time and estimating, budgeting, and controlling resources.
- *Risk management* is the continual assessment of threat or opportunity in terms of time, cost, technical feasibility, and customer satisfaction.

System Development Tools and Techniques

System development tools and techniques focus on the development or redesign of systems. These tools target invention and improvement of a system. The system may be the product and/or service itself or many systems within an organization. These tools are fundamental in customer-driven project management. All traditional project management–type projects require the use of the system development tools and techniques. The system development tools are outlined in Chap. 13, and a brief description of each tool follows:

- *Concurrent engineering* is a systematic approach to the integrated concurrent design of products and their related processes, including manufacture and support. This approach is intended to cause the developers, from the outset, to consider all elements of the product life cycle from conception through disposal, including quality, cost, schedule, and user/customer requirements.
- *Quality function deployment* is a disciplined approach for transforming customer requirements, the voice of the customer, into product development requirements.

- *Robust design* is a technique for designing a product that has minimal quality losses.
- *Design of experiments* is an experimental tool used to establish both parametric relationships and a product/process model in the early (applied research) stages of the design process.
- *Taguchi approach* includes several techniques for reducing variations in product or process performance to minimize loss.
- *Cost of quality techniques* are techniques to identify cost of conformance and nonconformance.
- *Cost of poor quality techniques* are techniques that focus on minimizing the cost of nonconformance.
- *Statistical process control* is a statistical tool for monitoring and controlling a process to maintain and possibly improve quality.

Use of Tools and Techniques within the CDPM Improvement Methodology

Customer-driven project management tools and techniques can be used in many places within the customer-driven project management improvement methodology. Customer-driven tools and techniques can be used to improve the project or processes. Specific tools and techniques are most beneficial during particular steps in the customer-driven improvement methodology. Figure 5.1 provides an overview of recommendations for using the CDPM tools and techniques. These recommendations for the use of the customer-driven project management tools within the project management phases and the customer-driven improvement methodology are outlined below.

Define quality (customer satisfaction) issue tools and techniques

- *Customer-driven project lead team.* The customer-driven lead team is formed to define the quality issue.
- *Focus setting.* Focus setting is used to establish the mission of the project. The mission provides guidance to the customer-driven lead team on achieving the vision.
- *Quality function deployment.* Quality function deployment is used to get the "voice of the customer." This gets the customer's requirements for the project. These requirements are translated into specific needs and expectations. The customer's needs and expectations are then converted into design requirements. This establishes the project processes.

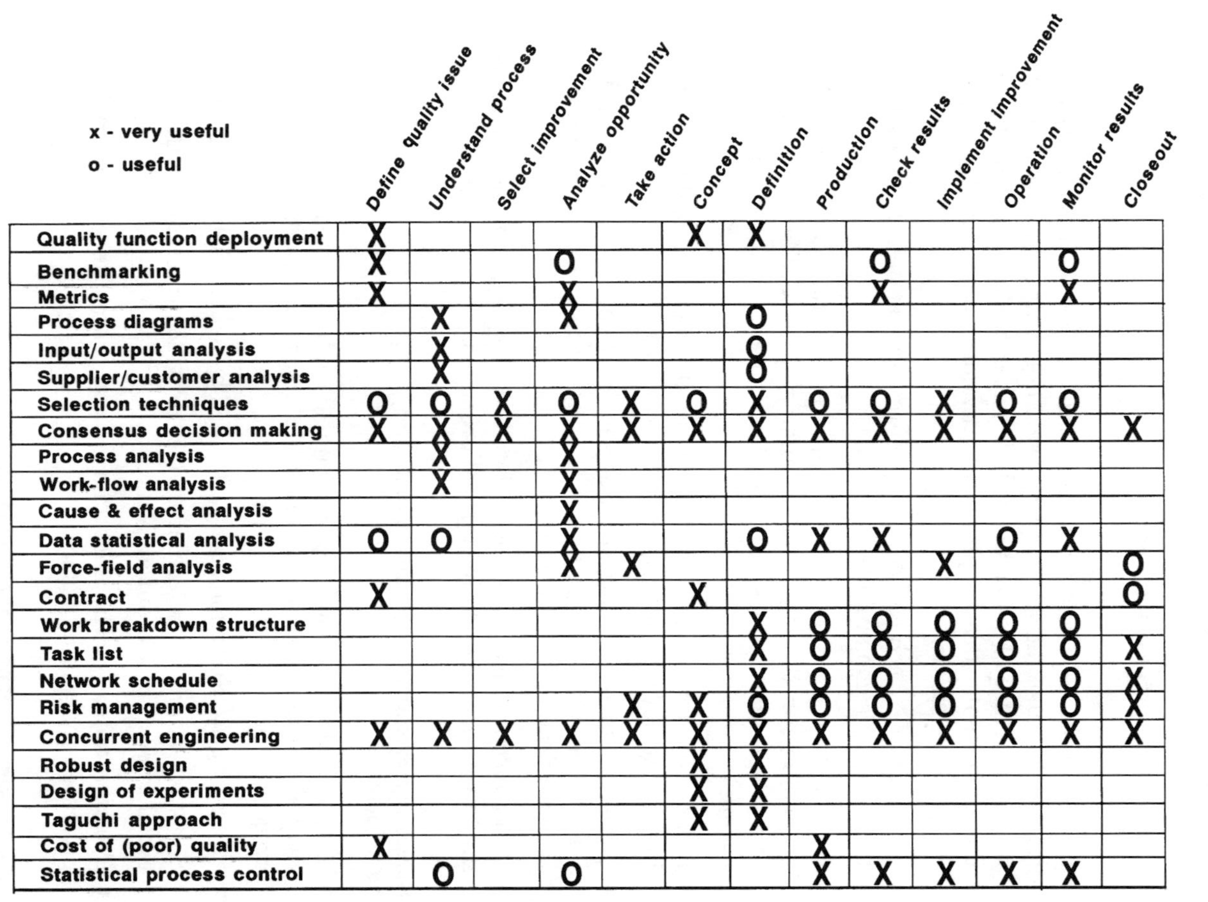

x - very useful
o - useful

	Define quality issue	Understand process	Select improvement	Analyze opportunity	Take action	Concept	Definition	Production	Check results	Implement improvement	Operation	Monitor results	Closeout
Quality function deployment	X					X	X						
Benchmarking	X			O					O			O	
Metrics	X			X					X			X	
Process diagrams		X		X			O						
Input/output analysis		X					O						
Supplier/customer analysis		X					O						
Selection techniques	O	O	X	O	X	O	X	O	O	X	O	O	
Consensus decision making	X	X	X	X	X	X	X	X	X	X	X	X	X
Process analysis		X		X									
Work-flow analysis		X		X									
Cause & effect analysis				X									
Data statistical analysis	O	O		X			O	X	X		O	X	
Force-field analysis				X	X					X			O
Contract	X					X							O
Work breakdown structure							X	O	O	O	O	O	
Task list							X	O	O	O	O	O	X
Network schedule							X	O	O	O	O	O	X
Risk management					X	X	O	O	O	O	O	O	X
Concurrent engineering	X	X	X	X	X	X	X	X	X	X	X	X	X
Robust design						X	X						
Design of experiments						X	X						
Taguchi approach						X	X						
Cost of (poor) quality	X							X					
Statistical process control		O		O				X	X	X	X	X	

Figure 5.1 CDPM improvement methodology with tools and techniques.

- *Benchmarking.* Benchmarking is used to determine process capability in relationship to the leaders of a specific process.
- *Metrics.* Metrics are used to measure the potential project processes.

Understand the process tools and techniques

- *Customer-driven project management/quality-improvement/work teams.* The customer-driven lead team determines project organizational structure. This could include other customer-driven project teams, customer-driven quality-improvement teams, and customer-driven work teams.
- *Focus setting.* Each customer-driven team must determine its focus to include mission and possible goals.
- *Process diagrams.* Each customer-driven team uses process diagrams to get a thorough understanding of its specific processes. At this stage, the top-level and top-down process diagrams are most useful.
- *Input/output analysis.* Input/output analysis is used to describe the inputs and outputs of each process.
- *Supplier/customer analysis.* Supplier/customer analysis provides information on the suppliers of the inputs to the process and customers of the process outputs.
- *Brainstorming.* All the issues related to a particular project process are brainstormed to define critical problems.

Select improvement opportunities tools and techniques

- *Voting.* Voting is used to get majority opinion on specific issues related to the information gathered during the understanding the process step.
- *Selection grid.* Selections can be limited by using a selection grid.
- *Selection matrix.* A selection matrix provides the team with additional information about an issue.
- *Decision making.* The customer-driven teams decide on critical processes and quality issues.

Analyze improvement opportunities tools and techniques

- *Process diagrams.* A detailed process diagram of any selected process is helpful at this stage.

- *Process analysis.* The detailed process diagram is used to perform a process analysis to make obvious process improvements.
- *Work-flow analysis.* The work-flow analysis provides a tool to make adjustments in the flow of the process to improve the process.
- *Cause-and-effect analysis.* The cause-and-effect analysis determines the underlying causes of the problems.
- *Data statistical analysis.* Data statistical analysis focuses the team on the critical or vital few items.
- *Force-field analysis.* Force-field analysis helps generate alternative solutions.

Take action tools and techniques

- *Brainstorming.* Brainstorming generates a list of appropriate actions.
- *Selection techniques.* Selection techniques are used to decide on actions to take.
- *Contract.* A contract establishes the specific action to take.
- *Work breakdown structure.* The work breakdown defines the coding and organization of the action deliverable. The work breakdown structure can be bridged from the quality function deployment matrix.
- *Task list.* The task list provides all the tasks involved in completing the project action.
- *Network schedule.* The tasks are planned, scheduled, and controlled using the network schedule.
- *Risk management.* Each task is continually assessed for risk in terms of cost, schedule, technical specification, and customer satisfaction.

Check results tools and techniques

- *Data statistical analysis.* Data statistical analysis provides information for focusing on a vital few, determining underlying causes and measuring process performance.
- *Metrics.* Metrics provide data on process performance.

Implement improvement tools and techniques

- *Presentation.* Presentation seeks approval and action.

Monitor results tools and techniques

- *Data statistical analysis.* Data statistical analysis provides information on the project.
- *Metrics.* Metrics focus on continuous process improvement.

Use of CDPM Tools and Techniques for Systems Development Project

The preceding recommendations relate to the customer-driven project management improvement methodology. These tools and techniques are sufficient for most applications of customer-driven project management. The CDPM tools and techniques outlined above still apply to a system development effort. However, in a complex systems complex development effort, the customer-driven project management improvement methodology requires the use of additional tools and techniques in the project management phases of concept, definition, production, operations, and closeout.

The following tools and techniques are useful in the concept phase:

- *Customer-driven teams.* A customer-driven lead team determines the specific concept for the deliverable.
- *Concurrent engineering.* Concurrent engineering is used to reduce the time required to get the deliverable to the customer.
- *Quality function deployment.* Quality function deployment takes the customer requirements and converts them into design requirements for product planning.
- *Robust design.* Robust design is used at this stage to establish both parametric relationships and a product/process model.

In the definition phase, the following tools and techniques help to define the project:

- *Quality function deployment.* Quality function deployment during this stage consists of turning design requirements into part characteristics through the parts deployment process. It also involves process planning the selection of manufacturing operations. Further, it includes establishing capable processes for production.
- *Metrics.* Metrics are used to monitor process performance.
- *Robust design.* Robust design at this stage is concerned with eliminating loss and reducing variation in product, parts, and processes.

The following tools and techniques are valuable during the production phase:

- *Statistical process control.* Statistical process control is used to monitor and manage processes.
- *Cost of poor quality.* Cost of poor quality assists in determining and reducing costs associated with inspection, rework, scrap, testing, change orders, errors, and inventory.

During operations, the following tools and techniques are useful to perform the project:

- *Cost of poor quality.* Cost of poor quality at this stage is concerned with litigation, warranty, lengthy cycle times, inventory, and customer complaints.
- *Statistical process control.* Statistical process control focuses on continuous improvement.

At closeout of the project, the following tools and techniques help to determine the exact fate of the project:

- *Brainstorming.* Brainstorming provides a tool to determine specific requirements for closeout.
- *Task list.* The task list documents all the tasks for closeout.
- *Network schedule.* The network schedule is used to plan, schedule, and control accomplishment of all the tasks.

Main Points

Customer-driven project management requires the appropriate use of many tools and techniques to be successful.

The specific tools used depends on the project.

Systems development projects require the use of additional development tools and techniques.

The categories of customer-driven project management tools and techniques are

- Customer-driven teams
- People involvement
- Definition
- Understanding
- Selection
- Analysis
- Program management
- System development

Customer-driven team tools and techniques target the effective use of teams within customer-driven project management.

People involvement tools and techniques aim at enhancing the productivity of the people resource.

The definition tools and techniques focus on the customer's needs and expectations and the critical processes required to meet the customer's requirements.

The understanding tools and techniques emphasize the performance of internal processes to meet the project's objective.

The selection tools and techniques move toward decision making.

The analysis tools and techniques seek the underlying issue, the vital few, the root cause.

The program management tools and techniques emphasize planning, monitoring, and controlling the project.

System improvement tools and techniques are aimed primarily at the quality of the design and prevention of defects.

Chapter

6

Customer-Driven Teams (CDTs)

Focus: This chapter details the application of customer-driven teams within the customer-driven project management system.

Introduction

Today, teams are the organizational structure of choice to meet the challenges of the global economic environment. People working together for a common goal in teams are absolutely essential to success. Teams maximize the use of human resources in the organization. In project management, functional, process, and multifunctional teams ensure that all aspects of the project are integrated to achieve the desired end result, satisfying the customer(s). Teams provide better decisions and the motivation to carry them out. Everyone can participate in a team. Relationships are nurtured for improved working coordination. Working together for a common goal leads to increased job satisfaction and rewards in the work itself. Teams foster freer contribution of information through more active communication. Further, the organization is thrust toward a common goal and an organizationwide perspective is fostered through teamwork. Teams provide the rapid, responsive organizational structure that is necessary for any organization to compete successfully in the ever-changing economic environment of today and tomorrow.

Traditional organizations in government and private industries usually are not geared toward using teams, especially project teams, to their full advantage. They do not structure project teams to both perform and improve projects. Normally, the project team is only responsible for completing the project. Thus there is much inefficiency and redundancy in the kind of organization that uses one kind of team to perform the project and in some of the more progressive organizations another kind of team to seek out quality improvements and build a customer-driven culture. In a traditional organization, management concepts typically focus on outputs and profits. The focus is on cost, schedule, and quality, usually in that order. Quality in the traditional sense means that the product or service meets technical acceptance. Traditional organizations do not empower teams with the responsibility, authority, and resources to continuously improve their processes geared toward customer satisfaction or "quality" in the total-quality sense. In these organizations, it is difficult to compete in the global marketplace because productivity and quality are separate and suppliers and the customer are isolated from each other.

Customer-driven teams are the preferred organizational structure for performing a project as efficiently and effectively as possible to satisfy customers' expectations. The customer-driven team focuses on both working a project and continuously improving the project. Customer-driven teams strive to marry productivity and quality through all phases of a project. In addition, customers and suppliers are linked from the external customer throughout the organization by supplier and customer relationships. Through customer-driven teams, empowered to constantly perform and improve all aspects of their process to satisfy the customer, an organization can use CDPM for success.

Traditional Project Teams

Traditional project management teams place accountability for results on one manager, the program manager who uses a disciplined planning, direction, control, and evaluation approach. Such a project manager is a member of the supplying organization, not the customer's organization. The basic function of the traditional project manager is to control quality, cost, and schedule; to manage the team's work; and to ensure the effective and efficient delivery of a project deliverable that is acceptable to the customer and within schedule, budget, and technical constraints. The traditional matrix project group borrows people from the functional departments of the organization, i.e., engineering, manufacturing, logistics/supportability, marketing, finance, human resources, and so forth, and pulls them into a project organization. The project manager focuses the team on the project specifications and requirements and keeps the customer informed as necessary.

This structure has presented major problems for many projects mainly because the customer is typically uninvolved in the key decisions in the project life cycle. Once a contract is signed, the supplying organization project manager is likely to feel "in charge" to the point that the team's creative and innovative efforts are too rapidly and too narrowly targeted on building the specification—instead of developing a fuller understanding of the customer's performance issues and expectations as the project "learning" progresses.

This tendency to isolate the project manager and the team from the customer too quickly is rooted in the propensity of project managers and project management organizations to drive projects by schedule and cost rather than by quality. This is due to the project manager's need to be "in control." Usually, the project manager can establish some control over cost and schedule within his or her organization. However, since quality is defined by customer satisfaction, quality can be a changing requirement. Therefore, project managers place extreme pressure on the customer to lock in on customer needs early in the concept phase of the project through quality standards and specifications, with little assurance that any adjustments can be made once the project proceeds into definition. In addition, customer satisfaction requires full communication between the supplier and the customer. This is difficult when the project manager emphasizes specifications, statement of work, schedules, and budgets instead of the customer's needs. Further, traditional project managers ignore the "learning curve" in both the supplier's and customer's organizations during the life cycle of the project. Without continuous improvement based on customer satisfaction, project deliverables often do not meet the customer's needs and expectations in a changing environment.

Customer-Driven Teams (CDTs)

Customer-driven teams are teams that are accountable for both project results and quality improvements, focusing on total customer satisfaction. They are empowered to do whatever is necessary to satisfy the customer. As shown in Fig. 6.1, the team consists of a customer project leader, a project facilitator, and process owners. Figure 6.1 shows some examples of process owners. They include the system engineer, process engineer, production manager, marketer, support lead, quality engineer, and contract officer. In addition, a program manager is an essential part of the team. The team is guided through the leadership of the customer. This can be through leadership directly from the customer or the customer's "voice." The customer or customer's "voice" acts as the team leader in a customer-driven team. The entire customer-driven team assumes ownership of all outcomes. The team members have the responsibility, authority, and resources to achieve expected objectives.

A customer-driven team is a functional or multifunctional team consisting of all the disciplines necessary to complete a project. For exam-

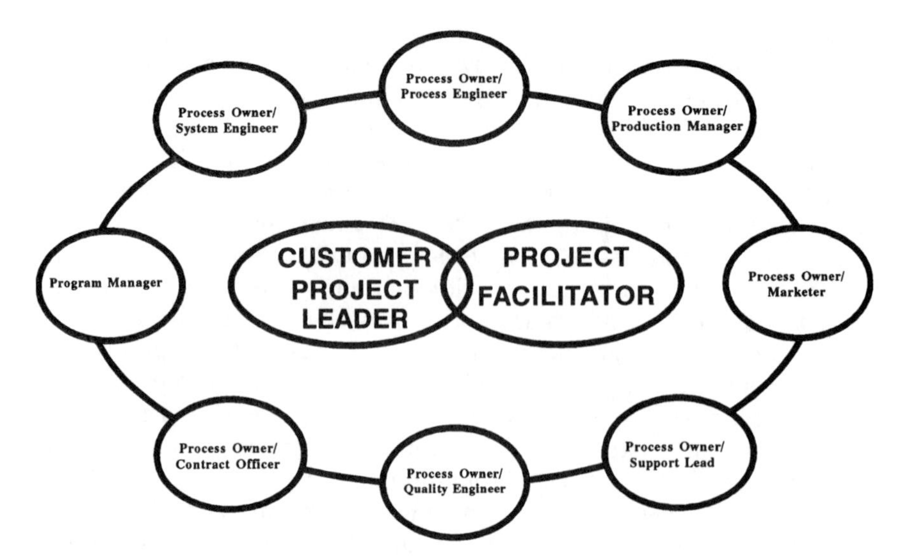

Figure 6.1 Customer-driven project team.

ple, a typical customer-driven team for a project to make a product consists of members from engineering, manufacturing, logistics/supportability, marketing, finance, and others as necessary. A customer-driven team for improving materials required for manufacturing might consist of members from suppliers, receiving, manufacturing, engineering, production control, and contracts. A customer-driven team for performing and improving a manufacturing process would consist of members from the work unit accomplishing the manufacturing process.

In addition to a team leader and members, the customer-driven team is supported by a project facilitator. Usually, the major project supplier or an external organization provides the customer-driven team facilitator. The customer-driven team facilitator helps the customer-driven team leader focus the team on desired outcomes and assists with establishing a collaborative union among all the members of the team.

An essential member of the team is the program manager. The program manager is vital to the integration of all program management functions. The program manager is normally the "manager of the project." In customer-driven project management, the program manager's duties are similar to those of traditional project management. The program manager is the team member who provides the liaison between the project and both the supplier and customer organizations.

In addition, there is a critical link between all the customer-driven teams. As shown in Fig. 6.2, the program manager becomes the critical link to other process managers throughout the many processes in

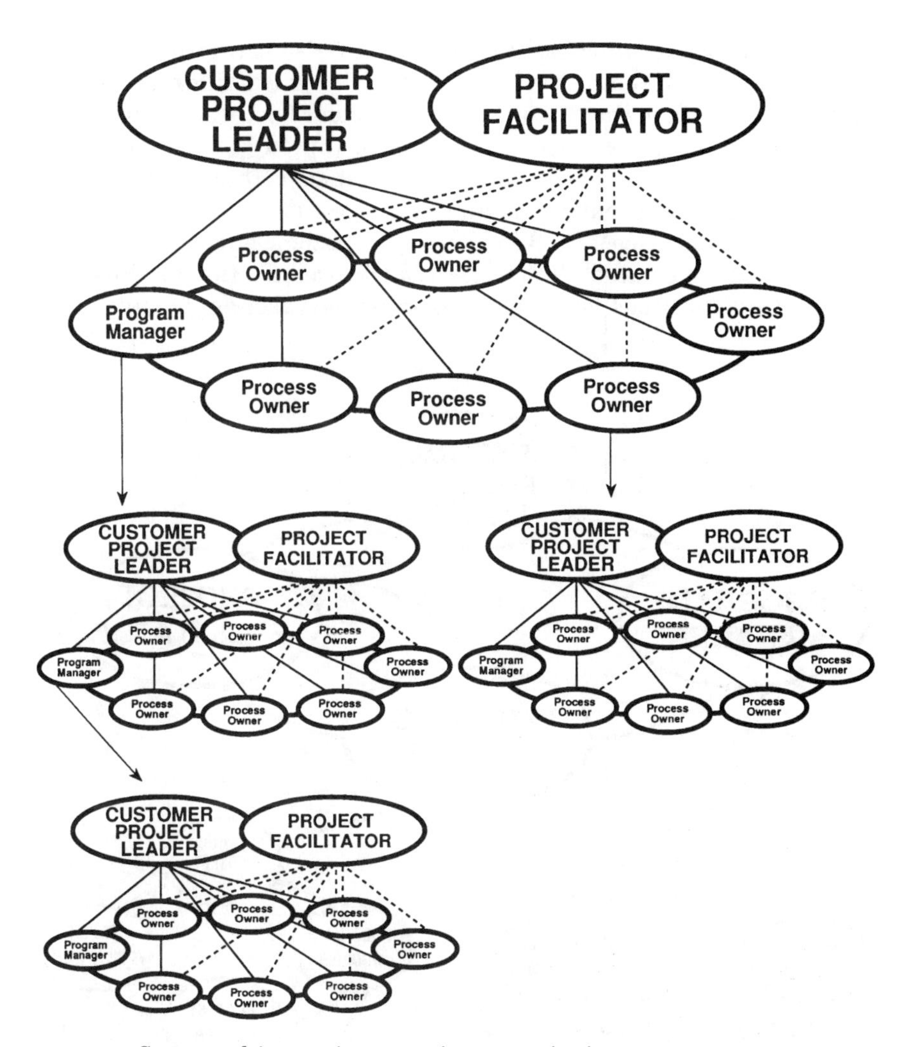

Figure 6.2 Customer-driven project teams in an organization.

the organization. The process managers can be linked in a project that has many layers of processes to accomplish the overall project. The program manager is a supplier as a member of one team. At the same time, the program manager is also the customer of processes in the project. For these activities, the project manager forms a customer-driven team to perform and improve these processes. The program manager becomes the customer project leader for this team.

Figure 6.2 also shows how the process owners can become program managers or, as they are called, process managers for their specific processes. Process managers are the traditional functional managers in today's organizations. Process owners are any person in the organization who owns a process. Process owners also can form additional customer-driven teams to perform and improve their processes. Again, they become the customer project leader for these teams.

Further, the project facilitators can be linked on their own customer-driven team whose purpose is to ensure that all teams are focused on the right outcomes, and teamwork is established over the complete project (see Fig. 6.3). Through customer-driven teams, all parts of the project and all processes in the organization are linked to ensure that the focus is on total customer satisfaction. Quality, cost,

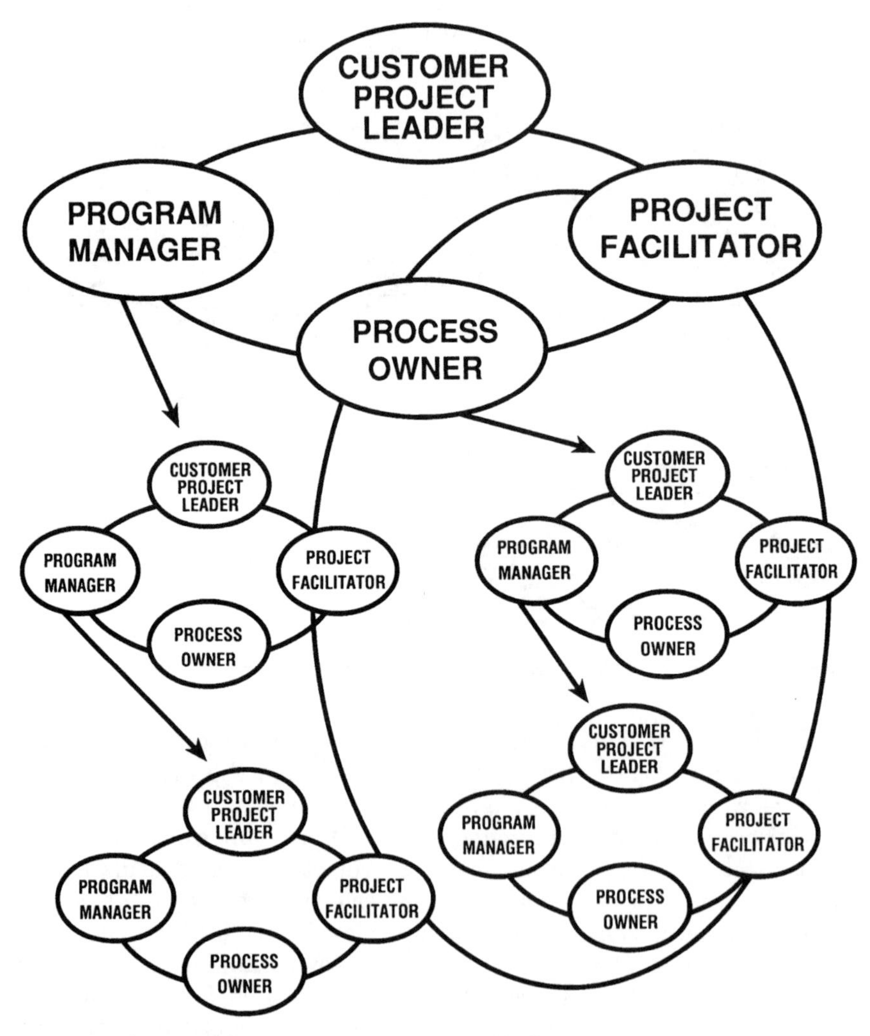

Figure 6.3 Customer-driven project organizational links.

and schedule are continually optimized through the synergy created by the integrated teamwork of customer-driven teams.

Customer-driven team philosophy

The customer-driven team approach stresses the following:

1. An organization is a dynamic system, not one that is static and locked into a vertical hierarchy or a structured, sequential approach to project management and/or quality improvement. It holds that an organization that encourages lateral as well as vertical communication will integrate its functions more effectively to achieve the objectives of each customer in the chain. Further, today's communications technology allows a free flow of information among customers, suppliers, and process owners. This provides the required support for the customer to lead the team in an effective way.
2. Customer-led teams will be successful by definition. Teams using a systems approach with the customer in the driver's seat throughout the life cycle of the project have more control over outcomes, focusing on total customer satisfaction. The customer must lead the critical decisions at every milestone in the project, and today's telecommunications and information technology makes this possible at a reasonable cost.
3. Tasking and accountability are best driven directly by the customer or customer's voice. The concept of responsibility, authority, and resources internally directed from the top down is not an overriding principle. The energy in a customer-driven team comes from the customer, not the internal pressure of the organization. The team accepts "ownership" with accompanying pride and commitment for the project and quality improvements.

Foundation for Establishing Customer-Driven Teams

Five basic principles provide the foundation for establishing customer-driven teams. These five basic principles are essential to starting any customer-driven team. The organization must buy into the following five principles prior to initiating any customer-driven team:

1. Top-management commitment exists through the willingness to take appropriate actions with a long-term view. This must be through active participation versus passive support.
2. Teams are the organizational structure of choice throughout the organization for project management, quality improvement, and process performance activities.

3. Accountability for the outcomes of projects can be shared among all members of the project team—the customer, process owners, and suppliers.
4. The customer or customer's voice is the leader of the project.
5. Continuous improvement focused on customer satisfaction is a primary pursuit within the organization.

Customer-driven teams require visible and continual actions from top management as support for the "way of life." Unless top management "sets the example," customer-driven teams will revert to the more familiar traditional project management. This requires management to make the commitment to take the actions necessary for the long term. In addition, top-management support is essential to ensure that team members cooperate within the team and across all functions of the organization, as well as with customers and suppliers, to meet the project's goal.

Customer-driven project management assumes that essentially all an organization's work that produces definable outputs should be structured into projects. These projects will be planned, accomplished, evaluated, and improved through customer-driven teams. These teams can respond rapidly to changing customer needs and expectations. "Ownership" for the project is shared by the team. The customer or customer's voice as team leader defines quality for the team. The team leader guides the team. The team facilitator keeps the team focused and establishes and maintains teamwork. The functional managers, as process owners, manage their processes. They ensure the technical competence of their people through continuous education, training, facilitating, coaching, and support.

In customer-driven teams, customer-driven project management philosophy, guiding principles, methodology, tools, and techniques are used through all eight phases of the customer-driven project management improvement methodology. Customer-driven teams stress the importance of using the customer-driven project management methodology throughout all project phases.

Establishing a Customer-Driven Team

A customer-driven team starts with a specific project. Once the project is identified, the customer-driven team is formed to include a customer project leader, a project facilitator, and team members. The focus of a customer-driven team can be on a project, quality improvement, or a process. The project must have a clearly defined customer and/or customers. The customer-driven team works together with the customer or customer's voice as the leader to complete the project.

While performing the work, the team looks for ways to continuously improve its processes to lower costs, reduce time, and increase quality. The customer-driven team progresses through all phases of the project using customer-driven project management to successful completion.

The key steps to establishing a customer-driven team include determining the focus and developing teamwork.

Determine the focus

The focus of the customer-driven team is to complete the specific project to total customer satisfaction. Within this overall focus of all customer-driven teams, each team must have a specific mission. This mission is the intended result of the project. This is included in a written statement that meets the following criteria:

- Must be customer-driven
- Includes the purpose of the team
- Sets a common direction
- Sets the expected result
- Involves all team members
- Opens communications
- Nurtures long-term relationships

When determining the focus of the customer-driven team, remember that besides completing the project within cost, schedule, and quality standards, the team focuses on the deliverable(s) satisfying the customer. The mission statement must consider the entire magnitude of the project in terms of this factor. In addition, resources must be committed, statements of authority and boundaries must be negotiated and clarified, and each team member needs to understand his or her specific role in accomplishing the mission. All these factors must be considered when the mission is set. The mission can be formed outside the team, but the mission statement must be written by the team. This completes the first step in forming a customer-driven team.

Develop teamwork in the customer-driven team

Developing teamwork starts with a group of people committed to a common purpose, i.e., project, quality improvement, or process. Next, roles and responsibilities must be identified and formalized.

Roles and responsibilities describe the specific results and expected outcomes from each team member. Every team member must understand what he or she needs to do. As part of performing their specific responsibilities, each team member should know exactly what he or she "owns" with the particular resources available. Further, the team members must define their boundaries of empowerment. Empowerment implies authority to act. This authority to act takes on an operational definition as agreed to by the specific organization and the team. Ideally, everyone has the authority to do what is necessary to meet customer needs and expectations, even in extraordinary ways. However, this is not always practical. Therefore, empowerment involves having the responsibility, authority, and resources to do whatever is required to satisfy the customer and achieve the mission within defined boundaries. The team member's specific authority to act, along with corresponding responsibility and resources, must be clearly understood by all team members. In addition, measures of performance ensure that quality standards focus on customer satisfaction.

Although roles and responsibilities will change over the life of a project, they must be identified in a "living" document. It is important for each member of the team to know what other members can contribute to the team. Only through this "living" document, which is updated as necessary throughout the project, can each member on the team know what the other members will supply toward the project.

The Customer Project Leader

The customer project leader, the customer or customer's voice, leads the team to completion of the project, satisfying the customer. This is a departure from traditional project management and a major innovation in the organizational structure. This unique arrangement requires the customer project leader, who is actually the customer or the customer's voice, to influence some of the work in various functional organizations usually outside the project manager's organization. This requires the explicit support of the outside organization(s) and the functional managers (process owners) to be successful.

Roles and responsibilities

The customer project leader or customer's voice project leader performs the following:

- Leads the team to successful project completion
- Ensures that the team is empowered to perform and improve all aspects of the project

- Is accountable for cost, schedule, and quality
- Drives total customer satisfaction as the primary consideration
- Encourages use of customer-driven project management
- Rewards and recognizes performance

Selection criteria

Besides a relentless pursuit of customer satisfaction, the criteria for selecting a customer project leader include

- Respect in customer's and supplier's organizations
- Developed communication skills, especially listening
- A high degree of interest in and commitment to project goals
- Demonstrated ability to use teams
- Ability to "set an example"
- Leadership and project management skills
- Systematic thinking
- Highest standards of integrity, ethics, and trust

Customer's voice project leader

In some situations the customer may not be able to function as a customer project leader. For instance, when there are many customers, it would not be practical to have many project leaders for the same project. When it is not possible for the customer to be the project leader, a customer's voice may be assigned to be the project leader. The customer's voice must fit the following criteria:

- Understands the customer(s) and the organization
- Capable of getting required management support
- Demonstrated ability to balance the needs and expectations of the customer with those of the organization
- "Sets the example" rather than blindly following organizational "norms"

The Project Facilitator

The project facilitator assists the team leader in focusing the team on the project, moving toward project completion, developing teamwork, and applying the customer-driven project management process. The facilitator can be from any organization. The facilitator can even be from an outside organization not specifically involved in the project. The facilitator acts primarily as the change agent for the application of customer-driven project management instead of traditional project management. The facilitator must be available to assist the team leader and the team in whatever capacity necessary. The facilitator helps to provide the right attitude, knowledge, and skills necessary for the team to perform and improve the project processes toward the end of customer satisfaction.

Roles and responsibilities

The project facilitator assists the customer project leader and the customer-driven team to do the right things right. The team facilitator is responsible for developing the team and applying the customer-driven project management techniques. As such, the facilitator helps the team concentrate on the project mission, gives on-the-job training, recommends formal training to use customer-driven project management tools and techniques, explains lessons learned from other team experiences, and assists with team dynamics. The team facilitator does whatever is necessary to make the team work as effectively and efficiently as possible while focusing on customer satisfaction. To accomplish these responsibilities, the team facilitator is usually expected to have strong organizational development and training capabilities, a positive feeling toward people and teams, and the ability to exercise the judgment necessary to ensure that the project team creates a highly motivated and productive team environment. At various times the team facilitator needs to be able to assume each of the following roles:

- *Coach*—Develop the total capabilities of the team through intensive application of instruction to enhance strengths and limit weaknesses.
- *Educator*—Provide specific knowledge.
- *Trainer*—Equip with specific skills.
- *Motivator*—Give encouragement, support, and recognition.
- *Mediator*—Promote a collaborative environment.
- *Negotiator*—Assist with finding a win/win solution.
- *Devil's advocate*—Stimulates open communication.
- *Public communicator*—Advertise team's successes.
- *Storyteller*—Communicates team's culture to others.
- *Confessor*—Listens to team and team members.
- *Cheerleader*—Provides boost to the team to induce progress.
- *Supporter*—Assists with obtaining resources.

Selection criteria

The project facilitator as the primary change agent on the customer-driven project team should have the following characteristics:

- Organizational development skills
- Ability to perform training, formal and on-the-job
- Appreciation of the value of people and their contributions in teams
- Demonstrated flexibility, creativity, adaptability, patience, and high energy
- Systematic thinker
- Sense of humor

Customer-Driven Team Members

The customer-driven team is made up of members from either one function or from many functions. The team consists of representatives from every discipline required to accomplish the mission, including customers, process owners, and suppliers involved in the project. Besides a team leader, a facilitator, and representatives from vital functions, a program manager is an essential team member. The customer-driven team membership may change over the life cycle of the project, but a steering team should remain as part of the structure from the cradle to the grave. The specific makeup of the steering team will depend on the particular project. Figure 6.4 shows the composition of one typical customer-driven team. This could be the core composition of a top-level customer-driven team for a new product development.

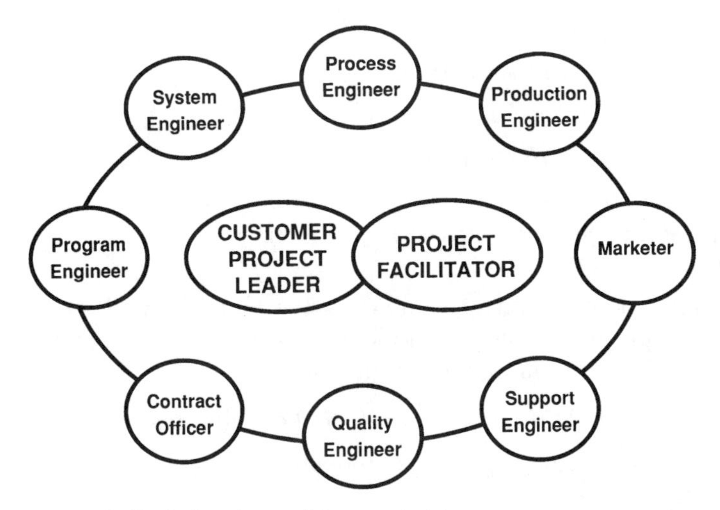

Figure 6.4 Typical customer-driven project team.

Roles and responsibilities

The team members share accountability for the performance and improvement of the project, striving for total customer satisfaction. Team members must feel that "they are all in it together." Therefore, they will do whatever is necessary to satisfy the customer. Specifically, team members responsibilities include

- Knowing customer needs and expectations
- Determining metrics
- Measuring performance
- Ensuring the performance and improvement of their process
- Making recommendations for improvements
- Being receptive to open communications
- Encouraging other team members
- Recognizing individual and team contributions
- Forming partnerships with suppliers
- Satisfying the customer (internal/external)

Selection criteria

A customer-driven team member needs to be willing to participate in a dynamic, flexible, responsive, and customer-driven organization. Generally, all customer-driven team members should display the following capabilities:

- Willingness to accept "ownership"
- Flexibility and adaptability
- Ability to do and follow at same time
- Strength to focus on a specific project
- Skill to make complex processes simple

Program manager team member roles and responsibilities

Besides the general roles and responsibilities shared by all team members, the program manager team member has the following additional responsibilities for his or her assigned program:

- Plans, organizes, staffs, directs, controls, and coordinates
- Recommends composition of own customer-driven team
- "Owns" program

- Guides the program
- Rewards and recognizes performance
- Is accountable for cost, schedule, and quality
- Maintains teamwork among customer, suppliers, and process owners

Process owner team member roles and responsibilities

Since each team member brings specific capabilities to the team to accomplish the project, team members have specific responsibilities geared to their particular expected contribution. Usually, a team member is a process owner or representative of a process owner. As such, the process owner team member has the following specific responsibilities for his or her particular process:

- Plans, organizes, staffs, directs, controls, and coordinates
- Rewards and recognizes performance
- "Owns" process
- Continuously improves process
- Empowers own customer-driven team
- Supports people to increase technical competence
- Is a systems thinker who statistically analyzes the process

The Customer-Driven Team in Project Management and Team Development

As with all project teams, a customer-driven team goes through the five stages of project development—concept, definition, production, operations, and closeout/continuous improvement. This is shown graphically in Fig. 6.5. As the team develops to full maturity, it moves through the stages of team development. The stages of team development are orientation, dissatisfaction, resolution, and production. Figure 6.5 also shows the stages of team development. Each team goes through all the stages as a normal course of development. In addition, each major change in a team may cause the team to revert to the orientation stage. In this section, the project management phases and stages in team growth are discussed together to show the relationship between project management phases and the stages of team growth. Since every team is different, the project management phases and team development stages usually do not progress at the same time. In fact, in a project consisting of many teams, the teams will be in vari-

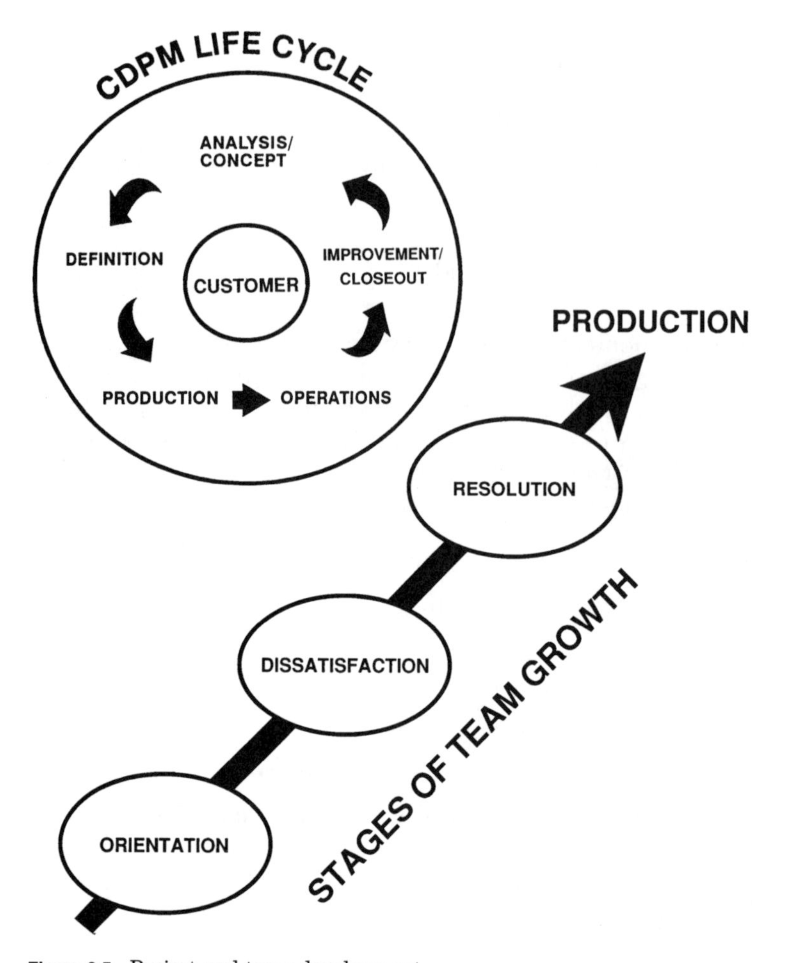

Figure 6.5 Project and team development.

ous stages of development during the project management stages. The faster the team develops, the more responsive the team can be to performance and improvement of the project. The most important point to remember is team development and the project must both receive appropriate attention so as to achieve total customer satisfaction.

The customer-driven team must be managed, facilitated, coached, and led through all the stages in team development. As the team moves through the stages in team development, the proper interventions must be applied to assist with transition of the team along the path from one stage to the next. The ultimate goal is to empower the customer-driven team to perform and improve all aspects of their project. This does not just happen. The team needs to be guided from one stage to the next. There is no known method of bypassing any of the stages, but transition from one stage to the next can be facilitated through application of the right intervention for each stage. The people must be led and the processes managed from orientation, through dissatisfaction and resolution, to production by truly empowered customer-driven teams. The team requires active managing during the orientation stage, facilitation during the dissatisfaction stage, coaching during the resolution stage, and shared leadership during the production stage. A support structure is helpful during all stages, at least for the first several years of developing customer-driven teams. Also, critical during all stages is training. This includes training in technical, business, and social skills. It is important that the team facilitator recognize the stages of team development and work with the customer-driven leader to apply the proper intervention to keep the team progressing toward completion of the project mission. The customer-driven team must understand that every team must go through all the stages. The length of each stage will vary with the team. In some cases, the team may not be able to progress from one stage to the next. In these cases, the team composition may have to be changed or the team disbanded. In most cases, though, the team will progress through all the stages and be successful if it is provided with the proper direction, leadership training, facilitating, coaching, and support.

Concept phase and orientation stage

For the customer-driven team, the concept phase of project management relates to the orientation stage of team development. During the concept phase of project management, the team determines the project scope, performance requirements, and customer needs. Also during this concept phase, the team uses the customer-driven project management improvement cycle to define the customer's initial quality issues and identify, select, and analyze improvement opportunities.

In the concept phase of project management the team is formulating the project; in the orientation stage of team development the team is forming. This orientation stage of team development involves establishing rapport, honesty, trust, and open communication. On the positive side, there is the usual enthusiasm associated with starting a new endeavor. The team members all want to contribute to accomplishing the overall goal of completing the project. However, the team has not yet learned how to work together as a team. Members may have hidden agendas. Their allegiance is normally to their functional group rather than to the team. To overcome some of the faltering that is common during this stage of team development, the team needs to be given information on the project and teamwork. Further, this stage requires the project manager and facilitator to display strong management of the team. Planning, organizing, staffing, controlling, directing, and coordinating are essential to ensure that the team knows exactly what needs to be done and how project goals and teamwork can be attained. With this direction, the team, through structured activities, can begin to further define the project. This focus on specific outcomes helps the team learn by doing.

Besides achieving outcomes, the team must be introduced to its expanded role of improvement. Each team member's paradigm must be broadened to aim at total customer satisfaction rather than the usual emphasis on optimizing their specific function in the project, which tends to suboptimize the entire project. Team members must be exposed to new possibilities, but this should be accomplished a little bit at a time. During this stage, relationship building is as important as task accomplishment. Therefore, it is not surprising that during this stage there may be little real progress toward achieving the project's mission.

All that should be expected from this phase is

- Setup of the team
- Definition of the focus of the project
- Formulation of the initial development of roles, responsibilities, and relationships

Definition phase and dissatisfaction stage

The definition phase of project management relates to the dissatisfaction with the start of resolution stages of team development. As the team defines the project in detail, specifying its requirements, determining performance standards, and identifying resources, team members are sometimes overwhelmed. During this phase the project is broken down into its work breakdown structure. Detailed plans for cost, schedule, and quality are established. This is when the team first realizes the magnitude of the project. Further, as administrative details, including policy and procedure documents, job descriptions, budgets, and tasks for team members, are developed, team members begin to realize the amount of work that lies ahead. This usually leads to dissatisfaction.

The dissatisfaction stage in team development is characterized by frustration and conflict. During this stage, team members usually rely on traditional management and interpersonal skills. Since they are comfortable with this style, they may forget any new technical, business, and social skills acquired earlier. They need structured support to help them discover that the new approaches are the way to success.

During this phase, the team project leader and the team facilitator need to provide any required support to move the team from dissatisfaction to resolution. They should focus the team on achieving some small specific action based on a defined need of the customer before attacking the whole project. This allows the team to experience success with a task as a team while still receiving support.

During this phase, the team should strive to achieve the following:

- Integrating all team members
- Achieving one success as a team no matter how small
- Applying the customer-driven project management methodology, tools, and techniques
- Defining the project through plans, specifications, policies, and procedures

Production phase and resolution stage

During the production phase of project management, the team must have progressed to at least the resolution stage of team development. This is the stage where the group has moved to being a team. The customer-driven project team must be truly working as a team during this phase to ensure the chances of successful project completion satisfying the customer. The team needs to work together to ready all detailed planning for the operation phase by making sure all processes are interrelated and interfaces ensured. Only through teamwork can all requirements be verified, feasibility tests completed, and support systems prepared.

At some point during this phase, the core team and other project teams must reach the production stage and should be functioning to accomplish and improve their processes. The team has moved from allegiance to functional group to sense of belonging to the customer-driven team. The team believes in its ability to perform. There is more concern for other team members. There is a focus on the issues. Personal conflict is avoided. Consensus decision making is the norm.

The team should focus on the following:

- Maintaining teamwork
- Preparing plans for action
- Ensuring that processes are in control
- Performing and improving the project

Operations phase and production stage

The operations phase of project management compares to the production stage of team development. The team is performing all the actions necessary to complete and improve the project aimed at satisfying the customer. During this phase, all customer-driven project team members and teams function as one unit. The team must be constantly striving to ensure that the deliverable(s) satisfy the customer. The emphasis is on ensuring that the customer is satisfied within an optimal cost and on schedule.

The team has developed to the point where using the customer-driven project management methodology is natural. The team is totally committed to the project. Team members are empowered. Pride is evident. The team is focused on actions. Each performs and improves the process as a natural course of action. Interaction between team members and other teams is commonplace to produce project outputs. The customer-driven project team responds to customer requests to ensure that the complete project satisfies the customer at the right cost and on schedule.

Closeout phase and disband stage

The closeout phase of project management is like the disband stage for the team. During closeout, the organization must decide to finish the project or continue to improve the deliverable(s) to expand its competitive position. If the decision is to stop the project, the customer-driven teams are disbanded or reduced to a support level. If continuous improvement is the choice, the organization can elect to continue the current customer-driven team(s) or start new teams.

Applications of Customer-Driven Teams

Customer-driven teams can be categorized as follows:

1. *Customer-driven project team.* The purpose of this category of customer-driven team is to complete a specific project, program, or task.
2. *Customer-driven quality-improvement team.* The customer-driven quality-improvement team focuses on improving a specific process.
3. *Customer-driven work teams.* Customer-driven work teams constantly perform and improve their particular process.

The basic customer-driven team philosophy, guiding principles, and concepts are the same for all these teams. They have many project management and total quality management processes in common. The major differences are in the application of the specific customer-driven project management methodology, tools, and techniques to achieve the specific focus of the team. The basic customer-driven team concepts apply to each type of team. Each team must develop its own mission, roles, and responsibilities to meet its specific purpose.

All the customer-driven teams have many items in common. First, they must develop collaborative relationships. Relationships must be developed within the function or functions represented on the team. In addition, the customer-driven team must develop relationships with many outside support functions, as well as suppliers. In customer-driven teams, the customer or customer's "voice" leads the team. Each customer-driven team must be empowered to achieve its defined mission. This includes all project management, continuous quality-improvement, and work activities.

Customer-driven project team

This is the customer-driven team that carries the same role as traditional project management. The purpose of this team is to complete a project. The project can be short or long term. In the customer-driven project team, the customer-driven project leader can be external or internal to the organization. For projects involving more than one customer-driven project team, the customer project leaders can be both external and internal. For example, a project involving the development of a new product would have the external customer as the project leader. It would also include internal customer project leaders to guide the performance and improvement of the many programs and processes. A project usually involves many tasks that must meet cost, schedule, and quality standards. Accountability for task performance is shared by the customer-driven project team.

A customer-driven project team can be a project team, program team, or construction team. For example, a customer-driven project team would be formed to

- Produce the B-2 aircraft.
- Develop a training program.
- Install a new computer system.
- Build a house.

Customer-driven quality improvement team

A customer-driven quality improvement team focuses on improving a process. Although a customer-driven quality improvement team may exist in the long term, it usually can complete its improvement activities within a short period of time. There can be many quality improvement teams in an organization. Customer-driven quality improvement teams are usually guided by a customer-driven steering team, just like a project-focused customer-driven project team. Since quality improvement focuses on internal processes, the customer-driven team leader is normally internal to the organization. However, there are cases where an external customer project leader would be appropriate.

Examples of customer-driven quality improvement teams include teams that

- Invent a manufacturing process.
- Reengineer a billing process.
- Increase the efficiency of responding to customers.
- Enhance supplier relations.
- Reduce waste.
- Raise productivity.
- Solve a specific problem.

Customer-driven work teams

The customer-driven work teams focus on constantly performing and improving their work processes. Normally, customer-driven work teams exist in the long term. They function as long as the work or processes are necessary. There can be many customer-driven work teams in an organization. They can exist in traditional blue-collar and white-collar work areas. In some cases, the customer-driven work team consists of both blue-collar and white-collar associates. The customer project leader in customer-driven work teams is usually internal to the organization. In self-directed customer-driven work teams, the team itself is the customer project leader. The customer-driven work team is empowered within defined boundaries to do whatever is necessary to perform and improve their work. Accountability is shared by all members of the customer-driven work team.

Customer-driven work teams are natural work groups. Some examples of customer-driven work teams are

- Design teams
- Shipping teams
- Assembly/manufacture teams
- Logistics support teams
- Customer service teams
- Accounting teams
- Procurement teams

Main Points

Teams, i.e., customer-driven project teams, are the organizational structure of choice to compete in today's global economic environment.

Teams can respond rapidly to changing customer needs and expectations.

A team is a group of people working together for a common goal.

Teams can be single-function, consisting of members from one discipline or organization, or multifunction, consisting of members across functional or organizational boundaries.

Customer-driven teams are accountable for the performance and improvement of a project.

In a customer-driven team, the customer or customer's voice is the team leader.

Top management must be willing to take appropriate actions over the long term for customer-driven teams to succeed.

The customer-driven team is established by determining the focus and forming the team.

The team must fully support the team mission.

The project facilitator helps the team establish and maintain teamwork.

The team must be able to do whatever is necessary to satisfy the customer within defined boundaries.

Process "owners" have pride of ownership of their processes.

A program/process manager is an essential part of the customer-driven team to help the team plan, organize, staff, direct, control, and coordinate the project.

The customer-driven team must be led through the project management phases and team development stages.

Customer-driven project management can be used for the following customer-driven teams:

- Customer-driven project management teams
- Customer-driven quality-improvement teams
- Customer-driven work teams

Chapter

7

People Involvement Tools and Techniques

Focus: This chapter describes tools and techniques to maximize the human resources contribution to a successful project.

Introduction

People are the key to success in customer-driven project management. It is people who perform and improve the project. Customer-driven project management aims to maximize the potential of the human resources in an organization by fostering both individual and team contributions to the organization. Customer-driven project management relies on individuals working intelligently and taking pride in their work. In addition, these individuals' contributions are multiplied through customer-driven teams. People-involvement tools and techniques include individual involvement, teamwork, communication (especially listening), focus setting, meetings, brainstorming, and presentations.

Individual Involvement

Individual involvement concerns each person's contributions to the organization. In customer-driven project management (CDPM), individuals strive to continually perform their work and improve the processes in the organization, focusing on total customer satisfaction.

CDPM seeks to benefit from each individual in the work force. All individuals are different, and each is unique and valuable. This diversity is a distinct advantage in today's economic environment; the organization that learns to use the diversity of its people to improve its competitive position is ahead of the game. People have a variety of attitudes, beliefs, perceptions, behaviors, opinions, and ideas that are potential sources of creativity. Innovation can be gained from different competencies, abilities, knowledge, and skills of the work force. Each person's culture, background, and personality foster an individuality that can be used for the good of the organization. Creativity, innovation, and individuality can be the edge needed for growth. Therefore, in a customer-driven project management organization, individual differences are valued as an important resource.

Although each person is different, people generally want some of the same basic things. They want to be safe and secure, trusted, and appreciated; they want to feel important, have pride in their work, be involved, and feel as if they belong; and they want to have opportunities for advancement and personal growth. The organization that provides a work environment where all these wants can be achieved by individuals will be rewarded with high productivity.

In a customer-driven project management environment, the goal is the actual empowerment of everyone in the organization. Empowerment does not just happen. The organization cannot simply announce that people are empowered and expect it to occur. Typically, empowerment comes in stages. First, people must trust the organization. Typically, most organizations have developed a number of internal adversarial relationships over the years. This internal conflict frequently leads to mistrust between management and workers, organizations and unions, and one department or function and another department or function. Such barriers must be removed before individuals can become involved in any extraordinary effort. Restoring trust may take some time, depending on the organization, and can only be accomplished by the actions of management working through structured activities that foster honest and open communication.

Once trust is restored, people will begin to assume more "ownership" of their work. At this point, the resources must be available to allow each person to take pride in his or her work. When pride in the work is the norm, people can be empowered to provide total customer satisfaction.

With the added emphasis on human resources, people must work more intelligently to perform and improve their work, with a focus on customer satisfaction. People have always known best how to do things right and do them better. However, neither the organization nor the people know how to tap this resource for the benefit of the organization, individuals, and customers. The organization must be transformed to provide an environment where individuals can maximize their potential. At the same time, people must be trained in a systematic process that provides them with the ability to influence their work. When this goal is met, individual involvement can reach its maximum potential.

Individual involvement is fostered by the following:

- *I*nstilling pride of accomplishment
- *N*urturing individual self-esteem
- *D*eveloping an atmosphere of trust and encouragement
- *I*nvolving everyone will make most of individual differences
- *V*isualizing a common purpose
- *I*mproving everything
- *D*emanding effective and open communications
- *U*sing rewards and recognition
- *A*llowing creativity and innovation
- *L*eading by example

The steps in individual involvement are

1. Establish a people-centered environment
2. Provide development opportunities
3. Provide experiences with expected behavior
4. Reward and recognize appropriate behavior

Teams

A *team* is a group of people working together for a common goal. Teams should not be confused with groups. A team shares responsibility, authority, and resources to achieve its collective mission. Team members feel empowered to do whatever is necessary within their defined boundaries. Action through cooperation is practiced both within the team and when the team seeks support. Problem solving and decision making are natural activities. Effective, open, and full communication, especially listening, is prolific. The team leader and team members possess a positive "can do" attitude even during difficult times. Team members motivate, respect, and support each other. Team members manage conflict, build self-esteem, and motivate other team members. They all contribute their technical competence in their speciality as well as all other skills. They acquire many skills so as to accomplish the mission and build and maintain teamwork. Effective teams realize that diversity, individuality, and creativity are their greatest advantages. Individual and team contributions are rewarded and recognized appropriately. The team accepts "ownership" of and takes pride in team performance. Everyone is totally committed to cost, schedule, and quality standards of excellence, with total customer satisfaction the primary focus of all team activities.

Types of Teams

Teams can be either functional or multifunctional. A *functional* team consists of members from the same discipline or organization. For example, in an engineering functional team, all the members would work in the engineering department. A *multifunctional* team would have members from engineering, manufacturing, marketing, and other areas as appropriate.

Teamwork

Teamwork is a technique whereby individual team members work together to achieve a common goal. It involves cooperative relationships, open communication, group problem solving, and consensus decision making. It can be effective only in an environment of honesty, trust, open communications, individual involvement, pride of accomplishment, and commitment. Specifically, effective teamwork involves

- *T*rust
- *E*ffective communication, especially listening
- *A* positive "can do" attitude
- *M*otivation to perform and improve
- "*W*e" mentality
- "*O*wnership" of work with pride
- *R*espect and consideration of others
- *K*eeping focus on total customer satisfaction

Benefits of teamwork

Teamwork provides the responsive work force required to survive in today's economic environment. Cooperation toward a common goal is essential for success. Some of the benefits of teamwork include

- *B*etter decisions and motivation
- *E*ncourage participation by everyone
- *N*urtures improved working relationships
- *E*ncourages rewards in the work itself
- *F*reer contribution of information
- *I*ncreased communication
- *T*hrusts an organizational focus
- *S*upports an organizationwide perspective

Principles of teamwork

In order to build and maintain teamwork, the team must follow some basic principles, such as these key principles:

- *K*eeping focused on the mission, not making it personal
- *E*ncouraging open communication and active listening
- *Y*earning for constructive relationships

In addition to these key principles, every team also must attempt to build and maintain teamwork over the long term. The team must be continuously maintaining and developing teamwork. Individual team members and the team as a whole must receive appropriate rewards and recognition so as to maintain an interest in the work of the team. Further, all members must be involved in team activities to maximize the true potential of the team. Team members must have enough self-esteem to contribute actively toward the team's goal. Communication is essential in any team activity. In addition, the strength of the team lies in the individuality of each of its members. Constructive cooperation is critical both within and outside the team. Relationships are important between team members and with customers, suppliers, and other teams. All the members, especially the team leader, must set the example. Team members can develop the behaviors necessary to work as a team through observation. Ideas are the power of the team. All team members must be encouraged to continually contribute innovative and creative ideas. Above all, the team must focus on the mission, not on the person. The work is not personal. Teamwork demands an unrelenting devotion to a common purpose. The basic principles of effective teamwork can be summarized as follows:

- *P*ursue a team environment.
- *R*eward and recognize the individual and the team.
- *I*nvolve all team members.
- *N*urture the self-esteem of all team members.
- *C*ommunicate freely and openly.
- *I*nclude individuality.
- *P*ursue constructive relationships.
- *L*ead by example.
- *E*ncourage all team members' ideas.
- *S*tay focused on the mission.

Building teamwork

Team building revolves around continuously diagnosing and improving the effectiveness of the team. In order to build team cohesiveness and effectiveness, it is important to pay particular attention to the roles and responsibilities, group dynamics, and interpersonal relationships within the team, as well as the team mission. The following actions are essential to building teamwork:

- Identify the team mission
- Establish roles and responsibilities
- Understand team dynamics
- Manage conflict
- Provide motivation
- Build individual self-esteem
- Critique teamwork

Identify the team mission. The team mission is the intended result, the focus for all team activities. The mission provides an indication of the magnitude and expected outcome(s) of the project. The mission should state the boundaries of the project and include specific process(es) involved. It is important for the mission to define the authority of the team. Further, the resources the team can employ to accomplish the mission must be identified. Normally, the mission originates from outside the team, coming in general terms from a variety of sources, such as management or the customer. This general mission must be negotiated and clarified by the team.

The mission should be specified in a written mission statement. Clarification of the mission should be the first outcome-related activity of the team. The mission statement must be clear and achievable, and the team must reach consensus on the meaning of the mission statement before doing any other team activity. Teamwork requires unrelenting devotion to a common purpose for success. The mission provides this common purpose.

Establish roles and responsibilities. Roles and responsibilities are the specific contributions expected from each team member in the attempt to accomplish the mission. These contributions can include any formal or informal offerings each team member brings to the team. Formal contributions include the expected roles and responsibilities of a specific discipline, function, or organization. Informal offerings are the contributions a team member can add as a result of personal strengths. Each team must develop its own unique roles and responsibilities based on the requirements of the mission and the capabilities of its members.

Roles and responsibilities must be defined in a "living document" developed by the team. Each team member must have distinct responsibilities and corresponding accountability. The roles and responsibilities change as the team develops and the project progresses. Developing an understanding of the initial roles and responsibilities should be the next team activity after agreeing to the mission statement. Team members' roles and responsibilities should include

- A clear understanding of the results and outcome(s) expected from each team member
- "Ownership" of the work, including the amount of control
- A grasp of the limits of resources, including funds, equipment, and people
- Empowerment and the amount of authority it carries
- Standards focusing on customer satisfaction

The roles and responsibilities should spell out the expected outcomes for each team member in terms relating to each member's contribution to the mission. If possible, the responsibilities should be stated in terms of metrics. In the initial stage of a project it many not be possible to include specific measurements, but performance measurements must be included as soon as possible. Each team member needs to know exactly what he or she needs to do.

Another part of roles and responsibilities involves "ownership." The roles and responsibilities must state which processes each team member owns, thus providing each team member a statement of what he or she does.

Critical to performance of roles and responsibilities is the amount of resources available. Again, these resources should be detailed so as to provide each team member with a statement of exactly what is available to accomplish the job.

Empowerment involves having the responsibility, authority, and resources to do whatever is required to satisfy the customer and achieve the mission within defined boundaries. The key to empowerment is defined boundaries. Each team member must know the boundaries, even though these boundaries will change as the team develops and the project progresses. In the beginning, team members usually are not fully empowered, but as they are trained and gain new experiences, each can assume more empowerment. Eventually, the team becomes fully empowered, maximizing the potential of the team through the creativity and innovation of team members.

Standards are an essential part of roles and responsibilities. They are the accepted norms for all team members, focusing on customer satisfaction. Standards must be a clear definition of what is acceptable performance.

Specific team roles and responsibilities. Each team consists of a team leader, team members, and sometimes a team facilitator. Each team player has a specific role. The team leader guides the team to mission accomplishment, the team members contribute toward achieving the mission, and the team facilitator assists the team with focus, teamwork, methodology, tools, and techniques. Detailed roles and responsibilities for each of these team players are discussed in Chap. 6.

The roles of the team leader and team members depend on the category of team. Figure 7.1 shows four categories of teams. The first category of team is the traditional *directive organization* with a manager. The role of the manager in this team is to get the task accomplished. The role of the team member is strictly to perform the directed job. The second category of team is a *participative organization.* The leader guides the team to a common goal through a process involving all team members. The team members provide their expertise and cooperation. The third category of team is a *collective self-led organization.* In this team, "ownership" is shared by all team members. A team facilitator creates and maintains teamwork. The fourth category is an *empowered organization.* An empowered team has the total responsibility, authority, and resources to perform and improve its process(es). In this category of team organization, a coach and/or resource person advises the teams.

Understand team dynamics. Each team must understand that although it is unique, all teams normally go through four distinct stages before they are truly performing as a team. As shown in Fig. 7.2, the four stages of team development are orientation, dissatisfaction, resolution, and production.

Each team must go through all four of the stages of team development before it reaches synergy. There is no shortcut. The duration and intensity of each stage vary with each team. It is important to maintain the focus and a positive attitude throughout all the stages—the team *will* achieve its mission.

A general description of each of the stages of team development follows. A detailed description of the stages of team development for a customer-driven team is presented in Chap. 6.

Stage 1: Orientation. During the first stage, team members spend their time becoming acquainted with each other and with the work of the team. Members attempt to build rapport, honesty, trust, and open

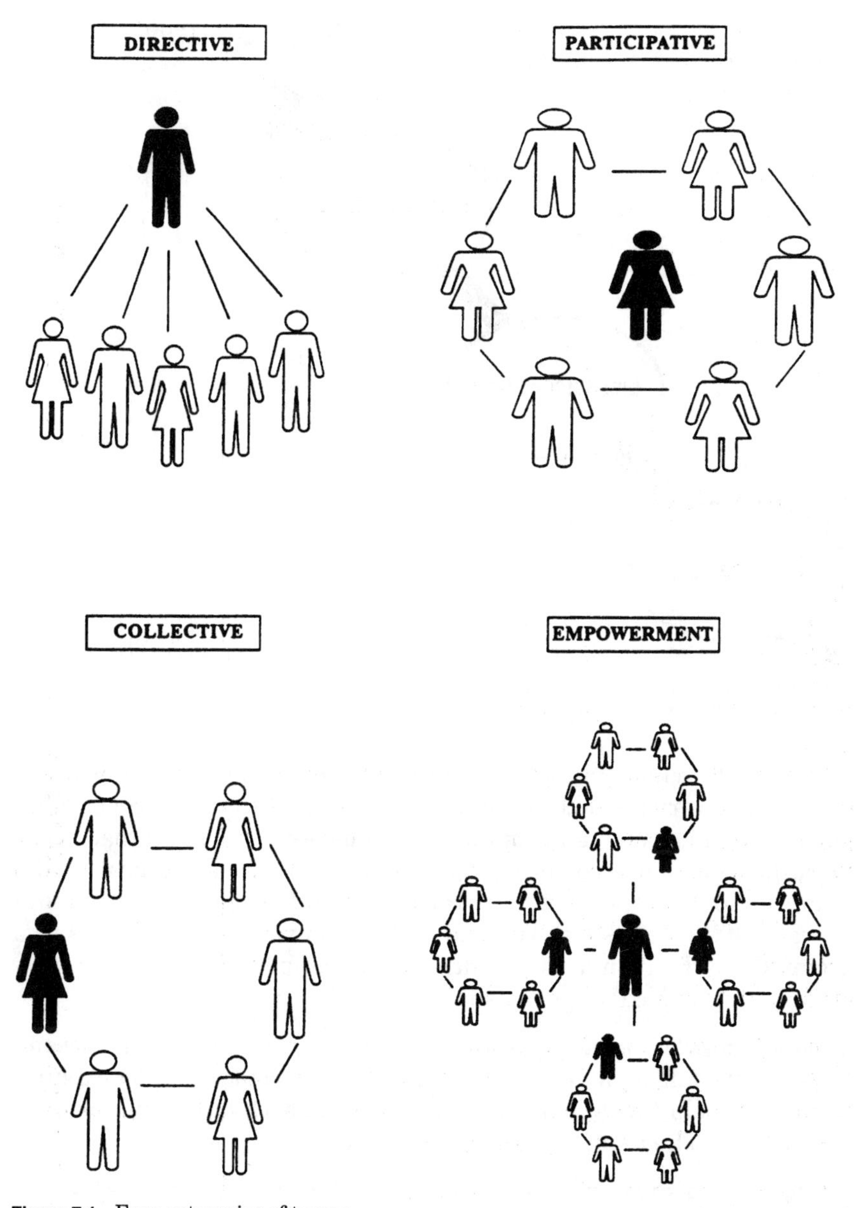

Figure 7.1 Four categories of teams.

communication. They try to determine what it takes to fit in. Team members frequently have great enthusiasm for the project, but they do not know how to work as a team to accomplish it. During this stage, the team is deciding what it needs to accomplish and who will be responsible for accomplishing it.

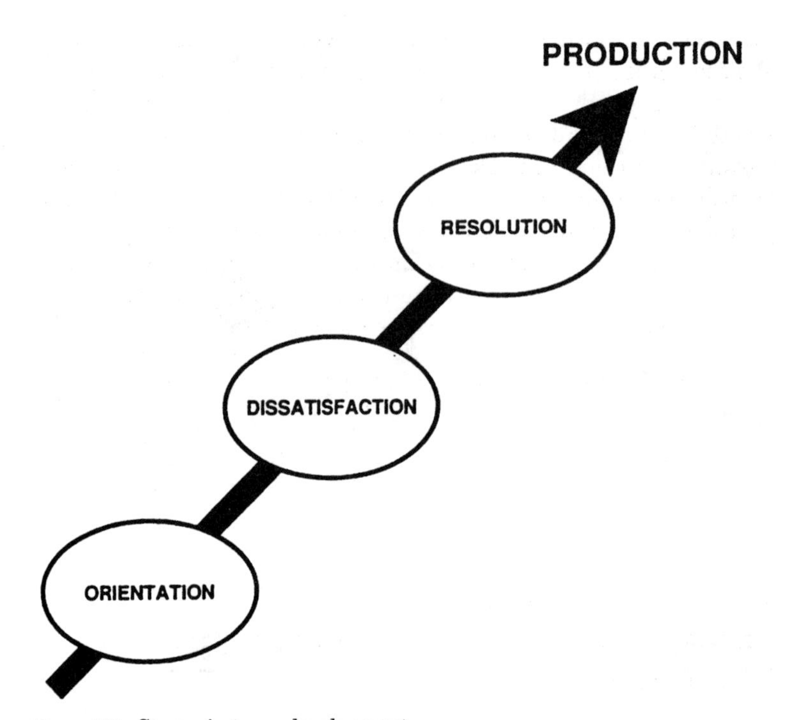

Figure 7.2 Stages in team development.

Stage 2: Dissatisfaction. Stage 2 is characterized by the team members being overwhelmed by the information and the task. Sometimes power struggles, emotions, and egos become evident. This stage is the most difficult to overcome, some teams never progressing past it. If a team cannot progress past this stage, it should be disbanded. To move forward to the next stage, the team must find some small success as a group. Once the team understands that it can perform as a team, progression to the next stage is usual.

Stage 3: Resolution. During stage 3, the team moves toward accomplishing its mission. In this stage, customer contact and measurements can help team members start to assist each other and focus on the mission. The team actually works as a team.

Stage 4: Production. Finally, in stage 4, the team becomes effective. The team members work together to achieve the mission. The team is using the full potential of all team members as efficiently and effectively as possible toward mission accomplishment.

Manage conflict. Conflict can exist whenever two or more people get together. Differences exist in every organization, and these differences are an advantage to any organization that has learned to manage conflict. Conflict can be positive, and agreement may be negative.

Differences exist in every organization. Diversity is one of an organization's major strengths. Teams must take advantage of their differences to be successful, and major benefits can be gained by paying attention to differences. For example, an organization can use its people of different cultures and backgrounds to research a potential new market or product targeted toward a specific culture or background. Further, an organization can gain new ideas from a diverse work force, and the new ideas can lead to improved operations, decreased cost, and/or time savings. The following is a list of potential sources of conflict that can be beneficial to an organization:

- *C*ultures and backgrounds
- *O*pinions
- *N*eeds and expectations
- *F*acts and perceptions
- *L*evels, departments, and organizations
- *I*nterests, personalities, and egos
- *C*ompetencies, knowledge, and skills
- *T*argets, missions, goals, and objectives

Conflict can be controlled. Cooperate rather than compete. Orient toward the issue, not the person. Negotiate win/win solutions. Take an organizationwide perspective. Recognize conflict as natural. Observe empathy with others' views. Limit perceived status differences.

Conflict can be positive. Conflict leads to the pursuit of win/win solutions. It allows team members to observe other team members' points of view. Conflict displays the team working through open communication. It forces the team to take an organizationwide view. When focused on mission, conflict takes personalities out of an issue. Conflict invites trust and involvement while accommodating a view of the entire issue, and it provides the opportunity to examine different sides of an issue. All this leads to effective consensus decision making, which establishes and maintains teamwork.

Agreement can be negative. This is commonly called *groupthink.* Groupthink is the tendency of groups to agree even though that agreement may have an adverse effect on the ability of the team to achieve its mission. Groupthink comes from many sources. Sometimes groupthink results from the good intention of maintaining the cohesiveness of the team. In other cases, groupthink stems from fear. Team members may be afraid of losing their jobs, losing face, or offending the leader, management, or other team members. Regardless of the source, groupthink must be identified and controlled. The following are some specific actions to overcome groupthink:

- *A*ppoint a devil's advocate.
- *G*et open discussion on all issues.
- *R*ecognize the impact of status differences.
- *E*xamine all agreement without resistance.
- *E*valuate all views/sides of the issue.

Conflict symptoms. The first step in managing conflict is recognizing that conflict exists. Everyone on the team, including the team leader, team members, and especially the team facilitator, must be alert constantly to the symptoms of conflict and groupthink. Some of the symptoms of conflict and groupthink are

- *S*topping open communication
- *Y*ielding to win/lose solutions
- *M*aking little movement toward solution
- *P*ressure to stop challenges
- *T*aking sides (we/they)
- *O*bserving no building on suggestions
- *M*embers silent
- *S*topping any resistance

Conflict management actions. Conflict can be managed during the day-to-day operations of the team. First, avoid face-saving situations. If honor and pride are at stake, people will defend their position even when they realize that they may not have the answer. Second, continuously self-examine attitudes. Sometimes a person may develop an attitude triggered by an emotional response. Such an attitude may be detrimental to teamwork. Focus on the mission, and maintain a positive attitude throughout all team activities. Third, target win/win solutions to allow the team to avoid we/they situations. Fourth, involve everyone in all team activities. People do not disagree with their own contributions. If all team members participate, each will support the decision. Fifth, observe the limits of arguing. Arguing is useless. It does not lead to positive solutions. Sixth, nurture differences of opinion. Everyone is right in their own mind. There are no right and wrong answers. Differences of opinion can be used to stimulate other ideas. Seventh, support constructive relationships. Relationships are the key to all teamwork. Build long-term relationships on a foundation of honesty and trust, thus fostering open and free communication—the real key to conflict management.

Provide motivation. Motivation is the behavior of an individual whose energy is selectively directed toward a goal. Performance is the result of having both the ability and the motivation to do a task. Motivation influences team members to certain behaviors. Motivation depends on satisfying the needs of individuals. Traditionally, motivation was equated with extrinsic rewards such as compensation, promotion, and additional benefits. The aim was to satisfy the basic needs of individuals for housing, food, and clothing. Today, people need to be motivated by a higher order of needs, such as a sense of belonging, a feeling of accomplishment, improved self-esteem, and opportunities for personal growth. Teamwork, especially that in customer-driven teams, provides intrinsic rewards.

Rewards and recognition for individual and team performance are essential to the promotion of teamwork. The intrinsic rewards are usually sufficient to start teams. Once a team is established, team members covet higher-level intrinsic rewards. An example of a reward that is effective in today's environment at this stage is inclusion in personal development workshops. During all the stages of team development, recognition is particularly effective in reinforcing positive behaviors. Praise and celebrations are necessary to maintain teamwork. Some examples of recognition include letters of appreciation, pizza parties, coffee and donuts, and public announcements. Particularly effective is a pat on the back with a "you did a good job" comment. In the early stages, extrinsic rewards have a short-term effect, and they may actually be a negative motivator for long-term teamwork. Extrinsic rewards are important for long-term teamwork, but they must be appropriate for the desired outcomes. Before any rewards are instituted, they must be analyzed thoroughly to ensure that everyone is treated fairly.

Besides rewards and recognition, the team can provide motivation to team members. The following are some specific actions the team can use to motivate team members:

- *M*ake it clear that the goal is shared.
- *O*rient, develop, and integrate team members.
- *T*hink and speak "we."
- *I*nstitute internal team rewards and recognitions.
- *V*alue individual contributions.
- *A*void frequent changes of team members.
- *T*ake time to develop relationships.
- *E*ncourage a sense of belonging.

Build individual self-esteem. An individual's self-esteem affects his or her performance of organizational tasks as well as his or her relationships with others on the team. There are actions that each team member can do to maintain and build the self-esteem of other team members. They are as follows:

- *E*stablish an environment in which an individual feels that his or her self-worth is important to performance.
- *S*tay focused on the mission; do not make it personal.
- *T*reat each person as you would want to be treated.
- *E*ncourage individual contributions.
- *E*nsure individual rewards and recognitions.
- *M*otivate, communicate, involve, and develop.

Critique teamwork. Periodically, the team should perform a self-assessment of team development. Each team should develop its own critique based on its criteria of a successful team. This critique should be completed individually, and the results should be tabulated, evaluated, and discussed as a team. The teamwork critique should be performed on a regular schedule. Figure 7.3 presents an example of a teamwork critique.

Communication

Communication is the most important tool in customer-driven project management. Communication involves exchanging information, and customer-driven project management demands a free flow of information. The success of customer-driven project management demands communicating with and among all team members. It also requires frequent and effective communications with people and groups outside the team. The customer-driven team needs information to understand the needs and expectations of the customer. The team also needs information from each of its members to complete and improve the project. Teams as a whole rely on information from their support teams. There must be constant communication between customers, process owners, program managers, suppliers, other support teams, and the functional organization. Communication coupled with the sharing of the right information is vital.

Communication of the right information is a complex process that includes verbal and nonverbal forms of communication such as speaking, listening, observing, writing, and reading. Because of this complexity, the information may not be communicated correctly or clearly. Even in the simplest communication model with just a sender, message, and receiver, there are many obstacles to effective communication. For communication to be effective, the sender must be credible, the message must be clear, and the receiver must interpret the

Teamwork Critique

Instructions Please rate the team based on the 5 point scales below. Circle the number on each scale that best states your opinion at this time. Discuss with team.

1. Trust

Is the level of trust among team members sufficient to allow open and honest communication without tension?

close/ tense	1	2	3	4	5	open/ relaxed

2. Effective Communication, especially listening

Do team members listen to each other?

members do not listen	1	2	3	4	5	members listen

Does everyone have a chance to express their ideas?

no ideas expressed	1	2	3	4	5	variety of ideas expressed

3. Attitude, positive "can do"

Do team members display a willingness to take risks?

avoid risk	1	2	3	4	5	take risk

4. Motivation

Are team members actively participating?

bored/ withdrawn	1	2	3	4	5	involved/ interested

5. "We" Mentality

Do team members demonstrate a togetherness in words and actions? Are decisions based on consensus?

individual contribution/ no consensus sought	1	2	3	4	5	team action/ consensus

6. Ownership

Do team members take the initiative to solve problems and/or improve their process as a natural course of action?

only do what told	1	2	3	4	5	take action to make things better

7. Respect, consideration of others

Do team members respect differences? Are people's differences managed to the team's advantage?

avoid others/ conflict	1	2	3	4	5	respect others/ manage conflict

8. Keeping Focused

Does the team remain targeted on vision, mission, and goals?

off target	1	2	3	4	5	on target

Figure 7.3 Teamwork critique.

message the way the sender intended. For example, if the sender is not trusted by the receiver, the receiver may not believe the sender. Regardless of the message, communication will be ineffective.

Communication gets even more complex if we add reality to the model. Rarely do we communicate with just a sender, message, and receiver. Normally, there are many distractions. We are influenced by our work environment, i.e., political pressure or fear. We are thinking about things at home or other things in the workplace. We have different values, cultural biases, perceptions, and so forth from the message sender. Communication can be improved by the following:

- *C*larify the message.
- *O*bserve body language.
- *M*aintain everyone's self-esteem.
- *M*ake your point short and simple.
- *U*nderstand others' points of view.
- *N*urture others' feelings.
- *I*nvolve yourself in the message.
- *C*omprehend the message.
- *A*ttend to the messages of others.
- *T*alk judiciously.
- *E*mphasize listening.

Because of the possibility of ineffective communication, it is critical to ensure through feedback that the right information is communicated. It is always the responsibility of the sender to ensure that effective communication has occurred. *Feedback* involves providing information back to the sender to verify the communication. Feedback can indicate agreement, disagreement, or indifference. Feedback, like communication, can be verbal or nonverbal. Some guidelines to effective feedback follow:

- *F*oster an environment conducive to sharing feedback.
- *E*ncourage feedback as a matter of routine.
- *E*stablish guidelines for providing feedback.
- *D*iscuss all unclear communications, paraphrase, and summarize.
- *B*e direct with feedback.
- *A*sk questions to get better understanding.
- *C*onsider "real" feelings of team members.
- *K*eep focused on the mission.

Listening

Listening is a technique for receiving and understanding information. Listening skills are critical to effective teamwork. Listening is one of our most important communication needs, but it is the least developed skill. Effective listening requires an effort to understand the ideas and feelings the other person is trying to communicate. An effective listener hears the content and the emotion behind the message. Expert listening requires active behavior and effort. It requires attention to the person and the message. An active listener attends not only to what the person is saying but also to gestures, posture, and vocal qualities. Thus, an active listener communicates that he or she is listening and trying to understand the other person. Active listening requires discipline, concentration, and practice. Effective listening requires the following:

- *L*etting others convey their message.
- *I*nvolving yourself in the message.
- *S*ummarizing and paraphrasing frequently.
- *T*alking only to clarify.
- *E*mpathizing with others' views.
- *N*urturing active listening skills.

Let the other person convey his or her message without interrupting or forcing your own views. Let the other person know that you are interested in what he or she is communicating without displaying an opinion or judgment.

Involve yourself in the message by actively listening to what the other person is communicating. Establish and maintain eye contact. Keep an alert posture. Look for verbal and nonverbal cues.

Summarize and paraphrase frequently to show an understanding of the message. By listening carefully and then rephrasing the content and feelings of the other person's message, the exact meaning of the message can be determined.

Ask questions to clarify points you do not understand. Points can be clarified by using open-ended questions. This type of question with an answer other than yes or no provides a more detailed explanation.

Understanding the other person's views is essential to effective listening. Set aside your opinions and judgments, and place yourself in the other person's place. Show the other person that you understand by requesting more information or by sharing a similar feeling or experience you have had and how you think it helps you understand the other person.

Nurture listening skills to improve communication. Listening skills must be practiced daily.

Benefits of active listening

Active listening is a skill that provides many benefits. These benefits make it in the best interest of any person to develop active listening skills. Here is a list of the benefits:

- A better understanding of the communication
- Effective communication
- Personal growth
- A position of trust in the organization
- Improved self-esteem
- An increase in competence
- Achievement of influence in the organization
- A savings in time

Focus Setting

Focus setting is a technique that allows one to move toward a specific outcome. The focus should be on the output of the process satisfying the customer. The focus is a vision, mission, and goals. The vision is the overall view of the organization for the future. The mission is the intended result. The goal is the specific desired outcome. To achieve results, customer-driven teams must focus on their specific purpose within the overall vision of the organization. Such teams must always strive for excellence while satisfying the customer. The purpose of the team is the focus, and the focus for the team is the mission. To achieve the mission, the team must accomplish goals.

Vision

A *vision* is the long-range focus for an organization or the team. The vision is usually the view of the future held by the organization's leadership. Although leadership creates the vision, it must be instituted throughout the organization. For this to occur, it must have meaning and be shared by everyone in the organization. It must be more than a slogan. It must foster some definite course of action that must be displayed by the organization's leadership. To have relevance to where the organization wants to go, the vision must be oriented toward the customer. The vision must communicate both inside and outside the organization a long-term picture for the organization. This vision must be constantly communicated to build the loyalty and trust necessary to develop a work force committed to its achievement.

The following are some guidelines for creating a vision:

- *V*iew the future.
- *I*nstitute the vision in the organization.
- *S*et the example through leadership.
- *I*nclude the direction in which the organization should go.
- *O*rient toward the customer.
- *N*urture through constant communication.

Vision examples. "GenCorp will be one of the most respected diversified companies in the world" (GenCorp). "To have the most effective fighting force" (U.S. Department of Defense). "Our vision is to become an organization that is internationally known as the premier regional quality and productivity center in the United States. Distinguished by its success in helping to create Total Quality organizations through assessment, training, networking and consulting services" (Maryland Center for Quality and Productivity).

Mission

The *mission* is the intended result. It should be stated in a results-oriented form. Specific details on missions for customer-driven teams are given in Chap. 6.

Mission examples. Figure 7.4 presents a mission statement for a customer-driven quality-improvement team. Other examples of mission statements for customer-driven teams are as follows:

- Provide a deliverable meeting the customer's total satisfaction.
- Continuously improve the deliverable's value to the customer.
- Eliminate errors in order processing.
- Decrease the cost of manufacturing.
- Improve assembly workmanship.
- Reduce failure rates of circuit boards.

Another example of mission statement follows: "The University Center for Quality and Productivity (MCQP) inspires positive organizational change to accelerate continuous performance improvement, supporting the long-term economic health of Maryland and the region."

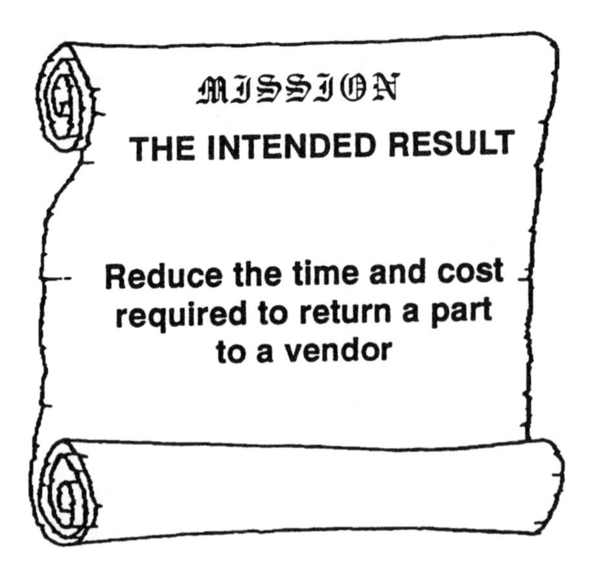

Figure 7.4 Mission statement example.

Goal

The goal is the specific desired outcome(s). It should be specific, measurable, attainable, results-oriented, and time-bound. Set a reasonable goal but do not set sights too low. Set a goal that will be a challenge. Orient goals to specific measurable results, and link goals to customer requirements.

*G*ear to specific results. Define within specific parameters.

*O*bserve by measurement. Check through identified system of measurement.

*A*ttain success. Challenge yourself, but include a high degree of success.

*L*imit to a specific time period. Define within a specific time period.

*S*et by an individual or the group. Determine by the people who make it happen.

Goal example. Figure 7.5 provides an example of a goal for customer-driven quality-improvement team. Some other examples of goals are as follows:

- Reduce manufacturing cycle time for assembly X from 6 to 2 hours within 1 month.
- Decrease errors in quantity required block on order processing sheet from 10 per month to 0 in 3 months.

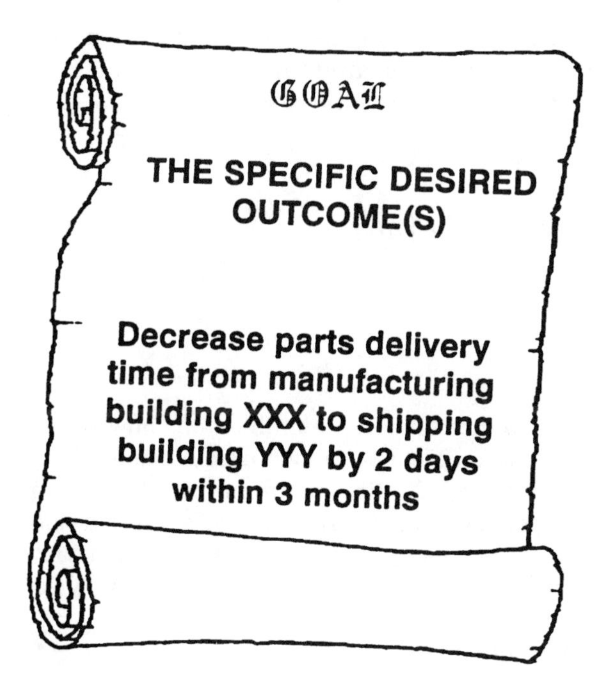

Figure 7.5 Goal statement example.

- Reduce rework on process A from 50 to 20 percent in 2 months.
- Produce a training program for customer-driven project management tools and techniques within 4 months.

Focus-setting tips

*F*ocus on the output of the process.

*O*rient toward the customer.

*C*ommit resources.

*U*nderstand the specific purpose.

*S*et by people who own the purpose.

Meetings

Meetings are a way of bringing a team together to work for a common goal. Effective meetings are an important aspect of customer-driven project management to get the team to develop improvements that an individual could not come up with. By bringing together people in a meeting to develop improvements for a common goal, better decisions can result. The key is making meetings effective. Effective meetings require an action-oriented focus. All team members must have a common focus and a common methodology geared toward specific actions.

Meetings can be made effective through the use of specific meeting tools, such as

- Rules of conduct
- Roles, responsibilities, and relationships
- A focus statement
- An agenda

Rules of conduct

Rules of conduct provide guidance for team members on "how" meetings will be conducted. Each team creates its own unique rules of conduct, which are determined during the first team meeting by consensus. The rules of conduct open communications for the team in a nonthreatening manner. They are posted during every team activity. Although they are established during the first team meeting, these rules can be changed at any time.

The rules of conduct consider the following:

- *Commitment of team members.* A rule on the amount of participation that might be appropriate for each team member.
- *"Owners" of meeting roles.* A rule to identify the specific meeting roles of the team leader, team members, team facilitator, and meeting recorder.
- *Negotiation process.* A rule for outlining the negotiation process.
- *Decision-making process.* A rule for delineating the process of decision making.
- *Unity issues.* Rules for maintaining the team's cohesiveness.
- *Communications procedures.* Procedures for allowing all members an opportunity to communicate on all issues.
- *Time management.* Rules for the start and end of meetings. Also, rules for conformance to the agenda may be needed by some teams.

Rules of conduct examples

- *R*ely on facts not opinions.
- *U*nderstand others' points of view.
- *L*isten actively to all ideas.
- *E*ncourage others.
- *S*ubmit assignments on time.

- *O*pen communication of all issues.
- *F*ocus.
- *C*ome to meetings on time.
- *O*rient toward customer satisfaction.
- *N*ever gossip about the meeting or team.
- *D*ecide everything by consensus.
- *U*se and build on everyone's ideas.
- *C*onduct the meeting using an agenda.
- *T*ake the time to self-critique the meeting.

Roles, responsibilities, and relationships

Besides normal team functions, team meetings involve additional roles, responsibilities, and relationships, that must be defined specifically. The team leader guides the team to mission accomplishment, and the team leader may guide the team during team meetings. Team members are expected to prepare for, participate in, and perform during team meetings. The team facilitator helps the team focus and apply methods, tools, and techniques during the meeting. In addition, for team meetings a recorder is needed. The recorder prepares all the administration documentation for the meeting, and could include such items as the agenda, minutes, assumptions, and a list of definitions.

In addition to roles and responsibilities, each team member must understand the relationships that exist. These relationships can have an effect on the meeting. The relationships involve the team as a whole, other team members, the organization as a whole, the functional organization, and self. A conflict in any of these relationships could cause a team meeting to be canceled or ineffective. These potential conflicts should be resolved as early as possible to ensure maximum participation by all team members.

Focus statement

A focus statement provides the purpose of a meeting. Each team meeting must have a written focus statement. If the team cannot write a focus statement, there is no need to hold a meeting. The focus statement should provide the following:

- *F*ocus for the entire meeting
- *O*utput expected from the meeting
- *C*lear, concise, and simple statement of direction
- *U*nderstanding for everyone on the team
- *S*tart for the agenda

Agenda

An agenda acts as a meeting guide. It gets the team to focus on the meeting's desired outcomes. An agenda encourages effective and efficient meetings because it provide a target for the meeting. It documents key team activities and acts to stimulate progress. An example of a meeting agenda is shown in Fig. 7.6. As you can see, the agenda

MEETING AGENDA

NAME OF TEAM: AS OF DATE:

START TIME: END TIME:

MEETING FOCUS STATEMENT

AGENDA ITEM	OWNER(S)	TIME

ACTION ITEMS	OWNER(S)	STATUS

COMMENTS

Figure 7.6 An agenda outline.

shows exactly what is expected during the meeting. It also provides desired outcome and action items with follow-up status.

Team meetings in action

In addition to the meeting tools just mentioned, actions must be taken before, during, and after each meeting to ensure proper preparation, conduct, and follow-up.

Initial meeting actions. It is important to get the team started correctly. During the first meeting, the following should be accomplished:

1. Establish rules of conduct.
2. Understand the mission of the team.
3. Establish roles and responsibilities.
4. Develop next agenda.

Before the meeting. The success of the team depends on the active involvement of all the team members. Team members should participate fully in all meetings. The following are some meaningful guidelines to assist the team in conducting an effective meeting:

- *Brainstorm ideas.* Review the focus statement, and write down your ideas on everything you know about the focus.
- *Evaluate what you know.* Start with ideas you brainstormed, and gather any addition information you may need. Analyze the information, trying to determine the specific opportunity, problem, or root cause.
- *Formulate alternatives.* Generate a list of alternatives to accomplish the focus.
- *Orient toward one alternative.* Determine one alternative you can support—your starting position based on the information you know. During the meeting, you may change your alternative based on additional information provided by other team members.
- *Review the agenda.* Do this to ensure that you are prepared with information, status, or assignments.
- *Ensure that you complete any assignments.* The team depends on you to accomplish your specific actions. Even if you cannot attend the meeting, try to make sure your assignments are on time.

During the meeting. During the meeting, speaking, listening, and cooperating are the key activities for all team members. Speak to make your point; present and clarify ideas. Listen actively. Cooperate with all other team members. During the meeting, do the following:

- *D*isplay teamwork.
- *U*nderstand the viewpoints of others.
- *R*emain focused.
- *I*nvolve yourself.
- *N*urture others' ideas.
- *G*o for win/win solutions.

During the meeting, *speak* to share information, but be short, simple, and concise. Plan what you are going to say before you say it. This will help you focus and save time. Encourage the building of ideas. This stimulates interest and involvement. Although you or others may not have anything to contribute initially, many people can add their ideas to others. Avoid personal remarks. Keep remarks focused on the mission, goal, problem, and issue. Also, avoid any words that may trigger an emotional reaction. These types of words may refer to race, sex, religion, politics, and the like.

Again, *listening* is essential during a meeting. Let other people convey their messages. Do not interrupt other people while they are speaking. Involve yourself in the message. Look for ideas you can support. Determine the central theme or concepts. Summarize and paraphrase frequently. This provides the speaker with feedback on the success of his or her communication. It also is the only way to confirm your understanding of the information. Further, there may be another team member who does not understand. All critical ideas must be repeated by another member and discussed to ensure the clarity necessary for consensus decision making. Talk only to clarify while you are listening. Effective listening requires your full concentration. Empathize with other people. In other words, put yourself in their shoes for awhile. You do not have to sympathize with them. Empathy helps you understand; sympathy may actually be a barrier. Nurture active listening skills. Active effective listening is not natural. It requires dedicated concentration of effort.

Cooperation makes a meeting work. Consider the self-esteem of others. This will give them the confidence to participate. Operate with the team, and give others a fair chance. Do not go outside the team to seek action or talk about other team members. Observe others' reactions to provide feedback on true reactions. Use this to find common ground for negotiations for win/win solutions. Pursue a common focus. As long as the team focuses on a common goal, the team can work. Many times a common focus overcomes conflicts, just as peer pressure to achieve shared results overshadows the personal needs of team members. Establish open communications, which are necessary for any cooperative effort. Recognize individual contributions, because this helps stimulate more participation. Allow positive conflict so as to lead to consensus decision making. The team will support a decision better if positive debate was endorsed during the meeting. Trade off ideas with the group so as to distribute "ownership" to the whole team. Encourage trust, the most important ingredient to developing and keeping cooperation on the team. Only with trust can there be real cooperation.

After the meeting. Once the meeting is over, the real team actions are performed. Team members act to perform their assignments and action items. Finding support and resources may be necessary after the meeting. A team member coordinates with management or a support function to ensure that the team can complete actions or implement a solution. All team members must "talk up" team activities to develop pride for their team in the organization. This gives all team members a feeling of belonging to a worthwhile team. It also help promote teamwork throughout the whole organization. Further, such actions are necessary to maintain team integrity. Finally, team members must review the agenda of the next meeting so as to begin preparing for that meeting.

Meeting critique

Some teams find it useful to perform a self-assessment at the end of each meeting. This is particularly beneficial when a team is just starting out. It provides a means to develop the skills required for effective team meetings while also fostering teamwork through finding success working on a non-mission-related activity. The more successes a team has as a team, the easier the team can develop and maintain teamwork. The team needs to design its own meeting critique. This critique should be completed as a team at the end of each meeting. It should take no longer than 5 to 10 minutes. The critique should address the following:

- *Communications.* Was communication effective? Was there discussion on all items?
- ***R**esults.* Was the focus statement accomplished?
- ***I**nvolvement.* Did everyone participate?
- ***T**raining.* Does the team require any specific training?
- ***I**ndividuals.* Were individual contributions recognized? Does any team member require more attention?
- ***Q**uestions.* Are there any items that require further research?
- ***U**nity.* Did the team work together? Was there any evidence of conflict? Groupthink?
- ***E**scalate.* Are there any issues requiring management resolution?

Brainstorming

Brainstorming is a group technique that encourages collective thinking to create ideas. The purpose of brainstorming is to stimulate the generation of ideas. It adds to the creative power of the team. The value of brainstorming lies in the fact that there may be more than one way to look at a problem or handle it. Through brainstorming, individual ideas or thoughts are not only brought out but may spark new ideas or thoughts from others, or improve on an idea already under consideration. The more ideas a team has, the greater is the probability of finding an opportunity or solution. Brainstorming accomplishes the following:

- *B*rings out the most ideas in the shortest time.
- *R*educes the need to give "right" answers.
- *A*llows the group to have fun.
- *I*ncreases involvement and participation.
- *N*urtures positive thinking.
- *S*olicits varying ideas and concepts.
- *T*empers negative attitudes.
- *O*mits criticism and evaluation of ideas.
- *R*esults in improved solutions.
- *M*aximizes the attainment of goals.

Brainstorming rules

For brainstorming to work effectively, the group leader must make sure that the principles of brainstorming are followed. Thus, each member must know the rules and follow them. It is a good idea to review the rules before each meeting until the group has established its brainstorming approach. The rules of brainstorming are as follows:

1. *R*ecord all ideas.
2. *U*se freewheeling ideas.
3. *L*imit judgment until later.
4. *E*ncourage participation by everyone.
5. *S*olicit quantity.

Let's look at each of these rules in more detail.

1. *Record all ideas.* Team members learn over and over the importance of recording things. This is the only way you can recapture what has happened. With brainstorming, it is easy in the excitement to be careless about recording ideas. Be sure someone is appointed to see that everything is written down. Remember, do not allow judgment on ideas during the recording process by letting the recorder omit any ideas. It is best to display every idea in full view of all members on a flipchart or whiteboard or similar device. After the brainstorming session, all ideas should be recorded on a sheet of paper so that the ideas can be preserved for use at a following meeting.

2. *Use freewheeling ideas.* Freewheeling has value in that while an idea may be unsuitable in itself, it serves as a stimulus for other members of the group. Even wild or exaggerated ideas have thought-provoking value that should never be underestimated.

3. *Limit judgment until later, with no criticism allowed.* Keep the ideas flowing. Criticism will shut off the flow. All ideas are encouraged and accepted. Remember, there are only "right" ideas. All ideas are right in each individual's mind.

4. *Encourage participation by everyone.* Good ideas are not necessarily in the minds of a few individuals. Give each member a turn to speak; don't miss anyone. It is important to give ample time for each member to speak. For example, solicit responses clockwise around the room. If a member has no idea at the moment, the member says "pass." By this remark, there is added assurance that no one is missed. Furthermore, a team member that passes on one round may very well have an idea on the next round.

5. *Solicit quantity.* Solicit a large number of ideas. Ideas build on ideas. They can be the combination or extension of other ideas. Ideas are thought-provoking and stimulating. Work toward a large number of ideas. Postpone judgment on ideas; that comes later.

Brainstorming steps. The steps to brainstorming are as follows:

1. *Generate ideas.* Follow the rules given above.

2. *Evaluate ideas.* During the evaluation step, the team examines each idea for value. This is the point at which to offer constructive criticism or analyze the ideas presented. Again, it is important that only the idea and not the generator of the idea be criticized. The ideas and alternative combinations of ideas are compared and examined. At this time, some ideas may be eliminated or combined with other ideas.

3. *Decide using consensus.* There are a number of ways to develop a consensus. Chapter 10, on decision-making tools and techniques, details methods to focus on consensus.

Brainstorming methods

There are three primary brainstorming methods:

1. Round-robin
2. Freewheeling
3. Slip

Each has advantages and disadvantages that the team or discussion leader will have to weigh before determining which one would be best to accomplish desired results. In some cases, the best method may be a combination of the various brainstorming methods. For instance, the brainstorming session may start with a round-robin or slip method and move into a freewheeling method to add more ideas.

Round-robin. Each group member, in turn, contributes an idea as it relates to the purpose of the discussion. Every idea is recorded on flipchart or board. When a group member has nothing to contribute, he or she simply says "pass." The next time around, this person may offer an idea or may pass again. Ideas are solicited until no one has anything to add.

Round-robin advantages include the fact that it is difficult for one person to dominate the discussion, and everyone is given an opportunity to participate fully. One of the disadvantages is that people feel frustration while waiting their turn.

Freewheeling. Each team member calls out ideas freely and in a random order. Every idea is recorded on a flipchart or board. The process continues until no one has anything else to add.

The advantage of freewheeling is that it is spontaneous and there are no restrictions. In terms of disadvantages, some individuals may dominate, quiet team members may be reluctant to speak, and the meeting may become chaotic if too many people talk at the same time.

Slip. Each team member writes down all his or her ideas on an issue, a problem, or an alternative on a "slip" of paper. He or she writes as many ideas as possible. Then the slips are collected, and all the ideas are written on the board. A variation to this method is the *Crawford slip method,* where each idea is written on a separate slip of paper. The slips are then put on a board and arranged in categories.

An advantage of the slip method is that all ideas are recorded and all contributions are anonymous. A disadvantage is that some creativity may be lost due to the inability of other team members to react to the contribution of others.

Figure 7.7 provides an example of a brainstorming session on the barriers to teamwork.

BARRIERS TO TEAMWORK

1. Personality conflicts
2. Egos
3. Management
4. Management styles
5. Language
6. Communications
7. Not listening
8. Shy person
9. Lack of motivation
10. Dominant person
11. Lack of interest
12. Lack of technical knowledge
13. Participation
14. Caste system
15. Not respecting others' individuality
16. Closed mind
17. Not a priority
18. Not familiar with the concept
19. Location
20. No focus

Figure 7.7 Brainstorming example.

Advanced brainstorming techniques

There are many advanced brainstorming techniques beyond the basic three mentioned above. Two of the most popular of these advanced techniques are nominal group techniques and affinity diagrams.

Nominal group technique. *Nominal group technique* is a refinement of brainstorming. It provides a more structured discussion and decision-making technique. The nominal group technique allows time for individual idea generation. This can be any amount of time. Sometimes, if the subject is not too complex, the team may have only 5 to 10 minutes. For a complex issue, the team may be asked to generate their ideas between team meetings. Once the ideas are generated, the nominal group technique then allows the leader to survey the opinions of the group about the ideas generated. Finally, nominal group technique leads the group to set priorities and focus on consensus. The steps of the nominal group technique can be summarized as follows:

1. Present the issue and give instructions.
2. Allow time for idea generation.
3. Gather ideas via round-robin, one idea at a time. Write each idea on a flipchart or board.
4. Process or clarify ideas. Focus on clarification of meaning, not on arguing points. Eliminate duplicate ideas, and combine similar ideas.
5. Set priorities.

Affinity diagram. The affinity diagram is another idea generator. It starts with the issue statement. Once the issue is presented, it continues like the nominal group technique, with some time for individual idea generation. The difference is that with an affinity diagram, each idea is written on an index card or piece of notepaper. All notes are then posted on a wall or put on a table. The team members arrange the cards into similar groupings. This is all done without discussion. Next, the team decides on a theme for each group of notes through discussion. The team creates a header card for each group of notes from the theme, and the cards are arranged under the header cards. Next, just as with the nominal group technique, the items are prioritized for action.

Presentations

Sometimes a presentation may be necessary to provide information, obtain approval, or request action. The presentation may be given formally or informally by the team. Involve as many team members as possible in the actual presentation. The presentation provides the team with an opportunity to inform management about team activities and accomplishments and recognize team members for their contributions.

Presentation steps

Step 1: Gain support. Gaining support requires identifying and involving key people early in the improvement process. Ensure support for recommendations from owners, suppliers, and customers by stressing the benefits to the organization.

Step 2: Prepare the presentation.

- Anticipate objections.
- Rehearse the presentation.
- Arrange the presentation.

Step 3: Give the presentation.

- Build rapport.
- Make the recommendation.
- Stress the benefits.
- Overcome objections.
- Seek action.

Step 4: Follow up on the presentation.

- Follow up to ensure that the recommended action is implemented.
- Reduce postdecision anxiety by repeating and summarizing benefits.
- Stress the benefits of early implementation.

Once the team knows that it has sufficient support for a recommended course of action requiring management approval, the team must prepare the presentation. Preparing the presentation involves the activities listed above. To accomplish these activities, the following processes must be performed:

- Develop presentation materials.
- Produce the presentation materials.
- Arrange for the presentation.
- Practice the presentation.

Development of presentation materials involves selecting a specific objective for the presentation and preparing a presentation outline to accomplish that objective. The presentation objective should state specifically the expected outcome of the presentation. The objective should be stated in terms of who, what, and when.

Presentation objective example. The organizational development and training manager will analyze within 3 months the specific needs of the organization to implement customer-driven quality-improvement teams throughout the organization. In this example, the "who" is the organizational development and training manager. The "what" is analyze the specific needs of the organization to implement customer-driven quality-improvement teams throughout the organization. The "when" is within 3 months.

The outline of the presentation should be geared to accomplish the objective. When preparing the presentation outline, consider the audience, understand how the recommendation affects others, and outline the organizationwide benefits. The audience may be supportive or unreceptive. Conduct a force-field analysis to determine the restraining forces and driving forces of the audience. At the same time, consider how the recommendation affects others. Anticipate objections. Again, conduct a force-field analysis to determine driving forces of any known objections to your proposal. Further, outline the organizationwide benefits through brainstorming and data collection.

Now you are ready to prepare the presentation outline. The presentation outline should contain an introduction, a body, and a conclusion. In the introduction, tell your audience what you are going to tell them. In the body, tell them. And in the conclusion tell them what you have just told them.

Presentation outline

In the *introduction,*

- Establish rapport through use of introductions.
- Get the audience's attention by answering what's in it for them.
- Tell the audience what you are going to tell them.

In the *body,*

- State your mission.
- Describe the process using a process diagram:
  - Significance of the process
  - Inputs with suppliers
  - Process itself
  - Output(s) with customer(s)
  - Owner(s)
  - Identify the underlying cause
  - Describe data collection
  - Discuss results
- Detail the action requested:
  - Alternatives considered
  - Solution selected
  - Plan for implementation

In the *conclusion,*

- Reinforce outcomes.
- Tell them what you told them.

- Get agreement on what you want.
- Summarize actions.

Prepare the presentation materials

Presentation materials can be as simple or complex as required to get the requested action from the audience. Presentation materials are used to attract and maintain attention on main ideas. They illustrate and support the team's recommendations. They focus on minimizing misunderstanding. Presentation materials include handouts, overhead transparencies, flipcharts, videotapes, and computer-based visuals. At a minimum, the presentation materials should consist of a handout for all participants. Normally, the presentation materials consist of a handout and some form of visual aid for the group to observe. In most organizations, this is either a flipchart or overhead transparencies. Specific tips for preparing the most common presentation aids follow.

Handouts. The handout supports the presentation by providing critical information and/or supplemental details. The handout should follow the same steps as the presentation, if given prior to the presentation. If the handout provides only supplemental or reinforcing information, it should not be given to the audience until an appropriate place is reached in the presentation or it should be given at the conclusion of the presentation.

Flipcharts. Flipcharts enhance presentations. They should emphasize the key points or graphically show concepts. Some specific tips for the design of flipcharts are as follows:

- List main points with bullets.
- Limit bullets to six or fewer per chart.
- Keep bullets short, i.e., around six words or less per bullet.
- The chart should be readable from every seat in the room.
- Multiple colors can be used to stress key words.
- Leave a blank sheet between pages of flipcharts.
- The chart should reflect the professional pride of the team.

Overhead transparencies. Overhead transparencies are used more frequently than any other information aid in a presentation. The overhead transparency is used in the same manner as a flipchart. However, with today's computer technology, especially graphic capabilities, overhead transparencies can be created easily and used to reinforce ideas graphically. Since people have different styles for understanding, it may be appropriate to use both words and pictures to convey your message. The words appeal to the more logical people, while the graphics focus on the creative people. The same tips for flipcharts apply to overhead transparencies. In addition, you should not attempt to show large amounts of data or complex processes on one transparency. Reduce the information to show trends, relationships, or overall processes. If detailed information is necessary, provide it as a handout.

Produce the presentation materials

Depending on the situation, the team may produce the presentation itself or may have to rely on various support services to produce the materials. If the team produces the materials, ensure that the presentation meets the standards of the audience. If the team has support services that make the presentation materials, planning and coordination are important. Many organizations have support services such as word processing, editing, graphics, and printing departments. When using these services, the team must plan enough time to allow for professional workmanship. This means determining all the tasks to be accomplished with an appropriate time period allowed to meet the team's scheduled presentation date. Ensure that there is enough time to use the presentation materials for a dry run before the actual presentation. Also, it is wise to coordinate periodically with support services people to ensure progress toward meeting the schedule.

Arrange for the presentation

Administrative details can have an effect on how the presentation is received. Ensure that the following administrative details are accomplished:

- Schedule the presentation time and place.
- Ensure that all the right participants can attend.
- Set up the room.
- Have the presentation materials on hand.

Practice the presentation

Rehearse the presentation prior to the actual presentation. If possible, practice the presentation once before an audience that provides a representation of the actual audience.

Give the presentation

Giving the presentation involves presenter preparation and the actual conduct of the presentation. The best way to ensure a successful presentation is by adequate preparation. This is enhanced through the development of some basic presentation skills. These skills can be grouped into the following categories:

- Presenter's preparation
- Presenter's style
- Presenter's delivery

Presenter's preparation. Depending on your experience, you will be more or less comfortable presenting to a group. Your level of comfort can be improved with time spent preparing. Specifically, the following techniques can be used to enhance your comfort level:

- *Practice.* Practice the presentation enough to get very familiar with the flow of ideas. Do not memorize the presentation. Practice enough to allow you to present the information naturally.
- *Plan for objections.* Perform a force-field analysis to determine objections and your response.
- *Visualize success.* Prior to the presentation, spend some time alone picturing yourself accomplishing a successful presentation.

Presenter's style. Although each presenter has his or her own style, the following guidelines will help any presenter become more successful:

- *Act naturally.* Make the presentation as natural as possible. Try to avoid doing anything that would appear faked, forced, or flaky.
- *Maintain a positive attitude.* Display a positive attitude by showing enthusiasm. Above all, be sincere in your commitment and support for the presentation goal.

Presenter's delivery. There are several nonverbal and verbal presentation tips to improve anyone's presentation. Most people concentrate on the verbal communication aspects of the presentation, although nonverbal behavior communicates much of the meaning. Nonverbal communication skills enhance your ability to communicate effectively with the audience. These nonverbal communication behaviors include eye contact, body movement, and gestures.

Eye contact shows interest in the audience. Look directly at your audience and include everyone equally. Good eye contact results in enhanced credibility.

Body movement is another important physical behavior for a presenter. It helps hold the audience's attention and puts the speaker at ease by working off excess energy that can cause nervousness. You can use body movement as punctuation to mark a change in your presentation. Moving from one spot to another tells the audience that you are changing the line of thought. Some body movement can be distracting. Pacing back and forth, rocking from side to side, or "dancing" serves no purpose and tells the audience that you are nervous.

Gestures can clarify, emphasize, or reinforce what is said. Make natural gestures by using your hands, arms, shoulders, and head. Fidgeting with your watch and scratching your ear are not gestures. This type of behavior usually distracts people from the presentation. Gestures take practice to use effectively.

Conduct the presentation

During the presentation, the team does the following:

- Builds rapport by developing a positive but professional relationship with the audience.
- Makes the recommendation, using the results of the customer-driven project management improvement methodology, and supports the recommendation with facts.
- Stresses the benefits of implementing the team's recommendation. The benefits should emphasize the tangible, measurable gains of the solution. In addition, show intangible advantages. Make sure you answer the question, What's in it for them?
- Overcomes objections by using the driving forces derived from the force-field analysis. Remember, focus on the issue, and never make it personal.
- Seeks action for implementation. The conclusion must provide a definite course of action.

Follow up on the presentation

After the presentation, the team should do the following:

- Follow up to ensure that the recommended action is implemented.
- Reduce postdecision anxiety by repeating and summarizing benefits.
- Stress the benefits of early implementation.

Main Points

Customer-driven project management aims to maximize the potential of the human resources in an organization through individual and team contributions.

People involvement tools and techniques include individual involvement, teamwork, communication (especially listening), focus setting, meetings, brainstorming, and presentation.

Individual involvement concerns each person's contributions to the organization. In customer-driven project management, individuals work to continually perform their work and improve the processes in the organization, focusing on total customer satisfaction.

Individual diversity is a distinct advantage in today's economic environment.

Open and honest communication leads to building the trust necessary for active people involvement.

Individual involvement is fostered by the following:

- *I*nstilling pride of accomplishment
- *N*urturing individual self-esteem
- *D*eveloping an atmosphere of trust and encouragement
- *I*nvolving everyone making the most of individual differences
- *V*isualizing a common purpose
- *I*mproving everything
- *D*emanding effective and open communication
- *U*sing rewards and recognition
- *A*llowing creativity and innovation
- *L*eading by example

Teamwork is a technique whereby individual team members work together to achieve a common goal.

Teamwork involves the following:

- *T*rust
- *E*ffective communication, especially listening
- *A*ttitude positive, "can do"
- *M*otivation to perform and improve
- "*W*e" mentality
- "*O*wnership" of work with pride
- *R*espect for and consideration of others
- *K*eeping focused on total customer satisfaction

The key principles of teamwork involve the following:

- *K*eeping focused on the mission; do not make it personal
- *E*ncouraging open communication and active listening
- *Y*earning for constructive relationships

The basic principles of teamwork can be summarized as follows:

- *P*ursue a team environment.
- *R*eward and recognize the individual and the team.
- *I*nvolve all team members.
- *N*urture the self-esteem of all team members.
- *C*ommunicate freely and openly.
- *I*nclude individuality.
- *P*ursue constructive relationships.
- *L*ead by example.
- *E*ncourage all team members' ideas.
- *S*tay focused on the mission.

The following are essential to building teamwork:

- Identify the team mission.
- Establish team roles and responsibilities.
- Understand team dynamics.
- Manage conflict.
- Provide motivation.
- Build individual self-esteem.
- Develop the team.

The mission is the intended result. The mission statement includes

- *M*agnitude of the improvement expected.
- *I*ndication of beginning process or problem
- *S*tatement of boundaries
- *S*tatement of authority
- *I*dentification of resources
- *O*rigination with management
- *N*egotiation and clarification

The roles and responsibilities should include

- *R*esults and expected outcome(s) from each team member
- "*O*wnership," including the amount of control
- *L*imits of resources—funds, equipment, and people
- *E*mpowerment, with the amount of authority specified
- *S*tandards focusing on customer satisfaction

The four stages of team development are orientation, dissatisfaction, resolution, and production.

Conflict can be managed.

Differences exist in every organization. These differences, as follows, are a major strength:

- *C*ultures and backgrounds
- *O*pinions
- *N*eeds and expectations
- *F*acts and perceptions
- *L*evels, departments, and organization
- *I*nterests, personalities, and egos
- *C*ompetencies, knowledge, and skills
- *T*argets, visions, missions, and goals

Conflict can be controlled as follows:

- *C*ooperate rather than compete.
- *O*rient toward the issue, not the person.
- *N*egotiate win/win solution(s).
- *T*ake an organizationwide perspective.
- *R*ecognize conflict as natural.
- *O*bserve empathy toward others' views.
- *L*imit perceived status differences.

Conflict can be positive; use it to

- *P*ursue win/win situations
- *O*bserve others' point of view
- *S*how open communication
- *I*nstill an organizationwide view
- *T*ake personalities out of issues
- *I*nvite trust and involvement
- *V*iew issues in their entirety
- *E*xamine different sides of an issue

Agreement can be negative. To avoid the negative aspects of agreement,

- *A*ppoint a "devil's advocate."
- *G*et open discussion on all issues.
- *R*ecognize the impact of status differences.
- *E*xamine all agreement without resistance.
- *E*valuate all views/sides of an issue.

Conflict symptoms are

- *S*topping open communication
- *Y*ielding to win/lose solutions
- *M*aking little movement toward a solution
- *P*ressure to stop challenges
- *T*aking sides (we/they)
- *O*bserving a lack of building on suggestions
- *M*embers silent
- *S*topping any resistance

Conflict management actions are as follows:

- *A*void "facesaving" situations.
- *C*ontinuously self-examine attitudes.
- *T*arget win/win solutions.
- *I*nvolve everyone.
- *O*bserve the limits of arguing.
- *N*urture differences of opinion.
- *S*upport constructive relationships.

Motivation is the behavior of an individual whose energy is selectively directed toward a goal.

Performance is the result of having both the ability and the motivation to accomplish a task.

Motivating actions are as follows:

- *M*ake clear that the goal is shared.
- *O*rient and integrate team members.
- *T*hink and speak "we."
- *I*nstitute team rewards and recognition.
- *V*alue individual contributions.
- *A*void frequent changes of team members.
- *T*ake time to exchange greetings.
- *E*ncourage a sense of belonging.

An individual's self-esteem affects his or her performance of organizational tasks as well as his or her relationship with others on the team.

Actions to maintain individual self-esteem are as follows:

- *E*stablish an environment in which an individual feels that his or her self-worth is important to performance.
- *S*tay focused on the issue.
- *T*reat each person as you would want to be treated.
- *E*ncourage individual contributions.
- *E*nsure that individual achievement is rewarded and recognized.
- *M*otivate, communicate, involve, and develop.

The success of customer-driven project management demands communicating with and among all team members.

Communication can be improved by the following:

- *C*larify the message.
- *O*bserve body language.
- *M*aintain everyone's self-esteem.
- *M*ake your point short and simple.
- *U*nderstand others' points of view.
- *N*urture others' feelings.
- *I*nvolve yourself in the message.
- *C*omprehend the message.
- *A*ttend to the messages of others.
- *T*alk judiciously.
- *E*mphasize listening.

Some guidelines on effective feedback are

- *F*oster an environment conducive to sharing feedback.
- *E*ncourage feedback as a matter of routine.
- *E*stablish guidelines for providing feedback.
- *D*iscuss all unclear communications.
- *B*e direct with feedback.
- *A*sk questions to achieve better understanding.
- *C*onsider "real" feelings of team members.
- *K*eep focused on the mission.

Effective listening requires the following:

- *L*et others convey their message.
- *I*nvolve yourself in the message.
- *S*ummarize and paraphrase frequently.
- *T*alk only to clarify.
- *E*mpathize with others' views.
- *N*urture active listening skills.

Focus setting is a technique to center actions on a specific outcome. The focus should be on the output of the process satisfying the customer.

The following are some guidelines for a vision:

- *V*iew the future.
- *I*nstitute the vision in the organization.
- *S*et the example through leadership.
- *I*nclude where you want the organization to go.
- *O*rient toward the customer.
- *N*urture through constant communication.

The mission is the intended result. It should be stated in a results-oriented form.

The goal is the specific desired outcome(s). It should be specific, measurable, attainable, results-oriented, and time-bound.

Meetings are a technique for bringing a team together to work for a common goal.

The meeting tools are

- Rules of conduct
- Meeting roles, responsibilities, and relationships
- Focus statement
- Agenda

Rules of conduct provide guidance for the team's conduct. The rules of conduct consider "how" meetings will be conducted.

Effective team meeting requires a leader, participants, a facilitator, and a recorder.

A focus statement provides the purpose for a meeting. Each team meeting must have a written focus statement.

An agenda acts as a meeting guide. It gets the team to focus on the meeting's desired outcomes.

Brainstorming is a technique used by a group of people that encourages their collective thinking power to create ideas.

Brainstorming rules are as follows:

- *R*ecord all ideas.
- *U*se freewheeling ideas.
- *L*imit judgment until later.
- *E*ncourage participation by everyone.
- *S*olicit quantity.

A presentation may be necessary to provide information, obtain approval, or request action.

Presentation steps are as follows:

Step 1: Gain support.
Step 2: Prepare the presentation.
Step 3: Give the presentation.
Step 4: Follow up on the presentation.

Presentation outline: In the introduction,

- Establish rapport with introductions.
- Get the audience's attention by listing the benefits.
- Tell them what you are going to tell them.

In the body,

- State your mission.
- Describe the process using a process diagram:
  - Significance of the process
  - Inputs with suppliers
  - Process itself
  - Output(s) with customer(s)
  - Owner(s)
  - Identify the underlying cause
  - Describe data collection
  - Discuss results
- Detail the action requested:
  - Alternatives considered
  - Solution selected
  - Plan for implementation

In the conclusion,

- Reinforce benefits.
- Tell them what you told them.
- Get agreement on what you want.
- Summarize actions.

When giving a presentation, act naturally and be sincere and positive.

Chapter

8

CDPM Definition Tools and Techniques

Focus: This chapter discusses the tools and techniques for defining the quality issue. The tools and techniques define the quality issue by determining customer's needs and expectations, the critical processes, and performance outcomes.

Introduction

Customers are the only people who can determine total customer satisfaction, and total customer satisfaction is the primary quality issue. In order to know if the customer is satisfied, intense observation is necessary. Only through observation, communication, especially listening, and measurement can the organization determine total customer satisfaction. The organization must use every means available to evaluate customer satisfaction. In this chapter, several tools and techniques are described to assist in defining the quality issue in terms of total customer satisfaction. The tools and techniques are quality function deployment (QFD), benchmarking, and metrics. The quality function deployment technique focuses on listening to the "voice of the customer." Benchmarking targets excellence. Metrics aim at meaningful measures of critical success factors forging continuous improvement actions.

All the tools and techniques in this chapter are especially useful during step 1 of the CDPM improvement methodology for defining the quality issues. In addition, these tools and techniques are helpful in other steps of the CDPM improvement methodology as follows:

- Quality function deployment assists in step 5, taking action to determine design requirements, part characteristics, process/manufacturing operations, and production requirements.

- Benchmarking is beneficial throughout the cycle. It helps one to understand and define processes in step 2, and it provides the targets for steps 3 to 8.
- Metrics provide a foundation for all steps in the CDPM improvement methodology. They define process performance, provide an indicator for selecting improvement opportunities, give a measure for analyzing the improvement opportunities, provide a target for action, and serve as a gauge to monitor results.

Before examining the specific tools and techniques, the customer-driven project team must understand that customers are the focus of all customer-driven project management (CDPM) efforts. Without customers, the organization ceases to exist. Therefore, every organization and everyone in the organization must strive constantly to satisfy current customers and to create new customers for the future.

Figure 8.1 shows what must be observed to achieve total customer satisfaction. The organization must know the customer, itself, its product, and its competition.

Know your customers

It is essential to get to know all internal and external customers. Customers are the leaders in customer-driven project management. The customer as leader is well-known in CDPM; however, other customers are not always obvious. Customers include all those touched by a product or service, internal and external to the organization. In order to be successful, the customer-driven project team must listen to all customers.

Next, customer needs and expectations must be determined. Customer expectations are dynamic; they increase and change continuously. This requires a continuous review of customer needs and expectations to ensure customer satisfaction. Like so much of customer-driven project management, this process, too, is forever ongoing.

Identify customer needs and expectations

The identification of customer needs and expectations requires systematic, thorough, and continuous communication. The most important aspect of this process is to *listen to the customer*. In this regard, the quality function deployment techniques described later in this chapter provide an excellent insight into customer needs and expectations. In addition, some common marketing research methods can be used to identify customer needs and expectations, including media

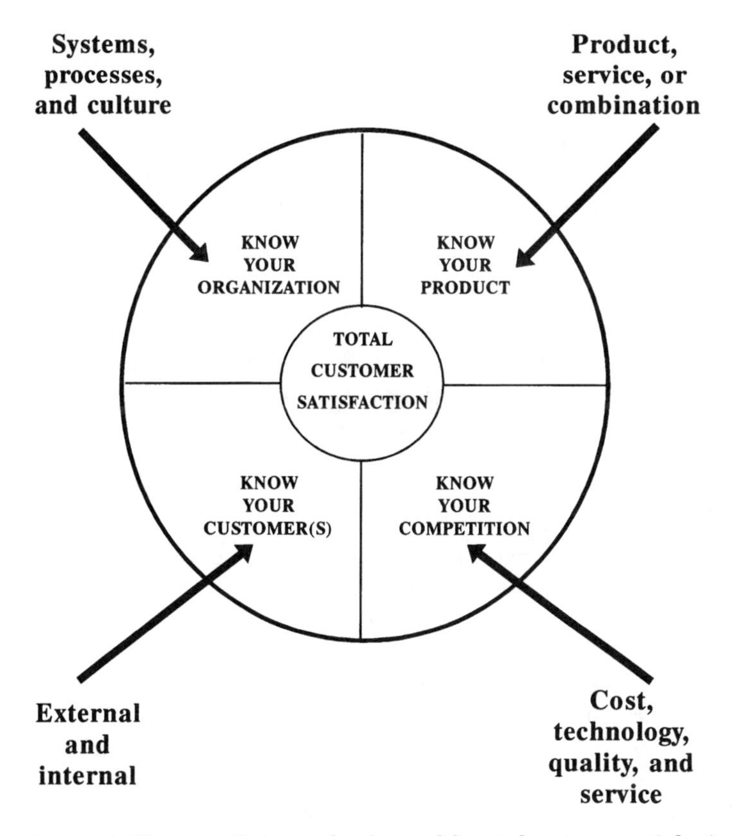

Figure 8.1 Elements that must be observed for total customer satisfaction.

research, test marketing, customer auditing, and customer focus groups. Interviews, group discussion, surveys, and direct observation offer other means to "listen" to the customer.

Customer needs are not static; they are always changing. Once customer needs are identified, these needs must be monitored continuously to ensure that the product and/or service still satisfies them. Customers have different needs ranging from basic survival needs, such as eating and sleeping, to safety needs, social needs, and esteem needs, to the total fulfillment of one's ultimate goal in life. Customers satisfy lower-level needs before higher-level ones. A need once satisfied is often no longer a need. Remember, needs are constantly replaced by other needs due to the changing world environment. Rapidly changing technology, differing tastes, and rising expectations due to past successes are some of the many factors that influence customer changes.

Develop customer relationships

Customer relationships are the core of customer-driven project management. This relationship ensures continuous customer satisfaction. The relationship determines whether the customer will continue or expand the business. The emphasis is on keeping current customers while seeking additional customers for the future.

Relationships demand continuous attention. Customer relationships require communication, support, and responsiveness. Communication, especially listening, is essential. The customer needs to be involved in as many aspects of the deliverable as possible. Support must be available to help the customer with the product after it is received. Responsiveness is the key to continuing the relationship. The organization must be able to respond to the needs of the customer in any situation.

Fundamental to all customer relationships is a foundation of integrity, ethics, and trust. Integrity implies honesty, morals, values, fairness, adherence to the facts, and sincerity. This characteristic is what anyone in the organization and the customer (internal/external) expect and deserve to receive. If the customer perceives that the organization is guilty of duplicity, customer satisfaction will not be achieved, and quite likely, the customer's business will be lost.

Ethics is the discipline concerned with good and bad in any situation. Ethics is a two-faceted subject represented by organizational and individual ethics. In the case of organization, most organizations have an established business code of ethics to which all employees are to adhere in the performance of their work. Individual ethics includes personal rights or wrongs. They are concerned with legal, moral, contractual, business, and individual dealings. A person should never do anything that goes against command media (policies, regulations, contracts, and so on) or that the person would not like done to himself or herself.

Trust is a by-product of integrity and ethical conduct. Trust is absolutely essential for any relationship and is necessary to ensure full participation of all members on the customer-driven team. Trust helps ensure that measurements focus on the areas critical to customer satisfaction. In CDPM, trust must be developed to remove the traditional conflicts between the project customer and the project supplier. Trust builds the cooperative environment essential for CDPM.

Know yourself

Knowing yourself is an important element in achieving total customer satisfaction. In the process of knowing itself, the organization looks inward to its culture, systems, and processes. In this process, internal quality is the focus, with total customer satisfaction still the target. The first aspect of knowing yourself is understanding your culture. The organizational culture must support the quest for total customer satisfaction. The current organizational vision, philosophy, principles, values, beliefs, assumptions, outcomes, and standards must provide the essential environment for prosperity.

In many organizations, the culture must be changed to focus on total customer satisfaction. For instance, the culture of many organizations is deeply rooted in the traditional management philosophy. In these organizations, management must assess its current state and determine the desired future state. This information provides the basis for a strategic plan to systematically adjust the organizational culture.

In addition to the organizational culture, all the systems and processes in the organization must target total customer satisfaction. Each internal customer in the organization must be satisfied. However, unlike external customers, internal customers may or may not be a person. An internal customer could be a person receiving the output of the job or the next process, the next task, the next activity, the next job, the next machine, or the next piece of equipment in the system.

To satisfy internal customers, all the processes in the organization must be understood, measured, and analyzed to determine existing performance. Once current performance is known, goals can be set to improve the internal organization. This continuous improvement focus steers the organization toward the future.

Know your product

The organization must know all there is to know about its product to achieve total customer satisfaction. This includes knowing all the aspects of the product. The product is an output of a process that is provided to a customer (internal/external) and includes goods, services, information, and so forth. The product is all aspects contributing to customer satisfaction, including such items as product quality, reliability, maintainability, availability, customer service, support services, supply support, support equipment, training, delivery, billing, and marketing. Again, every one of these elements of the product and/or service must focus on total customer satisfaction. The goods or services may be the best in the marketplace, but it is the entire product that contributes to total customer satisfaction. If the product does not provide total customer satisfaction, the customer will not be satisfied and most likely will go elsewhere to find a product that gives total customer satisfaction.

When striving for total customer satisfaction, the deliverable, at a minimum, must be comparable with the product and/or service offered by the competition. Obviously, a competitive advantage is gained by improving the deliverable. The deliverable may be differentiated by raising the level of customer satisfaction. A thorough analysis should always accompany any targeting of a product for customer satisfaction.

The goal is always to optimize customer satisfaction with the resources at hand. Raising the level too far beyond the current range of customer satisfaction is risky because of two factors. First, the cost factor could affect perceived value. Second, the customer may not be ready for the enhanced product or service. In both cases, the product or service may not be sold because it did not satisfy the customer.

Know the competition

The organization must know the competition to establish targets for its products and services and internal improvement efforts. The organization must establish product and service targets in relation to its competition. Typically, organizations compete on the major areas of technology, cost, product quality, and service quality. Frequently, product competition progresses from one area to the next area. It starts with technology, then cost, then product quality, and then quality service and other areas.

The organization must know in which major area of competition its deliverables are competing. This will show the organization where to target its product and services. In addition, the organization should always attempt to achieve the competitive advantage. Therefore, it is important to look for ways to differentiate within the major areas of competition.

The organization must know where it stacks up in relation to the competition to establish internal improvement effort targets. To determine its position in the competitive market, the organization should benchmark itself against its top competitor and the best in the field. Once benchmarks are established, the organization can set internal targets for improvement efforts.

As shown in the preceding subsections, knowing the customer, your organization, the deliverable, and the competition is the root of defining the quality/customer satisfaction issue. The tools and techniques useful in defining the customer issue are quality function deployment, benchmarking, and metrics.

Quality Function Deployment

Quality function deployment (QFD) is a disciplined approach to transforming customer requirements, the "voice" of the customer, into product development requirements. QFD is a tool for making plans visible and then determining the impact of those plans. QFD involves all activities of everyone at all stages from development through production with a customer focus.

QFD involves four phases: (1) product planning, (2) parts deployment, (3) process planning, and (4) production planning. The output from each phase is the input for the next phase. During phase 1, customer requirements are transformed into design requirements. In phase 2, design requirements are converted into a system (part) or concept design. Phase 3 examines candidate processes and selects one. Phase 4 looks at making capable production processes. All the phases of QFD are useful in customer-driven project management. An overview of each of the four phases is presented in Chap. 13. This chapter uses the first phase of quality function deployment for focusing on the quality issue. The first phase in QFD does the following:

- Defines customer needs and expectations
- Establishes initial design requirements
- Provides primary product planning information
- Initiates competitive assessments

QFD "house of quality"

The results of QFD planning are included on a chart called the "house of quality." The basic "house of quality" planning chart, which is for phase 1, is shown in Fig. 8.2. From this basic chart, many other useful charts can be generated to assist in moving from general customer requirements to specific production processes in other steps of the CDPM improvement methodology. The number and kinds of charts vary with the complexity of the project. QFD can be applied to the complete product, the system, the subsystem, and/or specific parts. At all stages of QFD application, prioritization is used to ensure that the overall analysis does not become excessively burdensome in terms of time and cost. For instance, many projects can be successful using just parts of the QFD chart. The QFD phase 1 chart is a combination of many matrix diagrams. It can consist of a matrix diagram of customer and design requirements, a priority rating, a correlation matrix, a customer competitive evaluation, a technical competitive evaluation, objective target values, technical difficulty ratings, and technical importance ratings. In most cases, complete detail for all the areas of the QFD chart is impractical and unnecessary. In summary, the customer-driven team uses only the charts or parts of charts appropriate for the specific project.

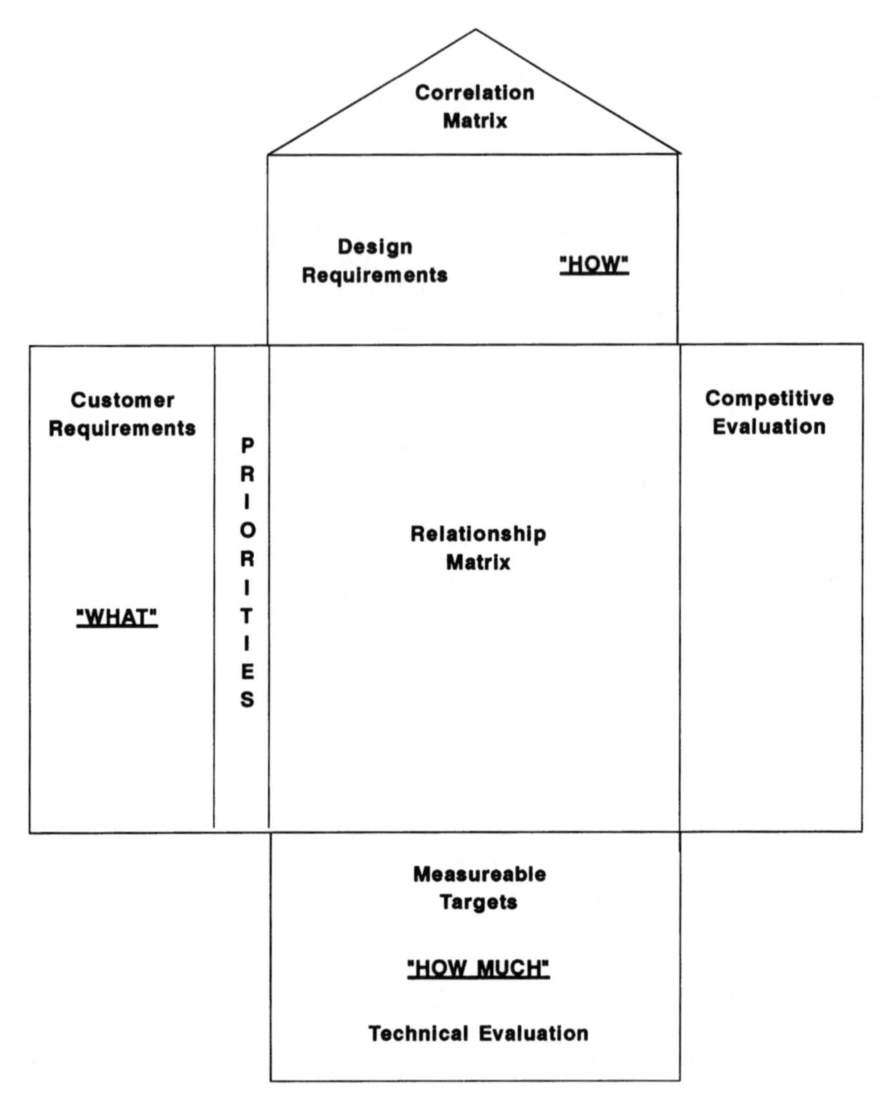

Figure 8.2 Quality function deployment phase 1 "house of quality."

QFD "house of quality" steps for phase 1

Step 1: Determine the "whats." These are the "voice" of the customer or customer needs and expectations.

Step 2: Transform the "whats" to "hows." The "hows" become the product design requirements or characteristics, which are measurable. These product design requirements establish the project deliverable, which forms the basis of the contract and work breakdown structure.

Step 3: Determine the nature of the relationships between the "whats" and the "hows" using the relationship matrix.

Step 4: Establish how much data are needed. This provides target values for design requirements.

Step 5: Correlate each "how" to each other "how." This is done in the correlation section, or the roof of the houselike matrix, and is used to aid in conflict resolution and tradeoff analysis.

Step 6: Complete the customer and technical competitive evaluation sections. These competitive evaluations rate the product under question against similar products produced by the competition. The customer evaluation relates the product features to customer satisfaction, and the technical evaluation assesses the product based on technical merits.

Step 7: Assign or calculate importance ratings. This helps to prioritize analysis efforts.

Step 8: Analyze results. This step includes a check and balance procedure to both identify planning gaps and point to wasteful activities.

Quality function deployment example

Figure 8.3 shows part of a QFD phase 1 chart for delivering an excellent cup of coffee. The customer requirements for an excellent cup of coffee are "hot," "eye-opener," "rich flavor," "good aroma," "low price," "generous amount," and "stays hot." These are the "whats." The "hows" are the product requirements listed across the top of the matrix. These include "serving temperature," "amount of caffeine," and so forth. The relationship matrix consists of the "whats" along the left column and the "hows" across the top. The evaluation of the relationship between the customer requirements and the product requirements is shown by the symbols depicting a weak, medium, or strong relationship. Next, the "how much" is shown on the bottom of the matrix. The "how much" is determined by examining the "what" to specify "how" to get the "how much." The "how much" items include "130°F," "____ ppm," and so on. In the figure, the "what" of "low price" is examined in relation to the "how" of "sale price" to get the "how much" of "$0.40." The roof of the "house of quality" is the correlation matrix. In the correlation matrix, tradeoff analysis is performed by comparing each "how" with the other "hows." These relationships can be strong positive, positive, negative, and strong negative, as indicated by the symbols.

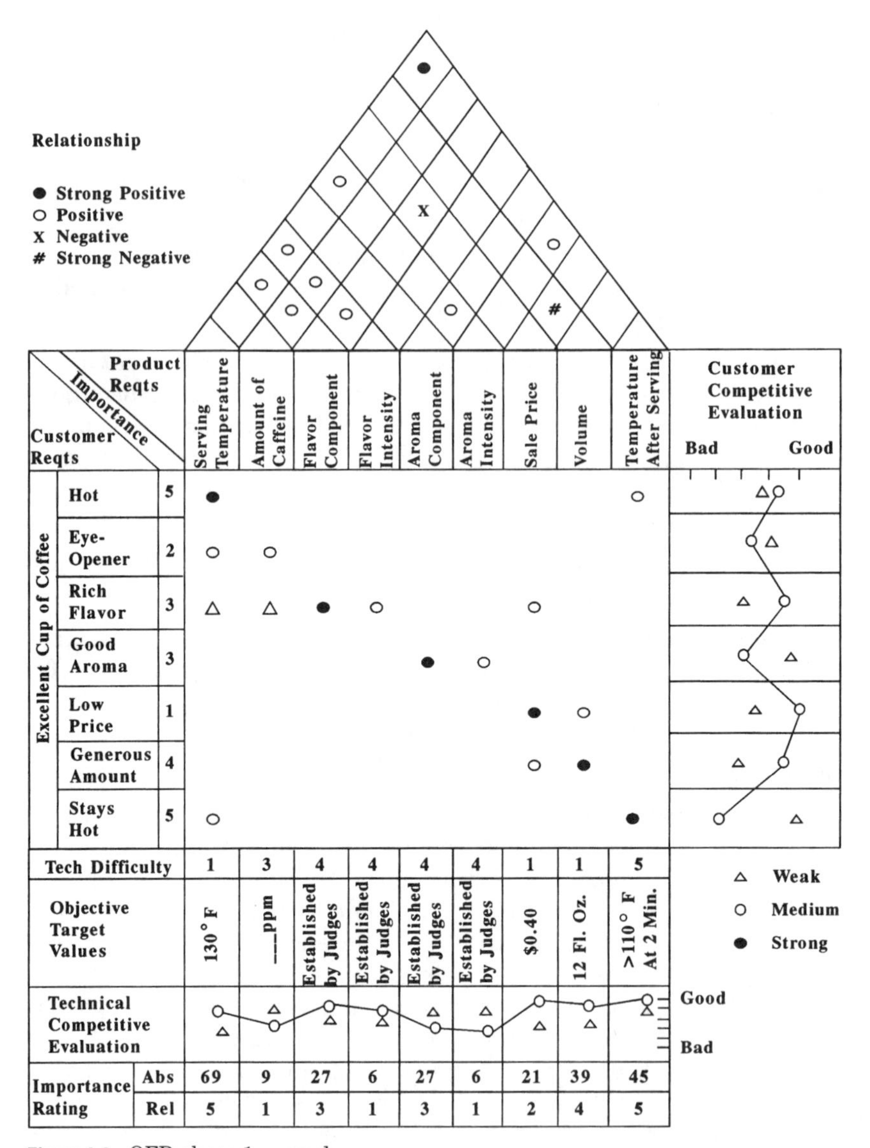

Figure 8.3 QFD phase 1 example.

The figure also shows the evaluation sections for customer and technical competitive assessments. The customer competitive evaluation is along the right side of the "house of quality," and the technical competitive evaluation is at the bottom portion.

In addition, the figure displays the ratings for customer requirements, technical difficulty, and technical importance. The customer importance rating is next to the "what." The technical difficulty rating is above the "how much." The overall importance rating is shown on the bottom of the "house of quality."

Benchmarking

Benchmarking is a method of measuring your organization against the recognized best performers in a certain industry, organization, function, system, or process. The purpose of benchmarking is to provide a target for improving the performance of your organization. The benchmark targets improvement of the process outputs or the performance of the actual process. Benchmarking focuses on customer-driven project management improvement efforts by emphasizing desired outcomes. It also nurtures wholesome competition by creating the desire to be the best. Benchmarking provides a common focus to hold the organization together by measuring areas and analyzing these areas against the best. This targeting of the best reinforces continuous improvement by keeping everyone centered on a long-term objective.

Figure 8.4 gives an example of steps toward "world class." The leaders are considered "world class." The organization starts with its current performance. This is the baseline. Through the implementation of continuous improvement, the organization moves toward improvement. As the organization institutionalizes continuous improvement, it progresses to competitive, best in class, and world class. With the help of benchmarking, this continuous improvement can be planned and implemented to meet the organization's specific objectives.

There are four methods of benchmarking: internal, competitive, functional, and generic. In each case, the type of benchmarking selected depends on the measures needed and the methods used to collect the data.

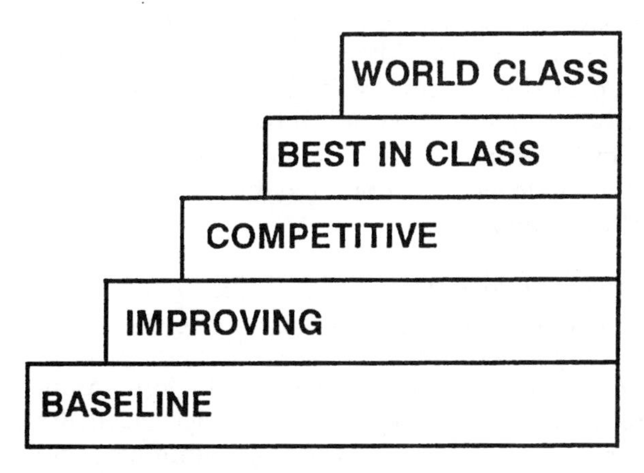

Figure 8.4 Steps toward "world class."

Internal benchmarking looks inside the organization for similar processes and units that seem to do it better.

Competitive benchmarking looks at competitors and examines their processes. This type of benchmarking seeks other institutions that are performing better than the customer-driven project management organization. When these processes are found, the competitors' performance is compared with that of the customer-driven project management organization.

Functional benchmarking looks at any outside or inside activity that is functionally exact to the process under review.

Generic benchmarking looks at any outside or inside activity that is generically the same as the one under review.

Benchmarking steps

1. Understand your organization.
2. Select critical areas for benchmarking.
3. Determine where to get benchmark information.
4. Collect and analyze data.
5. Select target benchmarks.
6. Determine your performance.
7. Set desired outcomes.
8. Use improvement methodology to achieve desired performance.

Benchmarking example

Benchmarking starts with the strategic intent of the organization performing the benchmarking process. There must be a commitment from the top leadership in the organization to pursue continuous improvement with benchmarking as a tool. With this in mind, the following provides an example of one cycle through the benchmarking process, beginning at the top leadership of an organizational development and training (OD&T) organization.

Step 1: Understand your organization.

Purpose: Define the focus of the benchmarks.

The benchmarking process begins as a result of strategic planning. This requires a complete understanding of all the areas needed to meet total customer satisfaction, as outlined in the beginning of this chapter. The mission outcome of the strategic planning provides the focal point for benchmarks for the organization. In our example, the mission is as follows:

> Our mission is to be the leading OD&T organization for individuals and organizations seeking continuous improvement focused on total customer satisfaction by delivering continuously improving, value-added, results-oriented, customer-satisfying organizational development and training products and services.

Step 2: Select critical areas for benchmarking.

Purpose: Determine what to benchmark.

The second step involves listing the areas considered significant for success of the mission. This involves listing the customer needs and expectations, the deliverable(s) to meet the customer's specification, and the internal processes to satisfy the requirements. If using QFD, this is part of phase 1. In Fig. 8.5, the top of the chart shows some of the areas for consideration by the OD&T organization. From this list, the organization selects the critical areas for benchmarking. The OD&T organization decides to benchmark all the areas critical to customer satisfaction at this stage. During other stages in the CDPM improvement methodology, the team may select other processes to benchmark.

The OD&T critical areas of customer satisfaction include the following:

- ***P**ersonal.* Ability to adapt the deliverable to specific customer needs and expectations. *Measure:* Percent of special request met.
- ***R**esponsive.* Ability to meet the needs and expectations of customers. *Measure:* Percent of total customer requests met.
- ***O**btainable.* Ability to provide deliverable within customer's affordability. *Measure:* Percent of customers lost as a result of cost.
- ***D**eliverable.* Provide deliverable when customer needs it. *Measure:* Percent on time delivery.
- ***U**seful.* Deliverable provides business results. *Measure:* Percent of customers reporting business results within 30 days.
- ***C**onvenient.* Provide deliverable where the customer wants it. *Measure:* Percent of requests for specific location met.
- ***T**imely.* Ability to provide OD&T solution at the time needed by the customer. *Measure:* Percent of times deliverable is just in time for customer's results.
- ***S**atisfaction.* Ability to totally satisfy the customer. *Measures:* Number of customer complaints; OD&T evaluation rating on

Major Processes

Customer Requirement	Deliverables	Processes
- Personal	- OD	- Training
- Responsive	- Training	- OD
- Obtainable	- Interventions	- Administration
- Deliverable	- Consulting	- Finance
- Useful	- Coaching	- Scheduling
- Convenient	- Facilitating	- Information systems
- Timely	- Assessing	- Support
- Satisfaction	- Support services	

Critical Areas for Success

	Measures of Success	Current Performance	Benchmark Target
Personal	% of special requests met	72%	98%
Responsive	% of total customer requests met	76%	100%
Obtainable	% of customers lost due to cost	25%	10%
Deliverable	% on time delivery	85%	100%
Useful	% of customers reporting business results	83%	100%
Convenient	% of requests for special location met	82%	98%
Timely	% of just in time deliverables	89%	100%
Satisfaction	Number of complaints	5 per month	5 per year
	Evaluation rating on 5 point scale	4.2	4.9
	Number of customer refunds	10 per year	0

Goals

	Year 1	Year 2	Year 3	Year 4
Personal	80%	90%	98%	100%
Responsive	80%	90%	100%	100%
Obtainable	20%	15%	10%	0%
Deliverable	90%	95%	100%	100%
Useful	90%	95%	100%	100%
Convenient	95%	95%	98%	100%
Timely	95%	98%	100%	100%
Satisfaction	1 per month	8 per year	5 per year	0
	4.5	4.8	4.9	5
	5 per year	1 per year	0	0

Figure 8.5 Benchmarking chart sample.

5-point scale; and number of customers requesting refunds from money-back guarantee.

Step 3: Determine where to get benchmark information.

Purpose: Find sources of benchmarking data.

Since the benchmarking information becomes the target, getting the right benchmarking information is the most important aspect of benchmarking. The sources of information for process performance measurements are numerous. The only real source of a benchmark for performing an actual process is the process performing organization. Some sources of benchmarking data include

- Computer databases
- Industry publications
- Professional society publications
- Company annual reports and publications
- Conferences, seminars, and workshops
- Other organizations within the same organization
- Consultants
- Site visits

For example, the OD&T organization selected the following sources of information:

- American Society of Training and Development
- *Training Magazine*
- Annual reports of leading organizations and training organizations

Step 4: Collect and analyze data.

Purpose: Get right and accurate data.

Once the sources of information are determined, the data are collected and analyzed. First, the data collected for benchmarking must be the right information. The information must fit the organization's requirement for benchmarking. It must truly reflect the leader. Second, the information must be analyzed to ensure that it is accurate. It is necessary to verify and validate any process performance measurement information as applicable to your specific operation before using it as a benchmark in the organization. In addition, it is imperative to verify and validate process operations by the direct observation of the process.

Step 5: Select target benchmarks.

Purpose: Establish a long-range focus.

During this step, the OD&T organization selects the target benchmarks to meet its mission. The middle section of Fig. 8.5 shows the critical areas for success, and sample target benchmarks are listed in the last column.

Step 6: Determine your performance.

Purpose: Know how the organization is currently performing.

During this step, the selected critical areas are measured. This involves the development of metrics. The metrics development process is described later in this chapter. The middle section of Fig. 8.5 lists sample current performance in the center column.

Step 7: Set desired outcomes.

Purpose: Establish short- and long-term goals to achieve targets.

The benchmarked target may take several years to achieve depending on the current performance and capability of the organization. The organization establishes a plan to achieve the benchmark. The plan may be set up to achieve the benchmark in stages, like the steps to world class shown in Fig. 8.4. The bottom section of Fig. 8.5 shows typical goals to reach the target. In year 1, the OD&T organization aims for improvement over its current performance. In year 2, it seeks to be competitive. In year 3, it targets best in class performance goals. In year 4, it strives for world class status.

Step 8: Use improvement methodology to achieve desired performance.

Purpose: Use a systematic approach to reach the benchmark.

Finally, the benchmarking and improvement process is continuous. The organization must establish a continuous improvement system to achieve the target.

Metrics

Once the organization knows the customer requirements and what it takes to be the best, the CDPM organization focuses on internal processes. One major technique to focus on continuous improvement of internal processes is metrics. *Metrics* are meaningful measures that target continuous process improvement actions. Metrics are differentiated from plain measurement by their specific focus on total customer satisfaction while supporting the organization's vision, mission, and objectives. Metrics are essential to assess all critical processes needed to accomplish the mission of the customer-driven project team. Metrics are necessary in defining the quality/customer satisfaction issue.

Metrics are a measure made over time, which communicates vital information about the quality of a process, activity, or resource. For a metric to be valid, it must be meaningful to the customer. The customer must agree that it is an accurate indicator of a customer value. The metric must stimulate appropriate action by the process owner. The metric tells how the process is performing. It distinguishes between acceptable and unacceptable actions. The metric must be repeatable over time. Continuous improvement actions require a measurement that focuses on the long term. The metric needs to indicate a trend. Again, the aim is constant improvement over time. The metric requires a clear operational definition that provides the exact description of the metric process. Also, the metric data must be simple to collect. The more complex, difficult, and time-consuming the data-collection effort, usually, the more probable it is that the data are incorrect. Therefore, the data for the metric must be simple, easy, and economical to collect.

Steps in the development of metrics

1. Define the purpose of the metric.
2. Develop operational definition.
3. Determine if measurements are available.
4. Generate new measurements if required.
5. Evaluate the validity of the metric.
6. Institute and baseline the metric.
7. Measure the process against baseline.
8. Prepare metric presentation.
9. Use the metric for continuous improvement action.

Metrics example

The following provides a step-by-step example of metric development for the customer-driven project team to improve its effectiveness.

Step 1: Define the purpose of the metric.

Purpose: Improve customer-driven team's effectiveness.

The purpose of the metric must be focused on a specific process geared to meeting the customer's needs and expectations. In this case, the process is project performance achieving total customer satisfaction.

Step 2: Develop operational definition.

Operational definition: The customer-driven project team will perform a monthly project performance review at the second team meeting of each month using an approved "customer satisfaction critique."

The operational definition provides the who, what, when, where, and how of the metric. In this example, the operational definition includes the following:

- Who: Customer-driven project team
- What: Perform "customer satisfaction critique"
- When: Monthly, second meeting of each month
- Where: Team meeting
- How: Presentation using approved "customer satisfaction critique"

Step 3: Determine if measurements are available.

Current measurements do not meet criteria for metric.

Although there are many existing measurements for project performance, they do not focus on customer satisfaction. They only target cost, schedule, and technical specifications. In this case, we have to generate new measurements. If at all possible, use existing measurements. Do not invent an unnecessary measurement.

Step 4: Generate new measurements if required.

New measurements include

- Deliverable satisfaction
- Project execution: schedule, cost, and technical
- Customer relationship
- Other working relationships: supplier, other CDTs, and so forth
- Team leader performance: team meetings, keeping focused, problem solving, decision making, team building, managing conflict

These metrics focus on the performance measurement of the key processes for project performance. If we measure these items that provide an indication of customer satisfaction, the customer-driven project team's performance will improve along with its processes.

Step 5: Evaluate the validity of the metric.

Rating of metric:

- Approved by customer-driven team? Yes
- Linked to organizational objectives? Yes
- Is simple and clear? Yes
- Can display a trend over time? Yes
- Is easy to collect data? Yes
- Can be used to drive process improvement? Yes

Step 6: Institute and baseline the metric.

Measurement tool: Customer satisfaction critique.

During this step, the measurement tool is selected and initial data are collected. The measurement tools can be any measurement tool described in Chap. 11 under "Data Statistical Analysis." For this example, a customer satisfaction critique was selected for collecting the data. Figure 8.6 shows a sample customer satisfaction critique. The results of the critique will be displayed on a line chart. Figure 8.7 shows the initial customer satisfaction charts.

Step 7: Measure the process against baseline.

Each month the customer satisfaction critique is distributed, collected, and charted to examine trends over time.

Step 8: Prepare metric presentation.

Each month the metric presentation is prepared. The metric presentation consists of the metric description and the metric graphic. Figure 8.8 displays the metric description for the example. The metric description consists of the metric operational definition, the measurement method, the desired outcome, the linkage to organizational objectives, and the process owner. An example of a metric graphic is shown in Fig. 8.9 for deliverable satisfaction. Besides the graphic data, a "stoplight" chart is integrated with the metric graphic to highlight status. A "stoplight" chart simply uses green, yellow, and red to indicate a particular assessment. For instance, green indicates no problems. Yellow means some potential problems exist. Red indicates a problem exists.

Step 9: Use the metric for continuous improvement action.

Based on the results of the metric analysis, the team leader initiates process improvement actions. These actions must be focused on continuous improvement.

Customer Satisfaction Critique

Instructions Please rate the customer-driven team based on the 5 point scales below. Circle the number on each scale that best states your opinion at this time.

1. Deliverable satisfaction

Is the customer satisfied with the deliverable to this point?

Very dissatisfied	1	2	3	4	5	Very satisfied

2. Relationships

Is the customer satisfied with the current customer relationship?

Very dissatisfied	1	2	3	4	5	Very satisfied

Is the customer satisfied with the team's relationships with suppliers, other teams, other leaders?

Very dissatisfied	1	2	3	4	5	Very satisfied

3. Project execution

Is the customer satisfied with schedule, cost, and technical performance?

Very dissatisfied	1	2	3	4	5	Very satisfied

4. Team leader performance

Is the customer satisfied with the team leader's performance?

Very dissatisfied	1	2	3	4	5	Very satisfied

Figure 8.6 Customer satisfaction critique.

DELIVERABLE SATISFACTION

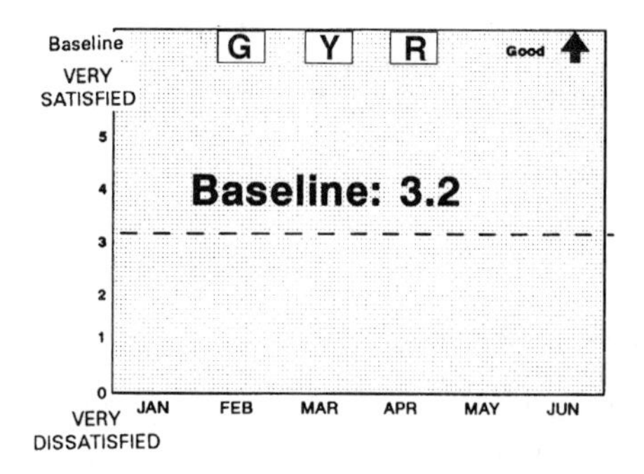

PROJECT EXECUTION

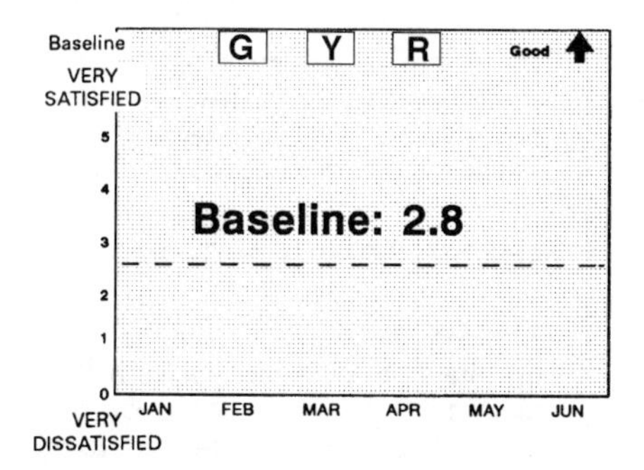

CUSTOMER RELATIONSHIP

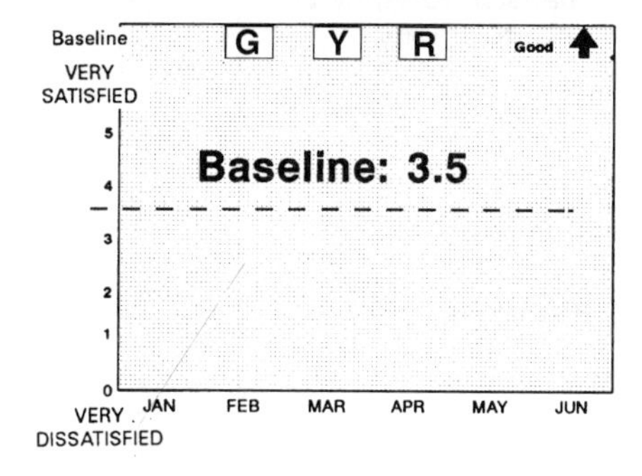

OTHER RELATIONSHIPS

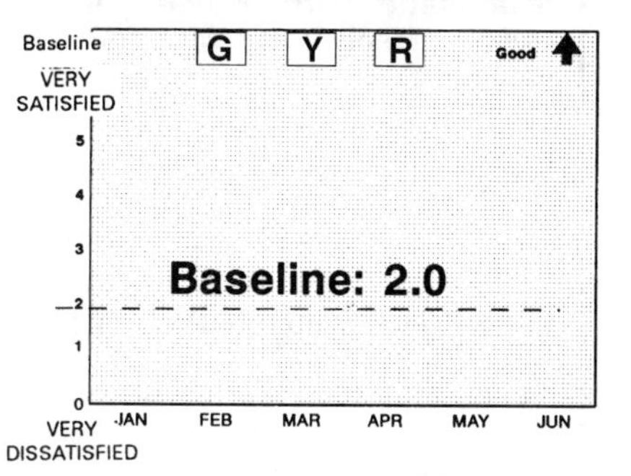

TEAM LEADER PERFORMANCE

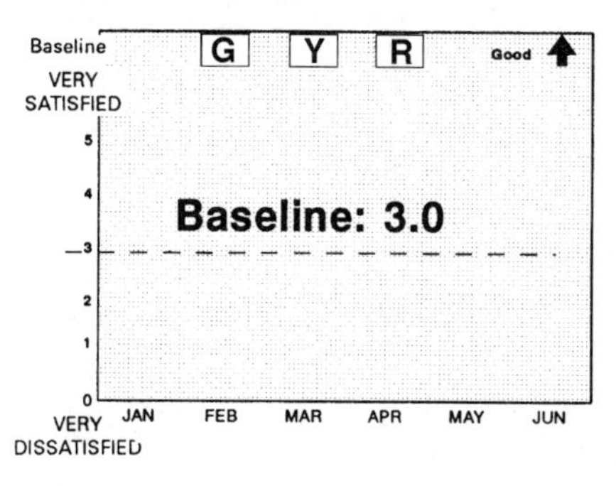

G Green

Y Yellow

R Red

Figure 8.7 Baseline customer satisfaction charts.

METRIC TITLE:

IMPROVE CDT's EFFECTIVENESS

Operational Definition:

The customer-driven project team will perform a monthly project performance review at the second team meeting of each month using the results of the approved customer satisfaction critique.

Measurement Method:

The measurement consists of a customer satisfaction index complied as an average from the approved customer satisfaction critique. The customer satisfaction index indicates the defined key customer's satisfaction with deliverable satisfaction, project execution, customer relationships, relationships with other key interfaces, and team leader performance. The customer satisfaction index uses a closed scale from 1(very dissatisfied) to 5(very satisfied).

Desired Outcome:

Improve customer satisfaction rating in all five areas.

Linkage to Organizational Objective:

Meets total customer satisfaction strategy.

Process Owner:

Customer-driven project team

Figure 8.8 Metric description example.

Main Points

Total customer satisfaction is the quality issue.

Customers are the only people who can determine total customer satisfaction.

To achieve total customer satisfaction, the organization must know the customer, itself, its product, and its competition.

Identifying and continuously responding to customers' needs and expectations are a never-ending challenge.

The key to total customer satisfaction is developing and maintaining customer relationships.

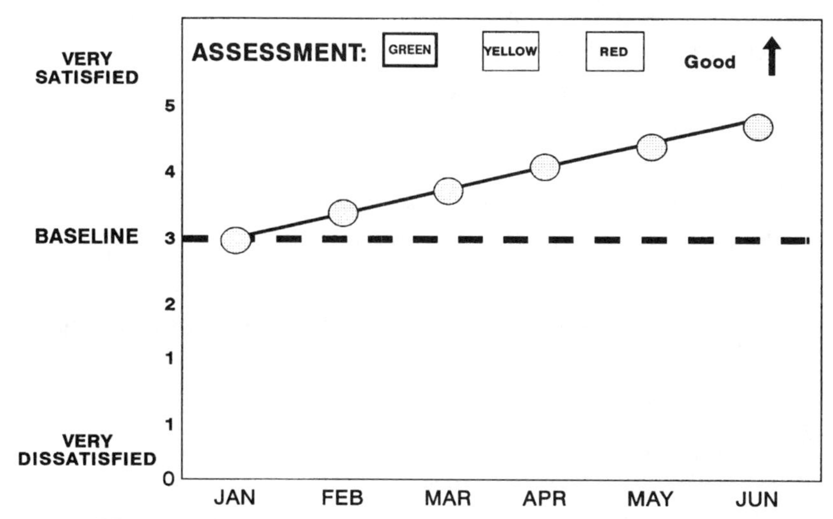

Figure 8.9 Metric graphic example.

Another key element in total customer satisfaction is a complete understanding of all internal processes.

Knowing the product is critical to total customer satisfaction.

Unless the organization can differentiate itself from the competition, the organization may never have a chance to totally satisfy the customer.

There are several tools and techniques that are particularly useful in defining the quality issue in terms of total customer satisfaction, such as quality function deployment (QFD), benchmarking, and metrics.

The quality function deployment (QFD) technique focuses on listening to the "voice of the customer."

In QFD phase 1, the customer requirements are transformed into design requirements. This provides both what the customer needs and expects and what the organization needs to do to satisfy the requirements.

Use appropriate parts of the QFD chart for projects for which a complete QFD chart is not practical or necessary.

Benchmarking targets excellence and focuses the internal process improvement effort.

Benchmarking

- *B*rings the focus on improvement efforts.
- *E*mphasizes desired outcomes.
- *N*urtures competitiveness.
- *C*reates the desire to be the best.
- *H*olds the organization together.
- *M*easures critical areas.
- *A*nalyzes critical areas against the best.
- *R*einforces continuous improvement.
- *K*eeps everyone on target.

Metrics aim at meaningful measures forging continuous improvement actions.

Metrics considerations include:

- *M*eaningful to the customer.
- *E*stablish appropriate actions.
- *T*ell how well the process is preforming.
- *R*epeatable over a period of time.
- *I*ndicate a trend.
- *C*lear operational definitions.
- *S*imple to collect.

Steps in the development of metrics:

1. Define the purpose of the metric.
2. Develop an operational definition.
3. Determine if measurements are available.
4. Generate new measurements if required.
5. Evaluate the validity of the metric.
6. Institute and baseline the metric.
7. Measure the process against baseline.
8. Prepare metric presentation.
9. Use the metric for continuous improvement action

An operational definition defines the who, what, where, when, and how of a metric.

A metric presentation consists of a metric description and metric graphic.

Chapter

9

CDPM Understanding Tools and Techniques

Focus: This chapter details the process understanding tools and techniques.

Introduction

Understanding the process is critical to any customer-driven project management (CDPM) effort. This step has a great impact on the effectiveness of the rest of the CDPM process. In CDPM, the process is the fundamental focus of all actions. A process can be a job, a task, a step, or an activity. A complete understanding of the project process and all the processes required to deliver the project is necessary before continuing to any other step in the CDPM improvement methodology.

To ensure that the organization is performing the right processes, satisfying the right customer, aiming toward the right target, and requesting accurate requirements, process understanding is critical. There are several tools and techniques to help in process understanding. Process diagrams define the process. Input/output analysis identifies interdependency problems. Supplier/customer analysis helps obtain and exchange information for conveying your requirements to suppliers and mutually determining the needs and expectations of your customers.

All the tools and techniques in this chapter are especially useful during step 2 of the CDPM improvement methodology for understanding and defining processes. In addition, these tools and techniques are helpful in other steps of CDPM improvement methodology as follows:

- Process diagrams are worthwhile during the step 4 while performing an analysis of the process. They provide a basis for process analysis. In addition, process diagrams provide insight into necessary processes to perform the project in step 5, taking action. Further, process diagrams assist in step 7, implementing the process. Process diagrams form an excellent tool for documenting and training a process.
- Input/output analysis and supplier/customer analysis provide information for step 5, taking action. Many of the actions needed are a result of the input/output analysis and supplier/customer analysis. They also should be used in steps 6 and 8 to check results and monitor performance.

The process

Before detailing process understanding tools and techniques, an understanding of basic process definitions, the nature of a process, the states of a process, process measurement, and process variation must exist.

Process definitions. The following are some basic definitions required to understand a process:

- A *process* is a series of activities that takes an input, modifies it (work takes place and/or value is added), and produces an output. Thus, a process is the job itself. Figure 9.1 presents a graphic representation of a process.
- An *input* is what you need to do the job.

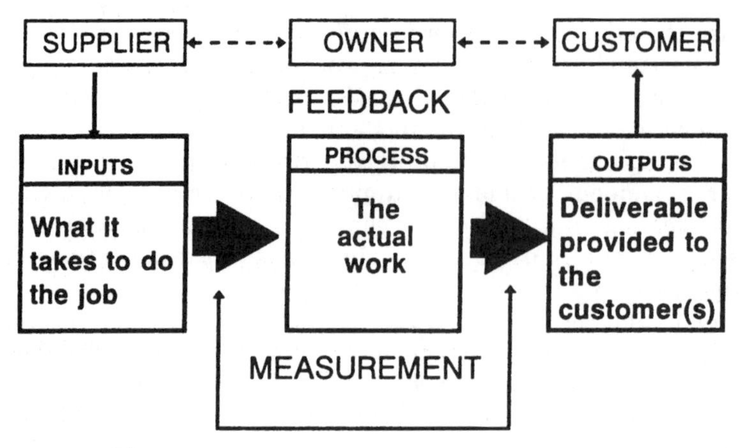

Figure 9.1 The process.

- An *output* is the product or service given to the customer.
- A *supplier* is the provider of the people, material, equipment, method, and/or environment for the input to the process.
- A *customer* (internal or external) is anyone affected by the product or service.
- The *process "owner"* is the person who can change the process.
- *Continuous process improvement* is the never-ending pursuit of excellence in process performance.
- A *measurement* of a process is the difference between the inputs and the outputs of the process as determined by the customer.
- *Variation* of a process is any deviation from the ultimate best target value.

The hierarchical nature of a process. An understanding of the hierarchical nature of processes, as displayed at the top of Fig. 9.2, is important when improving a process. There are many levels of processes. At the top are the major processes. These top-level processes can be broken down into subprocesses. Each subprocess consists of many tasks.

The bottom of Fig. 9.2 shows the nature of a process. At the top of the figure, there are the three major processes in producing a part. A part is engineered, it is manufactured, and it is tested. The manufacturing process breaks down into its subprocesses. This is another level of processes. When a part is manufactured, a shop order is prepared, material kits are provided, the part is fabricated, and the part is inspected. The kitting subprocess consists of several tasks. The building of a material kit subprocess requires pulling parts, preparing the kit, and releasing the kit to the shop.

Because of the various levels of processes, determining the boundaries of a specific part or parts of a process is a necessity. This means that the start and finish must be defined. Further, processes affect other processes on the same level or on different levels above or below. Therefore, it is important to know the impact of any improvement effort on other processes before the improvement is implemented. For instance, if an improvement recommends making a square hole, a further process may have to be changed to a square peg.

Process states. A process can be in one of several states depending on variation and capability. Figure 9.3 lists the states of a process. State 1 is the unknown state. In this state, the process performance has not been measured. There is no target. State 2 shows the process out of control. There is a target, but the performance cannot be pre-

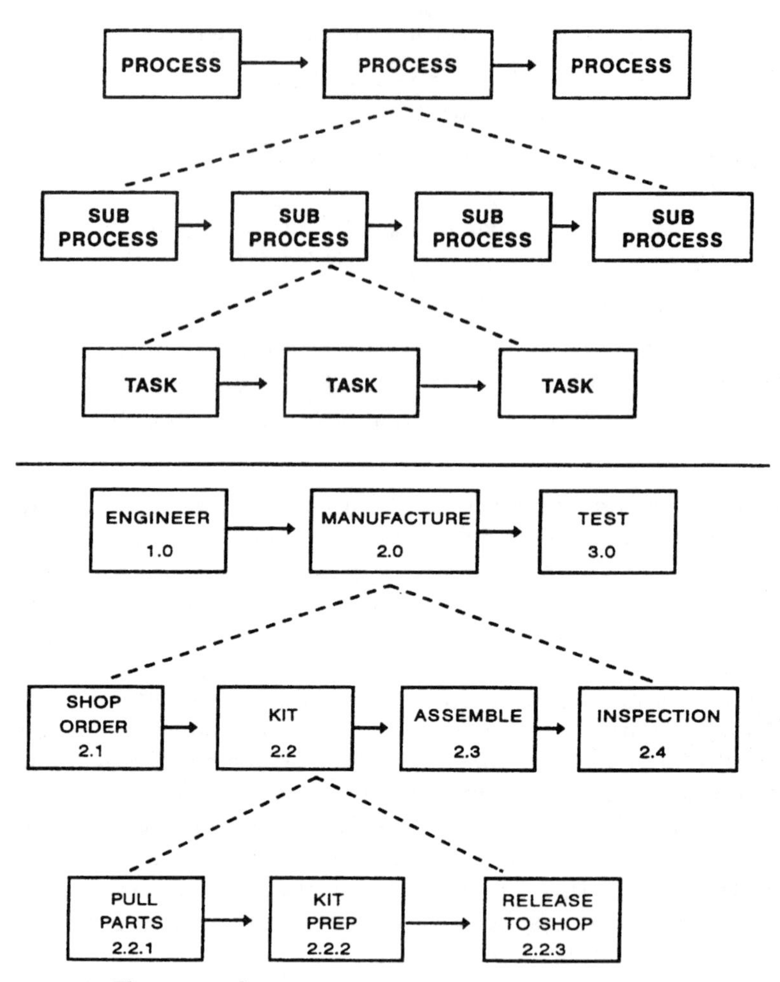

Figure 9.2 The nature of a process.

dicted. In this state, the process performance is an element of chance. State 3 displays a process in control, but the process is not capable. Performance can be predicted, but it will not always hit the target. In this state, the process is not within limits. State 4 is a process in control and capable. Process performance can be predicted within the target. State 5 is process improvement. In this state, the process is improved to reduce variability to the target value. The aim is to consistently hit the bull's-eye or center of the target. State 6 is continuous improvement. In this state, the process is constantly improved to its best possible performance. The target keeps getting smaller and smaller while still continuously hitting the bull's-eye.

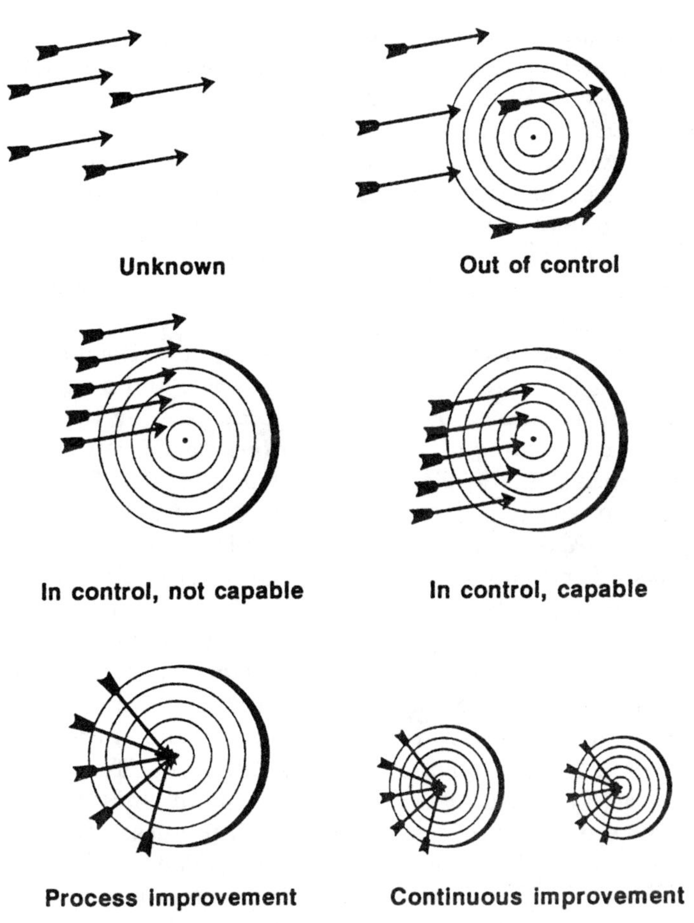

Figure 9.3 Process states.

The continuous improvement system moves the process from one state to another with the ultimate aim of consistently performing all processes at their ultimate best without variation.

Process measurement. All process performance can be measured through process indicators. The major process indicators focus on quality, cost, quantity, time, accuracy, reliability, flexibility, effectiveness, efficiency, and customer satisfaction. Usually, these process performance indicators are either the difference between the input and the output of the process or the output of the process from a customer's viewpoint. In manufacturing processes, the process performance indicators are called *quality characteristics.* These quality characteristics are such items as weight, height, thickness, strength, color, temperature, and density. Besides these quality characteristics, process performance indicators exist in every aspect of an organization. The possible performance indicators are limited only by the view of the organization. Each organization and process owner must determine their own process performance indicators. Some other examples of process performance indicators are errors, time to deliver to the customer, orders filled, number of repairs, number of skilled personnel, response times, number of changes, customer complaints, spares used, mean time between failures, and time available for operational use.

At a minimum, the organization needs to measure its critical areas for business success, key internal business processes, and customer satisfaction. The critical areas for business success are the three to ten most significant indicators of an organization's performance. For instance, the organizational development and training (OD&T) organization used as an example in Chap. 8 lists the VICTORY-C elements as its critical areas for success. Each one of the elements in the VICTORY-C model is measured continuously in a never-ending quest. IBM Rochester formulated its plans based on six critical success factors when it won the 1990 Malcolm Baldrige National Quality Award. The six areas were improved products and services definition, enhanced product strategy, six-sigma defect elimination strategy, further cycle time reductions, improved education, and increased employee involvement and ownership. Cadillac uses five "targets of excellence." Federal Express focuses on 12 service quality indicators.

Besides the key areas for business success, each process within an organization needs a process performance indicator. These indicators need to be developed by the process owners and should aim toward continuous improvement. They should not be used to drive or control the people in the organization. In this regard, the process performance indicators are not aimed at any particular person. They are focused on the process. Some typical process performance indicators of internal processes include cycle time, cost, schedule, number of items, amount of rework, number of errors, number of failures, delivery time, and so forth.

In addition to the preceding areas for measurement, customer satisfaction is an essential measurement area. The development of measurements in this area requires communication with the customer to determine the exact measures of total customer satisfaction. Usually the customer wants the deliverable to satisfy his or her specific need, at the time it is needed, whenever it is needed, for as long as it is needed, and at a cost the customer can afford. Throughout this book are many examples of customer satisfaction indicators. Customer satisfaction indicators aim at performance, response time, cost, availability, reliability, value, service, use, appearance, and so on.

The results areas mentioned in the Malcolm Baldrige National Quality Award provide an excellent starting point for process measurement. Each organization determines specific process measurements based on the following key areas of business performance:

1. Customer satisfaction/retention
2. Market share
3. Product and service quality
4. Productivity and operational effectiveness
5. Human resources performance/development
6. Supplier performance/development
7. Public responsibility

Specifically, the quality and operational results categories consisting of the following areas provide further insight.

- Product and service quality results
- Company operational results
- Business process and support service results
- Supplier quality results

Process variation. There is variation in every process. Variation results from common and special/assignable causes. Common causes are the normal variation in the established process. Such variation is always part of the process. Special/assignable causes are abnormal variation in the process. This type of variation arises from some particular circumstance. It is important to understand the impact of both causes of variation. A variation from a special assignable cause should be solved as a specific problem attributed to something outside the normal process. A variation from a common cause can only be improved by a fundamental change in the process itself. If a common cause is mistaken for a special cause, the adjustment could result in increased variation and frequently will make the process worse. For example, a test failure is encountered when performing a final test of an assembly. The test failure is determined to be attributed to the "A" board. If the failure's root cause is further blamed on the special cause of operator error in fabrication of the board when, in fact, the procedures in the process are incorrect, which would be a common cause, the test failure will be repeated in the future.

Process Diagram

A *process diagram* is a tool for defining the process, which is a major focus of CDPM activity. An initial step in any CDPM activity should be to define and understand the processes needed to accomplish the project. Each organization, function, and person should define their specific processes and understand how their processes satisfy customer needs and expectations (both internal and external customers). Each process is a customer of the preceding process, and each process has a customer for its process. Everyone must strive constantly to improve their processes both as suppliers and for customers.

A process diagram uses symbols and words to describe a process. It provides an indication of improvement opportunities, non-value-added tasks, and where simplification is possible. A process diagram identifies graphically the interrelationships of the process to show the roles and relationships between processes. It shows which elements affect process performance. Finally, it indicates where the process should be measured.

There are three types of process diagrams:

- Top-level process diagrams
- Top-down process diagrams
- Detailed process diagrams

Top-level process diagrams

A *top-level process diagram* is a picture of the entire process. This type of process diagram shows the input(s), the process, and the output(s). The top-level process diagram should focus on satisfying customer needs and expectations. The expected results must be determined, and the process must be measured to determine if it is achieving results. Figure 9.4 shows a top-level process diagram of a system development process.

Top-level process diagram steps

1. Define the specific outcome of the process, focusing on the customer.
2. State the process in terms of what work has to be done to meet the outcome.
3. Determine the input(s) required to satisfy the customer.
4. Measure the process. This measurement is usually the difference between inputs and outputs.

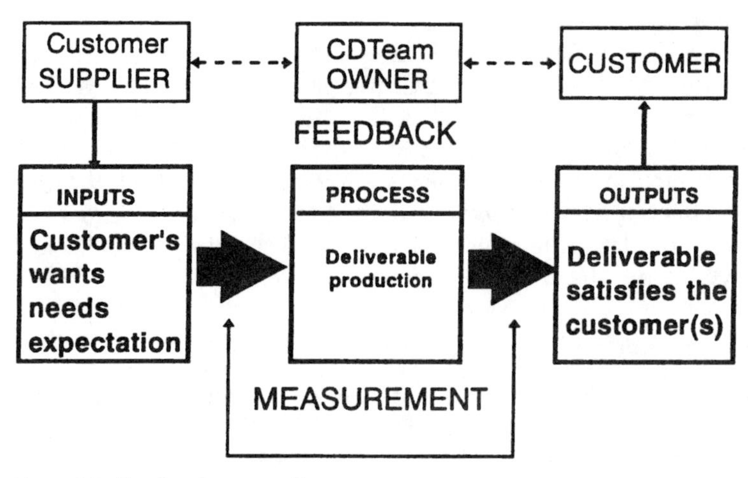

Figure 9.4 Top-level process diagram.

5. Use the CDPM improvement methodology to improve, invent, or reengineer the process.

Top-level process diagram example. As we analyzed our system development process, we discovered that the customer expects the product to be repaired in minimum time. This requires us to have trained technicians to service our product. Looking at Fig. 9.5, the

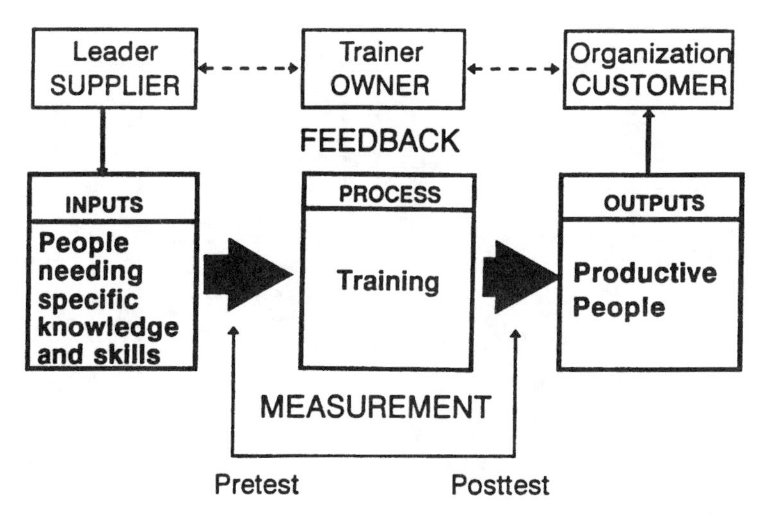

Figure 9.5 Top-level process diagram example for a training process.

top-level process diagram shows the training process. The customer expects skilled people as the outcome of the process. Entering the process are unskilled people. The training process takes the unskilled people and transforms them into skilled people. Before the people enter training, a pretest is administered to determine entry-level skills. At completion of the training, a posttest is given to determine the overall effectiveness of the training. This is the measurement of the process. This training process can then be continually checked and improved using the CDPM improvement cycle.

Top-down process diagrams

A *top-down process diagram* is a chart of the major steps and substeps in a process. By examining the major steps, the opportunities for improvement are focused on the essential steps in the process.

Top-down process diagram steps

1. List the major steps in the process. Keep it to no more than seven steps.
2. List the major substeps. Keep it to no more than seven steps.

Top-down process diagram example. Still using the training example, the top-down process diagram would list the major processes in the

MAJOR PROCESSES	ANALYSIS	DESIGN	DEVELOPMENT	IMPLEMENTATION	EVALUATION	
SUBPROCESS 1	Performance analysis	Objectives	Lesson plan	Presentation techniques	Evaluation plan	
SUBPROCESS 2	Process analysis	Lessons	Training materials development	Presentation preparation	Evaluation instruments	
SUBPROCESS 3	Task analysis	Measurements	Production of training materials	Presentation delivery	Conduct analysis	
SUBPROCESS 4		Course outline		Training administration	Analyze results of evaluation	
SUBPROCESS 5		Training plan				
SUBPROCESS 6						

Figure 9.6 Top-down process diagram worksheet.

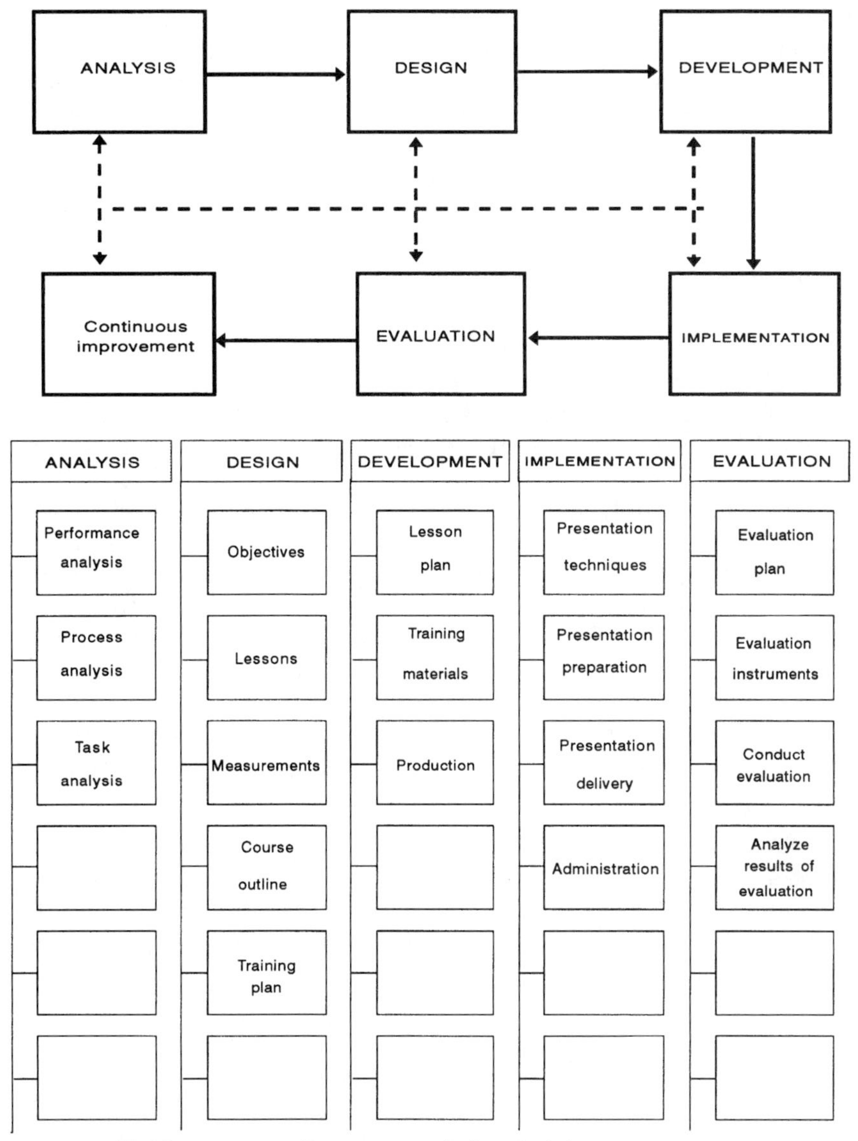

Figure 9.7 Top-down process diagram example for a training process.

top-level process. When working with a team, each team member completes a process worksheet, as shown in Fig. 9.6. Then the team develops the top-down process flow diagram, as shown in Fig. 9.7. In this figure, the major processes are analysis, design, development, implementation, and evaluation. The subprocesses under each major process are the next items listed. For instance, under the major process "Analysis," the following subprocesses would be written: (1)

training analysis (2) job/process analysis, and (3) task analysis. Under "Design," the subprocesses are (1) objective, (2) lessons, (3) measurements, (4) course outline, and (5) training plan. Under "Development," the subprocesses are (1) lesson plan, (2) training materials, and (3) training production. Under "Implementation," the subprocesses are (1) presentation techniques, (2) presentation preparation, (3) presentation delivery, and (4) administration. Under "Evaluation," the subprocesses are (1) plan evaluations, (2) produce evaluation instruments, (3) conduct evaluations, (4) analyze results, and (5) continuous improvement.

Detailed process diagrams

A *detailed process diagram* is a flowchart consisting of symbols and words that completely describe a process. This type of diagram provides information indicating improvement opportunities, identifying areas for data analysis, determining which elements affect process performance, and documenting and standardizing the process. This type of diagram is helpful in identifying non-value-added tasks and areas for simplification. Further, complex activities and unnecessary loops are visualized. This type of process diagram is useful for training, documenting, and explaining the process to others.

Before deciding to do a detailed process diagram, decide on the specific detail and boundaries of the process diagram. Detailed process diagrams are time-consuming. Therefore, specific boundaries are important to ensure progress on achieving improvements.

Detailed process diagram basic symbols. There are many detailed process diagram symbols. To keep it simple, four basic symbols are recommended. These symbols, as shown in Fig. 9.8, are enough for most process diagramming needs. The four basic detailed process diagram symbols are as follows:

Start or end is shown by squares with rounded sides.

Action statements are written as rectangles.

Decision statements are written as diamonds. A decision statement asks a yes or no question. If the answer is "yes," the path labeled "yes" is followed; otherwise, the other path is followed.

Wait/hold is illustrated by circles.

Detailed process diagram steps. The steps in completing a detailed process diagram are

1. Decide on specific detail and boundaries of the detailed process diagram.

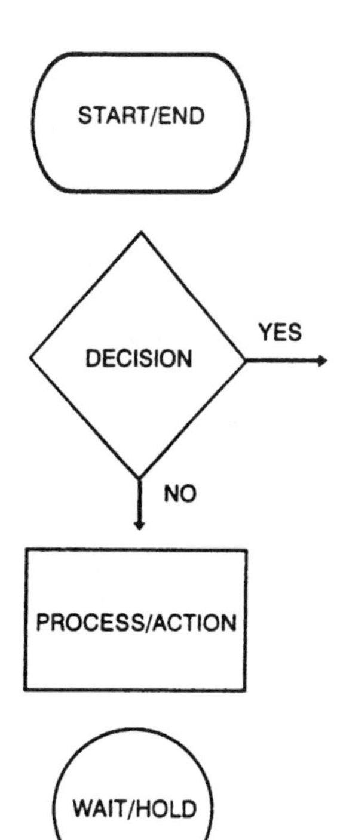

Figure 9.8 Basic symbols for detailed process diagrams.

2. List all the steps required in the process within the boundaries.
3. Construct a process flow diagram.
4. Determine times and costs of each activity, if appropriate.

Detailed process diagram example. A detailed process diagram example is shown in Fig. 9.9. It consists of the following activities:

1. A subcontractor data requirements list (SDRL) is received by materials. (*Input*)
2. Materials forwards the SDRL to data management. (*Action*)
3. Data management logs the SDRL. (*Action*)
4. Data management determines if an internal review is required before distribution. (*Decision*)
5. If an internal review is not required, the SDRL is distributed. (*No*)
6. If an internal review is required, data management forwards the SDRL to reviewer(s). (*Yes*)
7. The reviewers review the SDRL and make comments. (*Action*)

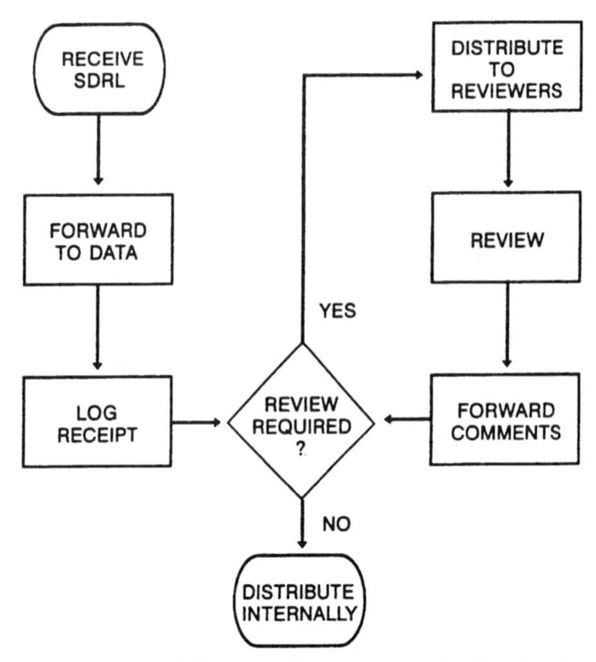

Figure 9.9 Detailed process diagram example (SDRL, subcontractor data requirement list).

8. The reviewer(s) forward the comments to data management. (*Action*)
9. Data management determines if further review is required. (*Decision*)
10. If no further review is required, the SDRL is distributed. (*Output*)

Input/Output Analysis

Input/output analysis is a technique for identifying interdependency problems. This identification is done by defining the process and listing inputs and outputs. Once the inputs and outputs are determined, the relationship of inputs to outputs is analyzed along with the roles of the organization. The top section of Fig. 9.10 shows the input/output diagram template. The bottom section shows the inputs and outputs from the logistics process.

Input analysis

The input analysis lists all the inputs of the process. These inputs are based on the requirements of the process. Once the inputs are known, the prime "owner" and support responsibilities for each of the inputs

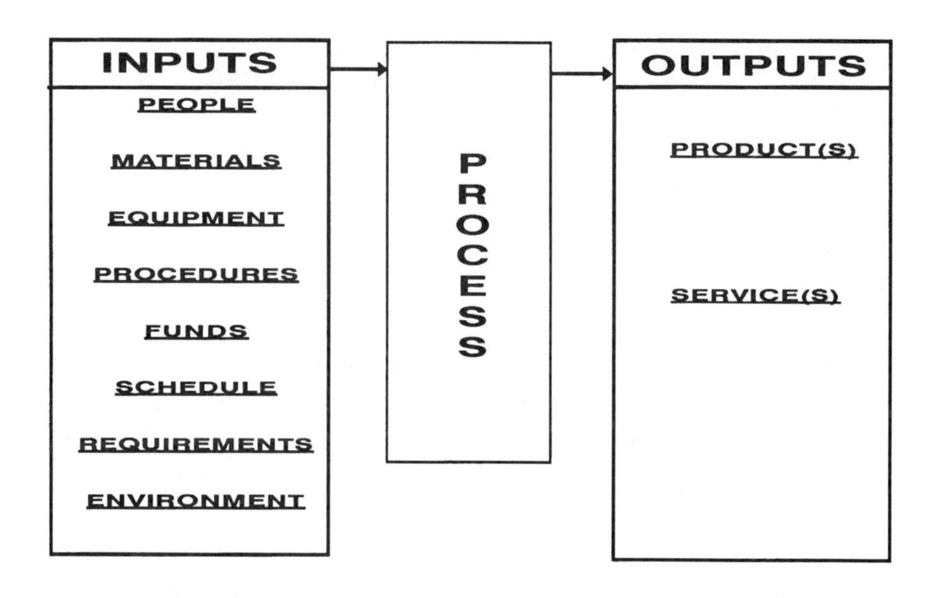

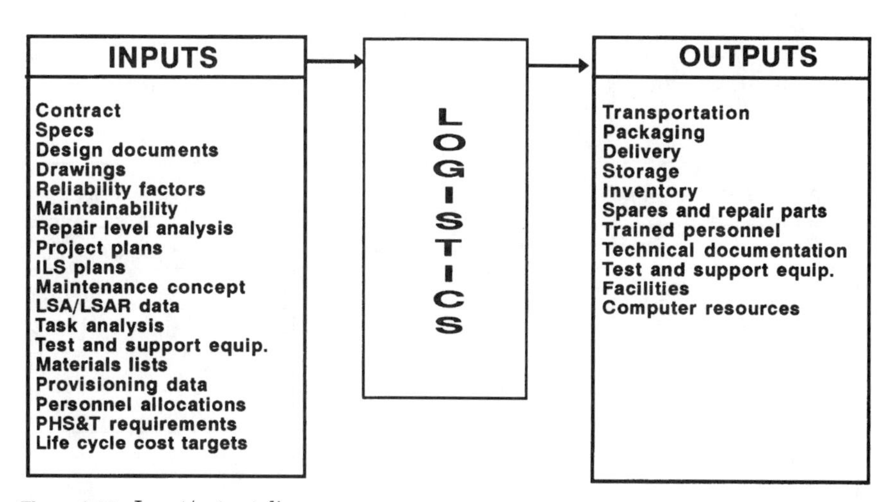

Figure 9.10 Input/output diagram.

are defined. This thorough analysis of all the inputs is used to match with outputs.

Output analysis

The output analysis lists all the outputs of the process. Again, the prime "owner" and support responsibilities are understood. This is accomplished by communicating with supplier(s), owners, and customer(s). The team needs to listen especially to the customer.

Input/output analysis steps

1. Define the actual process.
2. List inputs and outputs of the process.
3. Determine prime (owner) and support (influencing) responsibilities.
4. Match inputs and outputs with organizations.
5. Define roles of the organizations.
6. Document on input/output analysis worksheet.
7. Use CDPM improvement methodology to improve, invent, or reengineer the process.

Input/output analysis example. The input/output analysis uses information from the input/output diagram to complete the input/output analysis worksheet. In the example, the input/output diagram of the logistics process shown in Fig. 9.10 is the basis for the input/output worksheet, as shown in Fig. 9.11. The figure shows an output of the

PROCESS	INPUT	OUTPUT	PRIME	SUPPORT	ROLE
Logistics	Drawings	Documentation	Tech. pubs.	Engineering	Inter-dependent

Figure 9.11 Input/output worksheet.

logistics process as documentation. An input to the documentation is drawings. This shows the engineering drawings as an input for the output documentation. The prime owner of the documentation is the technical documentation section of the logistics organization. The engineering function has the main support role of providing the drawings for the documents. The roles are interdependent; without the drawings, the document cannot be produced. This relationship is shown in Fig. 9.11. This form can be used to look at relationships of output to input or input to output. Knowing this information, improvement opportunities can be identified and implemented using the continuous improvement cycle.

Supplier/Customer Analysis

Supplier/customer analysis is a technique that involves your suppliers in the development of your requirements and their conformance to them. In addition, it provides insight into your customers' needs and expectations and your meeting of those expectations.

It is important to develop a partnership with your suppliers and a relationship with the customers you want to keep or gain. Use surveys and interviews to ensure a mutual agreement on supplier requirements and customer expectations. The supplier/customer analysis worksheet can be used to document results. It is important to communicate, listen, and thoroughly analyze supplier and customer perceptions to continuously improve supplier performance, the process, and customer satisfaction.

Surveys and interviews

Surveys and interviews are the major methods for getting information from suppliers and customers. Surveys involve the following:

Set a survey strategy. The team develops the survey strategy to target the expected results. In addition, the survey strategy focuses on providing information that is valid, consistent, and free of bias.

Use simple, concise, and clear questions. Each question must be easily understood by all respondents.

Run a pilot. This is the only method to ensure an accurate survey. The pilot should be run on a representative sample of the true population for the survey.

Use the most effective and efficient format. This involves deciding on the best type of questions, either open or closed. Use closed questions to get specific answers. For instance, if you want to know if the customer agrees or disagrees with a particular item, ask a closed-ended question with a response scale of either agree or disagree. If you want to know how a customer feels about a particular feature, ask an open-ended question. In addition, the questions should be limited to 12 to 15 words. These types of questions are understandable. Further, put priority questions first. The respondent may become tired, bored, or disinterested as the survey questions progress. Also, cluster related questions. This keeps the respondent's mind on the subject and avoids confusion. In addition, provide a response scale with a through description. Finally, develop a scoring system. This allows for charting a detailed analysis.

Ensure room for comments. Typically, respondents' comments are valuable indicators of real information.

Yield to comments. If a respondent takes the time to make comments, the surveyor must pay particular attention to this information. Often it is worthwhile to schedule an interview with a respondent who offers comments on a survey.

Interviews involve the same considerations as surveys. In addition, the interviewer should do the following:

Instill an atmosphere of openness, honesty, and trust. This builds the rapport for the interview.

Nurture the self-esteem of the interviewee. This is essential to allow the interviewee to communicate freely.

Trust the interviewing process. In order to maintain the consistency of the interview, the interviewer must stay within the strategy of the interviewing process.

Empathize with the interviewee. The interviewer needs to put himself or herself in the shoes of the interviewee during the interview.

Respond to the interviewee frequently. Show the interviewee that you are sincere by nonverbal gestures such as nodding approval, smiling, and so on. In addition, show that you understand by paraphrasing and summarizing often.

Wait for the interviewee to respond. Resist the temptation to fill in gaps of silence. Silence in an interview should not be viewed as wasting time. Allow the interviewee time to collect his or her thoughts and formulate opinions.

Invite the interviewee to build on ideas. Ask open-ended questions to get additional information. Use close-ended question to get specific answers.

Ensure that the interviewee gets feedback on the results of the interview. This makes the interviewee feel that he or she has contributed to an outcome.

Write down all information. Documentation of the interview is the only means to analyze results. Ensure that the information is accurate and thorough.

Supplier analysis

The supplier analysis consist of answering the following questions:

Did you survey supplier(s) to ensure that requirements are known?

Is there a mutual understanding of requirements?

Have you established a partnership with key suppliers?

What are suppliers' perceptions of your requirements?

Did you listen to suppliers' concerns?

Did suppliers listen to your concerns?

Were interviews conducted to determine supplier perceptions?

Were customer expectations translated into supplier requirements?

Are your suppliers satisfying your requirements?

Customer analysis

Customer analysis seeks to answer the following questions:

Are you communicating to ensure that you are satisfying customers?

Do you understand customers' needs and expectations?

Have you conducted a survey to determine if you are satisfying your customers?

Has a thorough analysis been completed to ensure that the entire process is focused on customers' needs and expectations?

Does the "owner" understand how the process affects the customer?

Are process outputs measured in relation to customer expectations?

Are you satisfying mutually agreed on customer expectations?

Have you developed a relationship with key customers?

Supplier/customer analysis steps

1. Identify the customers (both internal and external customers) of the process.
2. Determine the needs and expectations of your customers.
3. Identify the products or services you provide to meet these needs and expectations.

4. Develop measures of your output that reflect customer expectations.
5. Determine if the customer expectations have been met or not.
6. Determine who "owns" or influences the product or service.
7. Identify your principal inputs (personnel, material, machine, method, environment).
8. Determine if suppliers know their requirements and their effect on your meeting customer expectations.
9. Involve your suppliers in the development of your requirements and their conformance to them.
10. Identify suppliers that are not meeting requirements.
11. Document results on the supplier/customer analysis worksheet.
12. Use CDPM, improvement methodology to improve, invent, or reengineer the process.

Supplier/customer analysis example. The supplier/customer analysis example continues using the engineering drawings required for documentation. First, transfer the input, output, and supplier information from the input/output worksheet to the supplier/customer diagram, as shown in Fig. 9.12. Second, transfer the information from the supplier/customer diagram to the supplier/customer worksheet, as shown in Fig. 9.13. In the example, the input of drawings provides the data for the maintenance documentation output. The customer expects a usable document for maintenance personnel. This is determined by communication with the customer. Third, a specific measurement must be determined. In this case, the customer expects the document to be 100 percent accurate. Fourth, measure performance. Fifth, determine if the document is meeting or not meeting customer expectations.

To complete the supplier side of the equation, engineering must be informed of the drawings' impact on providing an accurate document to maintenance personnel. The requirement for complete and accurate drawings must be measured to determine if engineering is meeting or not meeting requirements.

All the information from the supplier/customer worksheet is used within the improvement methodology to improve supplier performance, the process, and customer satisfaction.

Main Points

Process understanding is improved through process diagramming.

Process diagram only to the desired level.

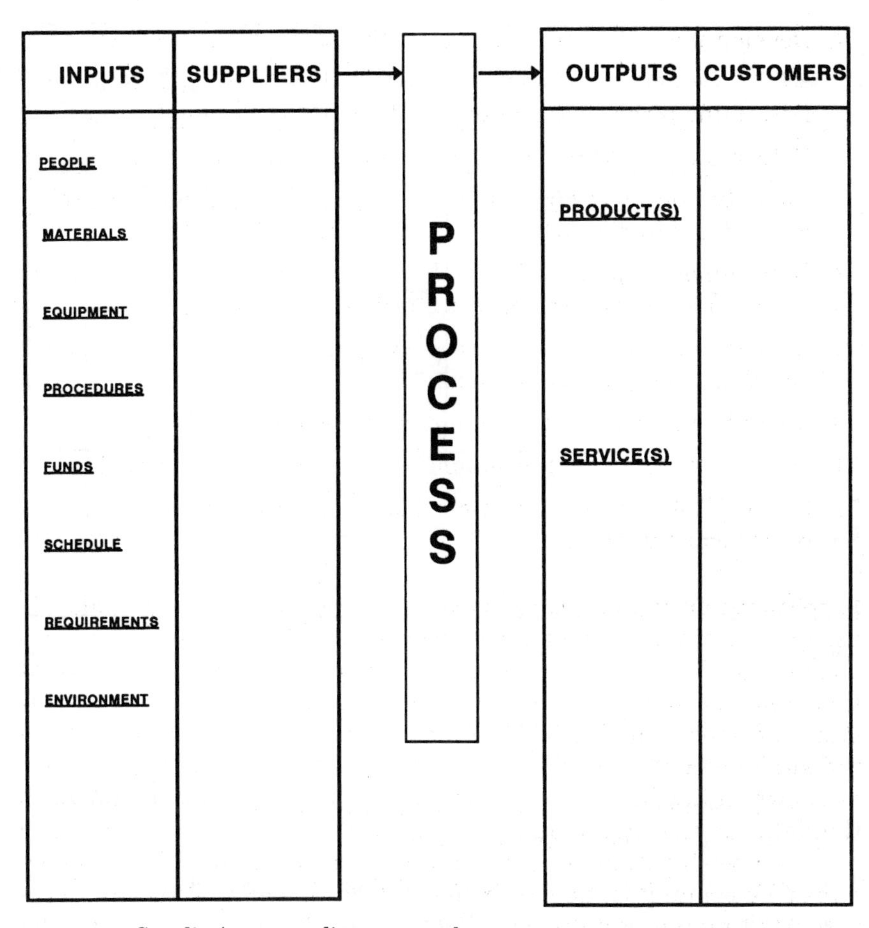

Figure 9.12 Supplier/customer diagram template.

Keep the process diagram within specified boundaries, but make it thorough and as complete as possible.

The relationship between inputs and outputs affects customer satisfaction.

Develop a partnership with your suppliers and a relationship with your customers, and constantly analyze these relationships.

Listening is most critical during customer/supplier analysis.

Surveys involve

- Setting a survey strategy.
- Using simple, concise, and clear questions.
- Running a pilot or test.
- Using the most effective and efficient format.

INPUT	SUPPLIER	REQUIREMENT	MET/ NOT MET	OUTPUT	CUSTOMER	NEED/ EXPECTATION	MET/ NOT MET
Drawing	Engineering	Complete and accurate	Met	Repair manuals	Maintenance people	100% accurate	Met

Figure 9.13 Supplier/customer worksheet.

- Ensuring room for comments.
- Yielding to comments.

Interviews involve the same considerations as surveys. In addition, the interviewer should do the following:

- Instill an atmosphere of openness, honesty, and trust.
- Nurture the self-esteem of the interviewee.
- Trust the interview process.
- Empathize with the interviewee.
- Respond to the interviewee frequently.
- Wait for the interviewee to respond.
- Invite the interviewee to build on ideas.
- Ensure that the interviewee gets feedback on results.
- Write down all information.

The process "owner" seeks to develop a partnership with suppliers and a relationship with its customers.

Supplier analysis aims at ensuring that the supplier provides the required inputs to the process.

Customer analysis focuses on ensuring that the process satisfies the customer.

Chapter

10

CDPM Selection Tools and Techniques

Focus: This chapter details selection tools and techniques to focus on consensus decision making.

Introduction

During customer-driven project management (CDPM) activities and decision points, the focus is on consensus decision making. Consensus decision making requires an orderly approach to making selections. There are many selection methods available to help the team focus on consensus. This allows the team to select from among several options. Normally, the selection methods funnel down the alternatives to the top few. From this point, the team aims for consensus on the "best" selection. Some of the most common selection methods are voting, the selection matrix, and the selection grid. These selection tools and techniques all assist a group in arriving at a decision through consensus.

The selection tools and techniques are useful throughout the CDPM improvement cycle to make decisions. Consensus decision making is the preferred method when the customer-driven team establishes the focus, selects an opportunity to work, and chooses an alternative. Specifically, this includes

- Making the selection of improvement opportunities during step 3.
- Selecting the project alternative during the "take action" phase, step 5.

Selection Techniques

Selection techniques are used to help clarify assumptions and focus on consensus. The first step to selection involves choosing a technique from among voting, selection matrix, and selection grid. Second, the selection criteria must be established by the team. Third, the team lists items to select. The list consists of issues, problems, or opportunities when selecting an item for the team to work or a list of alternatives when selecting a solution. Fourth, the team must communicate to get an understanding of each item for selection. This includes discussing tradeoffs of ideas with everyone involved. Fifth, the team needs to obtain consensus on the selection, which must meet the needs of the team for a problem to solve, an opportunity to improve, or a solution to implement. Finally, the team nurtures the decision by communicating the benefits to key people. The selection process can be summarized as follows:

- *S*elect a technique.
- *E*stablish criteria.
- *L*ist issues, problems, opportunities, or alternatives.
- *E*valuate the list using the criteria.
- *C*ommunicate until the team reaches an understanding.
- *T*rade off ideas.
- *I*nvolve everyone affected.
- *O*btain consensus.
- *N*urture the decision.

Selection technique steps

1. List issues, problems, opportunities, and alternatives.
2. Determine criteria.
3. Select method:
   a. Vote
   b. Matrix
   c. Grid
4. Make a decision.

Selection technique lists. The list of issues, problems, and opportunities is usually the result of a brainstorming session, process diagrams, input/output analysis, and customer/supplier analysis. Alternatives are generated from brainstorming sessions or force-field analysis.

Selection technique criteria. The team must determine realistic criteria to make the selection. The solution criteria equate to guidelines, requirements, and specifications for the selection. There are many factors that could be considered for selection criteria. Although the team can select any criteria, the following criteria are common:

- *C*ost
- *R*esources
- *I*mportance
- *T*ime
- *E*ffect
- *R*isk
- *I*ntegration with organization's objectives
- *A*uthority

In some instances, cost may not be a factor in the selection. In most cases, however, cost is a primary consideration. The specific cost factor should be stated in the criteria. For example, less than $1000 may be a selection criterion.

Resources involve many individual factors, including capital, labor, equipment, natural resources, etc. Resources are usually a major consideration in any selection. Again, the specific criteria must be determined by the team. Many times the selection must be based on no increase in resources of any kind.

Importance to the organization or team may be another criterion. Normally, a team does not want to waste time on unimportant projects. The team should always strive to focus on the vital few versus the trivial many. Again, there could be much debate on the importance of any single item. The specific criteria may have to be stated to avoid confusion.

Time is always a necessary criterion in any selection. Without a time criterion, almost anything is possible. Furthermore, a time criterion targets a selection geared to short- or long-term results.

Risk is a consideration many organizations use as a criterion. The amount of risk the team or the organization is willing to assume should be stated for a selection criterion.

Another selection criterion often overlooked is integration with the organization's objectives. Sometimes a solution is selected beyond the scope of the organization's focus.

Another criterion offered is authority. The authority should always be a selection criterion. Frequently, teams select a problem or a solution beyond their authority. Then they become frustrated with efforts to get others to implement their solution. The team needs the wisdom to work on the issues and solutions that the team "owns." Again, authority is a consideration for a selection criterion that the team decides.

Voting

Voting is a selection technique used to determine majority opinion. This technique may be useful in narrowing down a list of problems, opportunities, or alternatives. Since this method often leads to a win/lose situation, it is not recommended as a final decision-making technique for selecting an issue, problem, opportunity, or solution. The objective of all selection techniques is to focus on reaching consensus on a win/win solution. Again, this is just a preliminary step to limit the number of items to get a consensus. There are three primary voting techniques. The first is simply having people raise their hand to indicate a vote. The item with the most votes is selected. This method is effective for voting on a small number of items. For voting on a large number of items, use rank-order voting and multivoting.

Rank-order voting. Rank-order voting is a quick method for ranking a list of issues, problems, opportunities, or alternatives to determine the top priorities.

Rank-order voting steps

1. Generate a list of items requiring a decision.
2. Combine similar items.
3. Number items.
4. Have each member rate each item on a scale of 1 to 5, with 5 being the high number.
5. Total the points for each item.
6. Rank the items from highest to lowest based on total points.
7. Reach consensus on the top priority.

Rank-order voting example. The following is an example of the rank-order method of selection. First, a list of items for selection as opportunities was generated using a brainstorming session on the people issues in a specific organization. Second, similar items were combined. Third, items were numbered. Fourth, each member rated each item. Fifth, the total points were listed:

1.	Training	24
2.	Morale	17
3.	Location of personnel	10
4.	Not interested in someone else's responsibility	21
5.	Staffing problems	23
6.	Bad attitudes	16
7.	Tendency to separate people	13
8.	People's dissatisfaction	19
9.	Poor workmanship	20
10.	Lack of right skills	15
11.	Lack of indirect resources	21
12.	Not right people on shift	16
13.	Lack of team spirit	20
14.	No "ownership"	23
15.	Understaffed	22
16.	No accountability	21
17.	Restricted advancement	19
18.	Aging work force	18
19.	Lack of pride in work	17
20.	Too many leaders, not enough workers	25

Sixth, the list was rearranged highest to lowest.

1.	Too many leaders, not enough workers	25
2.	Training	24
3.	Staffing problems	23
4.	No ownership	23
5.	Understaffed	22
6.	Lack of indirect resources	21
7.	No accountability	21
8.	Not interested in someone else's responsibility	21
9.	Poor workmanship	20
10.	Lack of team spirit	20
11.	People's dissatisfaction	19
12.	Restricted advancement	19
13.	Aging work force	18
14.	Morale	17
15.	Lack of pride in work	17
16.	Bad attitudes	16

17. Not right people on shift	16
18. Lack of right skills	15
19. Tendency to separate people	13
20. Location of personnel	10

Seventh, the team reached a consensus on the selection. In this particular example, the highest number of points was "Too many leaders, not enough workers." However, the team decided through consensus to select "Training." This was the issue the team could have an impact on.

Multivoting. Multivoting is a technique used to reduce a large list of issues, problems, opportunities, or alternatives to a smaller number of items. Typically, during the understanding stage a team has a large list of concerns with the process. This list must be reduced to focus the team on a consensus about the most critical issues to work.

1. Generate a list of items requiring a decision.
2. Combine similar items.
3. Number items.
4. Have each member select a number of the items from the list. This is accomplished by each person writing the numbers of the items on separate sheets of paper. For instance, if the total number of items on the list is 50, they are numbered 1 to 50. Each person then may be asked to select the top 20 items from the list of 50.
5. Total the number of votes for each item.
6. Create a new list of the top 20 vote getters. Number the items 1 through 20.
7. Have each member select a lower number of items from the list. For instance, from the list of 20 items, each member may be asked to select the top 5 items.
8. Reach consensus on the top priority.

Multivoting example. Another team was looking into the issues related to methods in their organization. Figure 10.1 shows the flow. First, a list of 20 items was brainstormed. This list was pared down to the top 4 items. Finally, the team reached consensus on "Too many signatures and approvals" as the issue to focus their improvement efforts.

Selection matrix

A *selection matrix* is a technique for rating issues, problems, opportunities, or alternatives based on specific criteria. The issues, problems, opportunities, or alternatives are listed on the left side of the matrix.

Issues relating to methods in the organization include:

1. Unclear administrative procedures
2. Too much paperwork
3. Lack of configuration management
4. Too many signatures and approvals
5. Numerous engineering changes
6. Outdated methods
7. Policies and procedures overkill
8. Lack of understanding customer requirements
9. Delayed work authorizations
10. No scheduling
11. Poor planning
12. Firefighting
13. No data retrieval system
14. Inadequate design reviews
15. Lack of coordination
16. Design intent versus manufacturing producibility
17. We have always done it that way
18. Distribution
19. Information availability
20. Turf protection

Select the four most critical issues. Your selections should consider the following criteria: (1) Would require no resources to solve. (2) The team can solve the issue. (3) The issue can be solved in 30 days.

1 Unclear administrative procedures

2 Too many signatures and approvals

3 Outdated methods

4 Lack of understanding customer requirements

Selection is:

Too many signatures and approvals

Figure 10.1 Multivoting example.

The team selects the criteria to be considered in evaluating the alternatives. These criteria are placed along the top of the matrix. Then the members individually rate the issues, problems, opportunities, or alternatives. The selection matrix should always be completed individually first. Then the group ratings are determined. The results of the selection matrix do not set the actual decision. The purpose of the matrix is to lead to a more focused discussion of each item. This process helps to clarify assumptions and focus consensus.

Selection matrix steps

1. The issues, problems, opportunities, or alternatives are listed on the left side of the matrix.
2. The team selects the criteria to be considered in evaluating the problems, opportunities, or alternatives.
3. The criteria are listed at the top of the matrix.
4. The members individually rate the problems, opportunities, or alternatives.
5. Each team member's highest total point item is tabulated.
6. Discussion of the issues, problems, opportunities, or alternatives ensues.
7. The group tries to reach a consensus.

Selection matrix example. The selection matrix example is shown in Fig. 10.2. In the example, the organization needs to select a method of training that ensures that test technicians can perform a certain test. The opportunities are listed down the right side of the figure. The opportunities are formal classroom, on-the-job, combination of both, or none. The criteria are listed across the top. For this example, the following criteria were selected: effect, cost, and time to implement. Each opportunity is rated against the criteria individually by each team member. The team member who completed the selection matrix in Fig. 10.2 rated the combination of both formal and on-the-job training as the highest solution. This solution received 22 points. The issue, problem, opportunity, or solution is then rated by the total number of team members giving it the highest number of total points. For instance. the team considering the method of training consists of 10 members. The results of the 10 team members' selection matrices were as follows:

Formal classroom	3 members rated this highest
On-the-job training	2 members rated this highest
Combination of both	5 members rated this highest
None	0 members rated this highest

The bottom of Fig. 10.2 shows a data-collection chart for team member totals. Once all team members' highest ratings are known,

Selection Matrix

Issue Opportunity Problem Alternative	EFFECT High- 10 Low - 1	COST High- 10 Low - 1	TIME High- 10 Low - 1	OTHER High- 10 Low - 1	TOTAL
Formal classroom	8	6	3		17
On-the-job training	4	8	7		19
Combination of both	9	7	6		22
None	1	7	10		18

Team Total

ITEM	HIGHEST ITEM	TOTAL
Formal classroom	| | |	3
On-the-job training	| |	2
Combination of both	| | | | |	5
None		0

Figure 10.2 Selection matrix example.

the group focuses on consensus. The final selection is based on a discussion of the items on the selection matrix.

Selection grid

A *selection grid* compares each issue, problem, opportunity, or alternative against others using the established criteria.

Selection grid steps

1. List problems, opportunities, or alternatives.
2. Determine criteria.
3. Compare each pair of issues, problems, opportunities, or alternatives against others using the criteria.
4. Try to reach a consensus as a group.

Selection grid example. This example continues the example of a team looking into the issues related to methods in their organization. Remember the team used multivoting to narrow the list of issues to the top four. The top four issues were

1. Unclear administrative procedures
2. Too many signatures and approvals
3. Outdated methods
4. Lack of understanding of customer requirements

In this example, the selection grid is used to focus on consensus. The selection grid is shown in Fig. 10.3. Again, the selection grid should be completed individually. First, each choice is given a number. Then each choice is compared against the other choices. The selection is made by circling the preferred choice. For instance, in block 1, item 1, "Unclear administrative procedures," is compared with item 2, "Too many signatures and approvals." This team member selected item 2, "Unclear administrative procedures," by circling item 2 on the selection grid. This procedure is repeated in each of the blocks. Once the selection grid is completed, the choices are totaled. The bottom left of Fig. 10.3 shows a data-collection chart for totaling the individual selections. The individual data-collection chart shows item 4, "Lack of understanding of customer requirements," as the highest selection. If the selection grid is completed individually, the team's selections must be totaled like the selection matrix. The bottom right of Fig. 10.3 shows the team data-collection chart. The team's selection also was item 4. Again, from this process, assump-

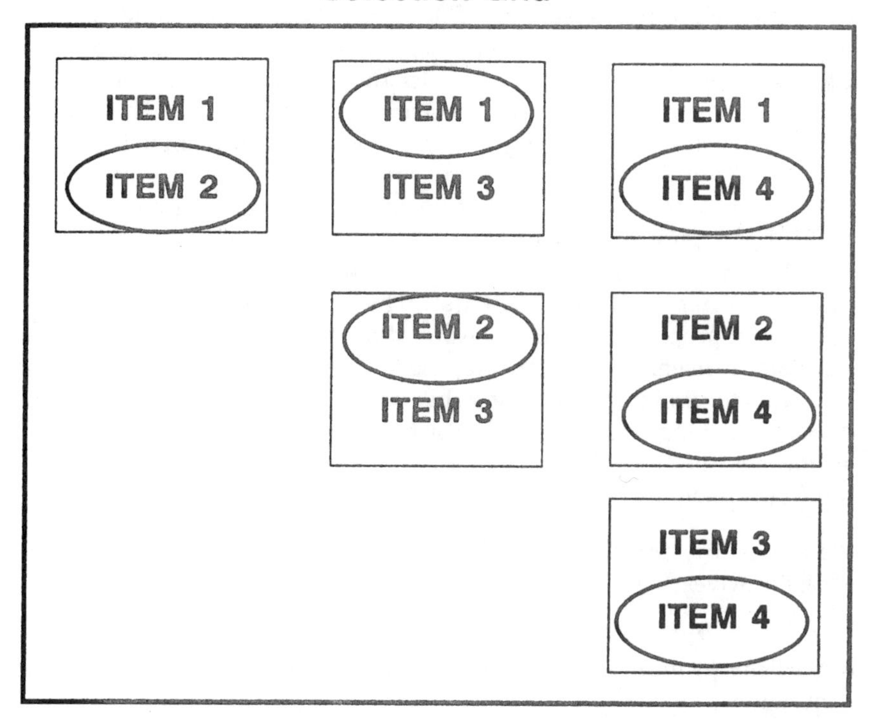

Individual Total

ITEM	CHOICE	TOTAL
ITEM 1	I	1
ITEM 2	I I	2
ITEM 3		0
ITEM 4	I I I	3

Team Total

ITEM	HIGHEST ITEM	TOTAL
ITEM 1	I I	2
ITEM 2	I I	2
ITEM 3	I	1
ITEM 4	I I I I I	5

Figure 10.3 Selection grid example.

tions are clarified and a consensus can be reached. Remember, the ultimate selection may differ from the selection grid result.

Decision Making

The decision-making process is the process of making the selection. In making a decision, the impact and the support of the outcome should be considered. A group will be more committed to success if the decision is reached by consensus. Therefore, consensus should be used when selecting an issue, problem, or opportunity to work as a team. Consensus is also necessary when deciding on a solution. It should be used at all major decision points in the CDPM methodology. Consensus decision making targets a win/win outcome. Decisions reached by any other method than consensus can lead to win/lose situations. A win/lose decision results in not having total commitment and support for the selection.

Although consensus is the recommended method for team decision making, other methods also exist. These types of decision making may be appropriate at times other than those mentioned above as essential consensus decision making times. In many instances, time constraints, insignificance, or other considerations make consensus decision making unrealistic, impractical, or inappropriate. In certain situations, the team determines an alternate method of decision making from the following methods:

- *Decision by majority.* This is a decision by more than half the representatives.
- *Decision by leader.* In some cases, the leader makes the decision.
- *Decision by management.* Management sometimes must make the decision.

Consensus

Consensus means everyone in the group accepts and supports the decision. This does not mean that everyone wants the selection but that everyone on the team agrees to the decision. Consensus equals support and commitment. This can only be reached by open and fair communication among all team members. Consensus requires understanding and discussion. Consensus is arrived at by understanding the process, mission, problem, and all the possible alternatives. Further, discussion of all the possible driving and restraining forces, causes and effects, and process interactions from all the viewpoints of the group is absolutely essential. Once understanding and discussion take place, the group can proceed with the process of arriving at a consensus.

Consensus involves communicating, especially listening to other's points of view. It means opening the team members' minds to new ideas. It requires nurturing the feelings and ideas of all team members. It allows the sharing of information, and it encourages participation. Consensus nurtures discussion by not voting or agreeing too quickly. It fosters the support of ideas that are best for everyone with the understanding that differences are a strength. Consensus decision making seeks a win/win solution.

Consensus decision steps. Consensus is reached by allowing everyone the opportunity to express their ideas about a decision. Instructions for developing consensus follow:

1. Present the decision to be made. The decision statement should include the what, when, and why. The decision statement should not be given as an either/or alternative.
2. Write the decision statement. The decision statement should be written on a flipchart or board that everyone can see. The team clarifies the decision statement at this time.
3. Review background information. Provide a common foundation of information to all team members. Facilitate a discussion of background information from all team members.
4. Decide why consensus would be the best decision-making method. Lead a discussion determining the best decision-making method for the situation.
5. Brainstorm selection criteria. First, each team member takes some time to write items for selection criteria. Second, the team conducts a round-robin brainstorming session to list items on a flipchart for consideration as selection criteria. Third, the list is added to by a freewheeling brainstorming session.
6. Clarify selection criteria items. Open discussion on the items on the list requiring further explanation.
7. Arrange list as nonnegotiables, primary needs, and secondary needs. First, delete all obvious items that the team agrees should not be used as selection criteria. Second, determine items that are nonnegotiables, that absolutely must be included as selection criteria. Third, evaluate the remaining items as primary or secondary needs for the selection criteria.
8. Brainstorm or use force-field analysis to determine alternatives. Write the alternatives on a flipchart or board.
9. Evaluate each alternative against selection criteria. Use the appropriate selection techniques outlined previously in this chapter to focus the alternatives to reach consensus.
10. Agree on a decision. Write the final decision on a flipchart or board.
11. Get each team member's personal commitment and support for the decision. Ask each team member individually to commit to the decision.

12. Implement the decision. Develop an action plan with the specific what, when, and who to implement the decision. Conduct periodic reviews to follow up.

Decision by the majority or leader

In many cases during routine group activities, decisions may be made by the majority of the membership or by the leader. Decisions made by the majority or the leader are usually reserved for relatively minor aspects of group activities. For example, the decision to use a specific tool or technique may be determined by the majority or the leader. In cases where it is feasible, the particular decision-making method should be determined by the team.

Decision by management

Management may make some decisions. Decisions by management are necessary in many cases. In a CDPM environment, management should always be aware of the impact of decisions on maintaining the environment. As with consensus, if decisions by management are accompanied by understanding and discussion, they frequently lead to support, even if they are unpopular.

Main Points

Selection techniques help to clarify assumptions and focus on consensus.

Consensus is important when selecting an opportunity and an alternative.

Consensus means everyone understands and supports the decision.

The criteria and method selected are critical to the desired outcome.

All decisions should focus on a win/win outcome that is best for the whole organization.

Considerations for reaching consensus are

- *C*ommunicate, especially listen to others' points of view.
- *O*pen team members' minds to new ideas.
- *N*urture the feelings and ideas of all team members.

- *S*hare information.
- *E*ncourage participation.
- *N*urture discussion; don't vote or agree too quickly.
- *S*upport ideas that are best for everyone.
- *U*nderstand that differences are a strength.
- *S*eek a win/win solution.

Chapter

11

CDPM Analysis Tools and Techniques

Focus: This chapter describes the basic tools and techniques for analysis.

Introduction

When using the customer-driven project management (CDPM) improvement methodology, thorough analysis is extremely important. The tools and techniques for analysis help to improve the process, enhance work-flow efficiency, determine underlying or root causes, identify the vital few, and look at both sides of an issue.

All the tools and techniques in this chapter are especially helpful during step 4 of the CDPM improvement methodology for analyzing improvement opportunities. In addition, these tools and techniques are helpful in other steps of CDPM improvement methodology as follows:

- *Process and work-flow analysis* assists in step 2, understanding and defining the process. When employed during this step, the process analysis is used to make some immediate changes to processes.
- *Cause-and-effect analysis* is beneficial during step 5, taking action, to determine the cause or effect of recommended actions.
- *Data statistical analysis* provides tools and techniques for all steps in the CDPM improvement methodology. This type of analysis assists in defining the process performance, providing an indicator for selecting improvement opportunities, giving a target to take action, furnishing the standard to check results, and serving as a gauge to monitor results.

- *Force-field analysis* is particularly useful anytime the team needs to look at both sides of an issue. It also is excellent for generating alternative solutions. In addition, a force-field analysis is worthwhile during step 7, implementing the improvement, to examine the forces for and against the solution prior to making a management presentation.

Process Analysis

Process analysis is a tool used to improve a process by eliminating non-value-added activities, waits, and/or simplifying the process. The focus of process analysis is on specific defined outcomes. These desired results usually aim at time and/or cost reduction. Process analysis is extremely useful for getting the output of the process to the customer as quick as possible at the lowest possible cost. The major goals of process analysis are elimination or reduction of high costs, non-value-added processes, activities, and tasks, and the waits between processes. This is an excellent initial focus for process improvement, invention, or reengineering.

High-cost areas are usually a primary area of focus. This is accomplished by adding cost figures to the process diagram to determine processes that are excessively costly. Then the team does a complete process analysis on each of the high-cost areas.

In addition to high-cost areas, many organizations lose money performing non-value-added processes. These non-value-added processes are another target for process analysis. The value and nonvalue of a particular process, activity, or task is a judgment based on facts within a specific environment. Each process, activity, or task deserves a thorough analysis to determine its value. The value is based on organizational and/or customer needs and expectations. The customer-driven team conducts the process analysis with inputs from all people affected by the process under review. As part of the process analysis, the customer-driven team conducts a risk assessment to reduce the probability of eliminating an essential task. Processes, activities, or tasks are not eliminated without the concurrence of the process "owner" and the consensus of everyone affected by the process improvement.

Once non-value-added tasks are evaluated for possible savings, the team can focus on reducing or eliminating waits. Many hours are wasted between the performance of processes. This down time affects the organization's ability to respond rapidly to customers. Rapid response time is a major differentiator in today's economic times. By concentrating on improving the waits between processes, the team can have a considerable impact on its cycle time.

During process analysis, the team first challenges the following:

- Excessive costs
- Inordinate waits
- Bureaucratic procedures
- Duplicate efforts
- Inspection or overseer operations
- Layers of approval
- Noncontributors to customer satisfaction

Once the preceding have been examined, process simplification becomes the next step. This involves probing the high-cost and high-time processes for simple, innovative, and creative improvements in accomplishing the process. Can the process be combined with another activity? Can the process be done less frequently? Can the process be automated to be accomplished more quickly? Can the process be done another way? These initial actions achieve quick results at little or no cost. This level of process analysis aims at process improvement to achieve increased financial performance, improved operating procedures, and greater customer satisfaction.

During this step, the team challenges the following:

- Complexity
- Unnecessary loops
- Frequency
- Methodology
- Use of technology
- Optimization of resources
- Innovative application of telecommunications and information systems

Process analysis steps

1. Construct a process diagram (top-down or detailed).
2. Ensure that waits between processes/activities are identified.
3. Determine the time and cost of each process/activity and time of waits.
4. Reduce or eliminate waits.
5. Select critical activities (high time or cost).
6. Eliminate non-value-added processes/activities.

7. Eliminate parts of the process.
8. Simplify value-added processes/activities. During this step, look to combine processes/activities, change the amount of time or frequency, do processes/activities in parallel with another process, and use another method to do process.
9. Use CDPM improvement methodology to improve, invent, or reengineer the process.

Process analysis example

This example uses the detailed subcontractor data requirements list (SDRL) process flow diagram from Chap. 9. Figure 11.1 shows the process flow diagram with times, and Fig. 11.2 shows the process flow diagram with costs. By examining the process flow diagrams, the team decides that the waits are extremely excessive. This would be the team's first area of focus for improvement. The team builds a time line of activities and waits. This is shown at the bottom of Fig. 11.1, which indicates that the process with an internal review takes 41 days, including 23 days of wait and 18 days of actual work. The team decides to aim at reducing the waits by 50 percent, which reduces the total time to 29.5 days with 11.5 days of wait. Next, the team targets high-time and high-cost processes. In the example, the high-time and high-cost process is the review process, as shown in Figs. 11.1 and 11.2. The team would next do a detailed process diagram on just the review process. This process diagram would be used to eliminate non-value-added activities and simplify value-added activities.

Work-Flow Analysis

A *work-flow analysis* looks at a picture of how the work actually flows through an organization or facility. Like the process analysis, focused on eliminating and simplifying the process, the work-flow analysis targets inefficiencies in the work motion. The work-flow analysis aims for identification and elimination of unnecessary steps and reduction of burdensome activities.

Work-flow analysis steps

1. Define the process in terms of purposes, objectives, and start and end points.
2. Identify functions of the organization or facility.
3. Identify activities within each function.

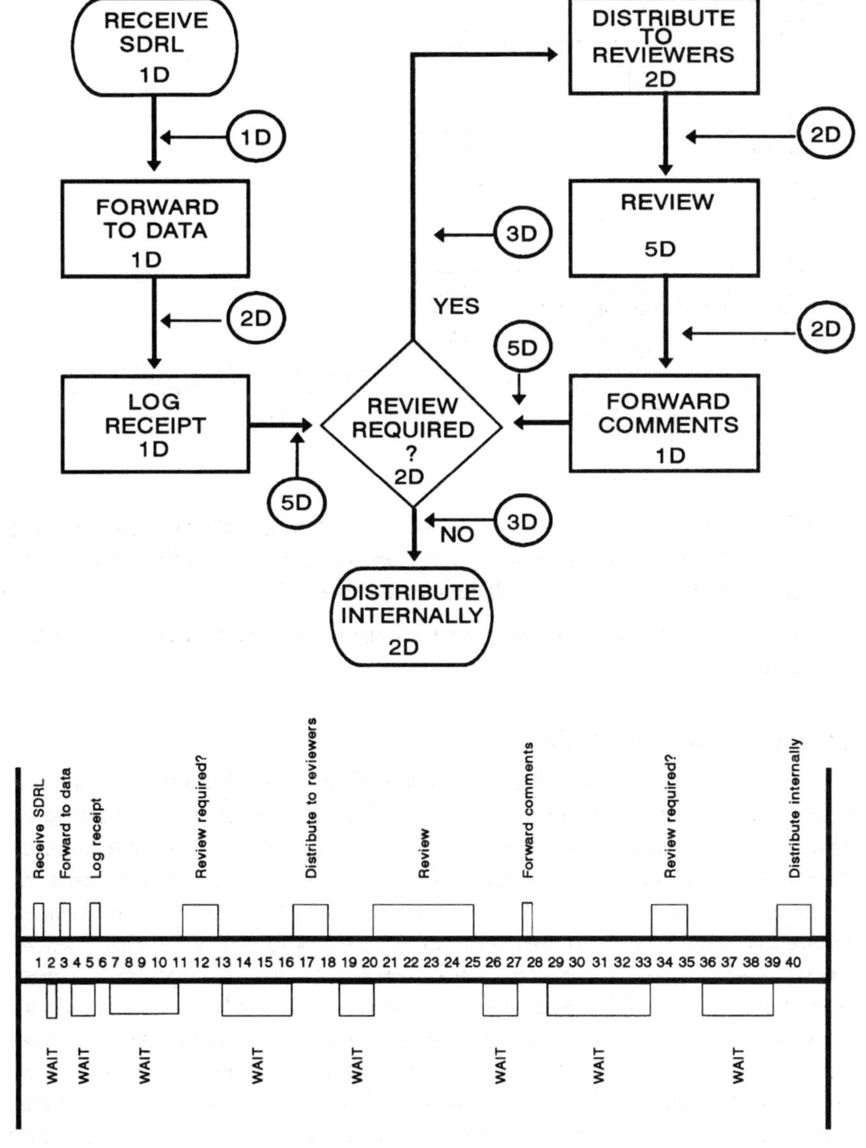

Figure 11.1 Process analysis time example (SDRL, subcontractor data requirement list).

4. Identify tasks or basic steps within each activity.
5. Using process diagram symbols or drawings of the organization or facility, graphically display the actual work flow.
6. Analyze the work flow by identifying major activities, lengthy or complex tasks, decision points, and duplicate or repetitive tasks.

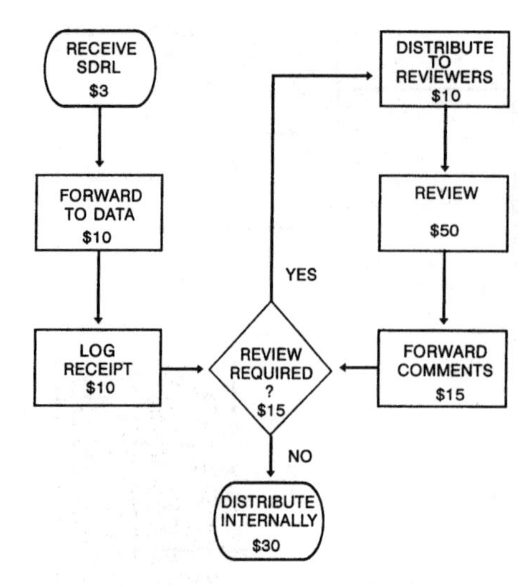

Figure 11.2 Process analysis cost example (SDRL, subcontractor data requirement list).

7. Check the logic of the work flow by following all possible routes through the organization and facility for all work activity to ensure that all possible alternatives are explored.
8. Determine improvement, invention, or reengineering opportunities.

Work-flow analysis example

The work flow for processing a customer's request for training within a learning center is shown in Fig. 11.3. The customer request is received by the administrative specialist at the desk. The administrative specialist goes to the file cabinet to get the person's training record. The training record is brought back to the desk to be checked to ensure that the customer's request meets the person's competency needs. Next, the administrative specialist makes a copy of the customer's request. The work-flow diagram shows the administrative specialist going to and from the copier. Once the administrative specialist determines that the customer's request requires enrollment in a training course, the course enrollment needs approval by the training manager. The administrative specialist takes the customer request with course enrollment information to the training manager for signature. The administrative specialist returns to the desk. Finally, after the training manager's signature is obtained, the customer's request is put in the course enrollment books in the bookcase. By examining this work flow, customer request processing can be improved to save time and motion. For instance, duplicate copies can be eliminated, saving time and motion. The file cabinet and the bookcase can be moved closer to the desk to reduce motion. In addition,

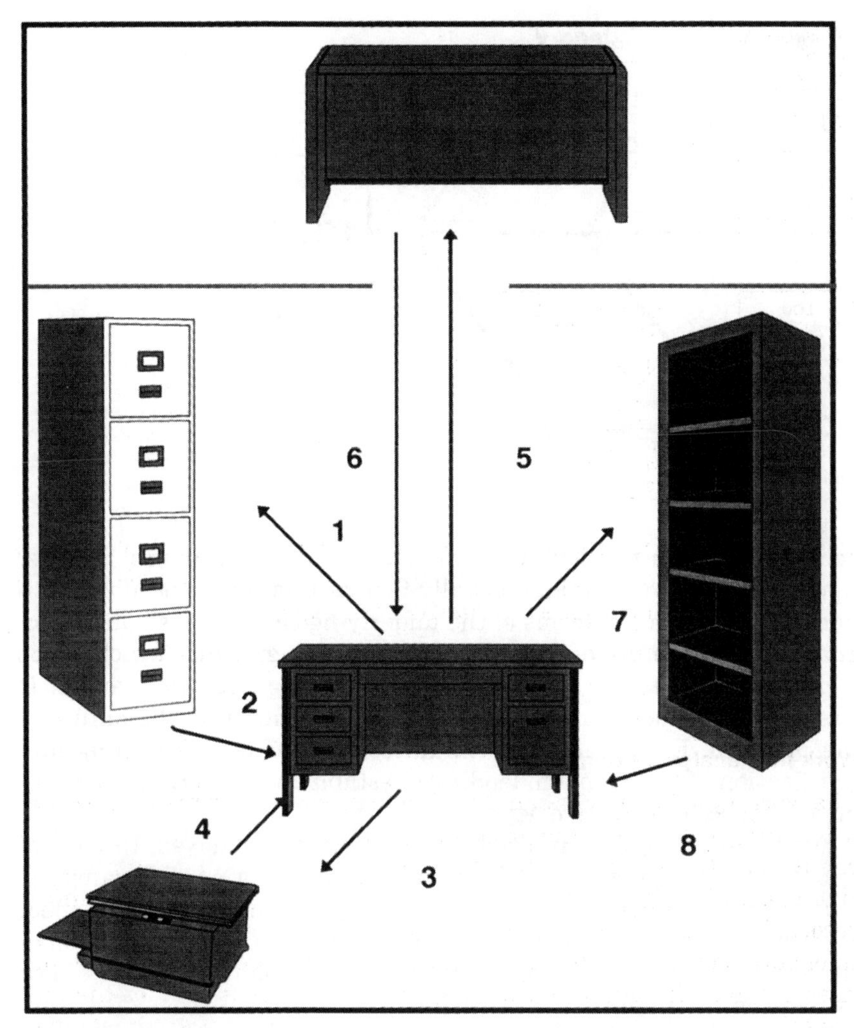

Figure 11.3 Work-flow analysis.

the training manager's signature could be eliminated by empowering the administrative specialist to make the approval.

Cause-and-Effect Analysis

Cause-and-effect analysis is a useful technique for helping a group to examine the underlying cause(s) of a problem. Figure 11.4 shows a basic cause-and-effect diagram, which is a graphic representation of the relationships among a list of issues, problems, or opportunities. Such a diagram is a useful tool in association with brainstorming

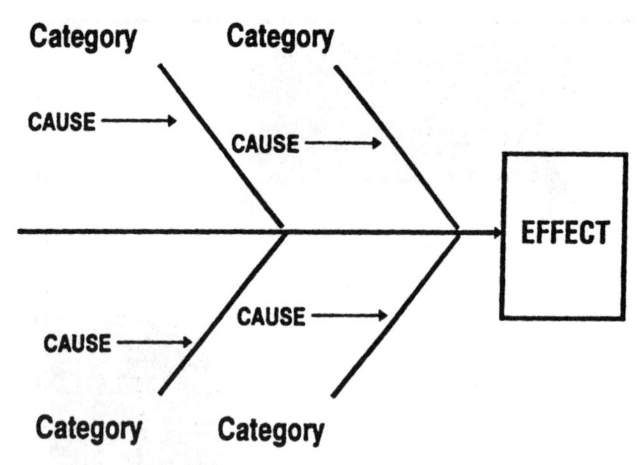

Figure 11.4 Cause-and-effect diagram.

because it takes the brainstorming to core issues or root causes. Application of the technique usually results in a more specific definition of the problem in terms of the underlying cause. Cause-and-effect analysis has the added benefit of being very graphic, which helps members see patterns and relationships among potential causes. It lets individual members express their interpretation of the nature of the problem. Frequently, it stimulates further brainstorming and clarification of the problem, leading to establishing priorities and taking appropriate corrective action.

For a process to be improved or a problem to be solved, the action taken must target the real issue, the underlying or root cause. It must attack the disease, not merely the symptoms. Cause-and-effect analysis aims at the cause of the issue or problem bringing the undesirable effect. The cause-and-effect diagram graphically displays how causes and their effects are associated.

The cause-and-effect diagram starts the "fishbone," as shown in Fig. 11.4. A box is drawn for the effect. Next, an arrow is drawn pointing to the box. From the arrow, slanting lines are drawn. Start with four slanting lines. As the categories are selected, lines can be added or not used.

Cause-and-effect analysis begins with the issue or problem as the effect. The effect is written in the box. Next, the team decides on categories. There can be as many major categories as necessary. Typically, three to five categories are selected. Potential categories include personnel (men and women), materials, methods, machines, environment, people, culture, systems, procedures, training, plans, facility, policies, technology, information systems, communications, structure, etc. The categories are written above the slanted lines.

Once the categories are selected, the team brainstorms all the possible causes of the effect within each category. The brainstorming rules outlined in Chap. 7 apply for cause-and-effect analysis as well. During identification of causes, the team builds on each other's ideas. The same cause can occur in more than one category. For instance, the cause "unavailable" may be appropriate for several categories, i.e., people, machines, methods, and materials. The causes are written as little branches connected to the major branch slanted line.

Once the causes are listed, the team interprets the diagram. The team looks for the underlying cause or causes by critically examining the pros and cons of each individual cause. In addition, the team should repeatedly ask why. This provides a basis for focusing on consensus of selecting the underlying or root cause, which is accomplished by using the selection techniques described in Chap. 10.

Next, the selected underlying cause or causes must be validated by collecting data on their occurrence. Again, the focus is on a cure for the root disease, not just the symptoms. Cause-and-effect analysis steps can be summarized as follows:

1. Define the problem.
2. Define the major categories.
3. Brainstorm possible causes.
4. Identify the most likely causes.
5. Verify the most likely cause.

Cause-and-effect analysis example

Figure 11.5 shows the cause-and-effect diagram for this example. The steps of the cause-and-effect analysis are described in the following paragraphs.

Define the problem. The team is asked to identify the problem, and the problem is the effect. In the example, the problem is a piece of equipment with a mean time between failure (MTBF) of 400 hours. This MTBF is too low. The desired goal is an MTBF of over 1000 hours. Thus the MTBF is the problem or effect.

Define the major categories. Next, the major categories of possible causes of the problem are identified. The most popular categories are the machines, methods, people, and materials. These categories were selected for this example. It is important to tailor the categories to the specific problem. Remember, you are not limited to these categories.

Brainstorm possible causes. The team then brainstorms possible causes. These causes are listed under the appropriate category. The

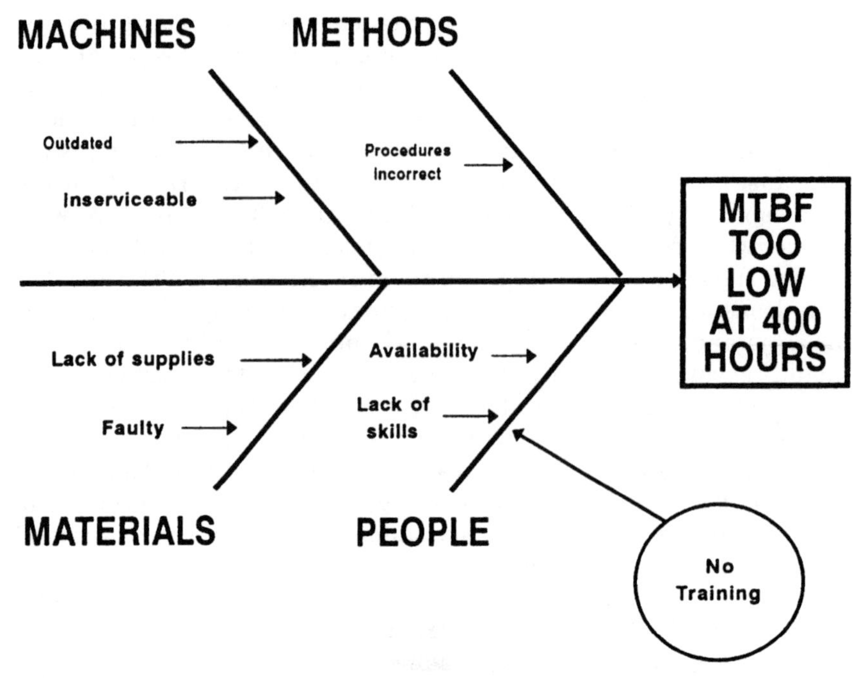

Figure 11.5 Sample cause-and-effect diagram (MTBF, mean time between failures).

brainstorming rules described in Chap. 7 apply in this step. It is sometimes helpful for the leader or facilitator to keep repeating the heading for the cause in relation to the effect. For instance, under "Method," what is a cause of low MTBF? What is another method causing the low MTBF? This questioning is continued in each category until all ideas are exhausted. If someone comes up with an idea that applies to more than one category, list that idea in each category. If someone comes up with an idea that falls into a category other than the one being brainstormed, list the idea in the appropriate category. If an idea is generated that the person cannot immediately categorize, list the idea on the side of the diagram. The idea can be categorized at the completion of the brainstorming session. If someone cannot generate an idea, the person can pass or build on other people's ideas. Continue the brainstorming session until the team is satisfied that it has completed the search for underlying causes for the problem.

Identify the most likely causes. The team looks for clues to the most likely causes. Once all the causes are examined, the team selects by consensus the most likely cause by using the selection techniques from Chap. 10.

Verify the most likely cause. The most likely cause is verified by data statistical analysis, a test, collecting more data on the problem, or communicating with customers to verify or reject the cause. When using data statistical analysis, histograms or Pareto analysis is usually very effective in verifying the most likely cause. Once your team has identified the underlying cause, the team can begin to generate alternative solutions and work toward improvement.

Data Statistical Analysis

Data statistical analysis is an essential element of any CDPM endeavor. Customer-driven project management uses quantitative methods to continuously improve project processes and all the processes in the organization aimed at total customer satisfaction. This involves monitoring, analyzing, correcting, and improving processes using rational decision making based on facts. Statistics are one method to establish factual data. Statistics are used for many purposes in a CDPM environment, including problem solving, process measurement, and pass/fail decisions. Statistics are useful in step 1 of the CDPM improvement methodology to define the quality issues. Statistics help quantify total customer satisfaction. In step 2, understanding and defining the process, statistics provide factual data on process performance. In step 4, analyzing the improvement opportunity, statistics aid in the verification of underlying or root causes, determining the vital few, evaluating variation, and establishing correlation. In step 5, taking action, statistics provide data for robust design and capable processes. In steps 6 and 8, checking and monitoring results, statistics furnish a yardstick for decision making.

Data statistical analysis includes tools for collecting, sorting, charting, and analyzing data to make decisions. A chart can make the process easier to understand by arranging the data so that comparisons can be made so as to focus on the right problems. Sorting and resorting the data can help the team focus on the most important problems and causes. Targeting smaller and smaller samples or categories funnels the data to the underlying or variation causes. Even small improvements on the right problems can yield significant benefits.

Data statistical analysis steps

1. Collect data.
2. Sort.
3. Chart data.
4. Analyze data.

Data collection

Data collection is the first step in data statistical analysis. It starts with determining what data are needed. Sometimes the data required are already available. In these cases, all that is required is to sort, chart, and analyze the data. However, in many cases, the specific data required for data statistical analysis are not available. If this is the case, the team needs to determine what data to collect, where to collect the data, and how to collect the data.

Data-collection plan

Data collection requires a plan. The data-collection plan establishes the purpose, strategy, and tactics to get the data for data statistical analysis. The data-collection plan answers the following questions:

Why does the team need to collect the data?

What data are needed?

What process provides the data?

Where in the process are the data available?

Are the data already being collected?

If the data are not already collected, how will they be collected?

Who will collect the data?

How long will the data be collected?

What data collection method will be used?

What sampling method is needed?

Who will chart the data?

How will the data be reported/presented?

Is a pilot or test necessary?

How will the pilot or test be conducted?

Who will participate in the pilot or test?

Are the data timely, accurate, and consistent?

Data-collection methods

Data must be collected to measure and analyze a process. There are many methods for data collection. The data-collection method must accomplish the purpose as stated by the customer-driven team in the data-collection plan. Data-collection methods include

- Observation
- Questionnaires
- Interviews
- Tests
- Work samples
- Checksheets

Observation. Observation is looking at actual performance or data. This could include reviewing documentation or watching the work process. This type of data collection is useful when the team wants to distinguish between perceived and actual outcomes or behaviors.

Questionnaires or surveys. A questionnaire requests in writing some particular information. A questionnaire is most useful when one needs to get information from a large number of people in a short time or when the people with the information are geographically distant. Questionnaires have some disadvantages in that respondents may misinterpret the questions, the returns are low, and there is no opportunity to probe deeper into responses.

Interviews. An interview involves communicating directly with the people with the information. Interviewing is used to collect the information needed to improve processes, solve problems, and involve those outside the group in generating and implementing potential solutions. Interviewing is also useful when evaluating implemented solutions. Interviewing is essential for supplier and customer analyses. The disadvantage of an interview is that many times the outcome depends on the skill of the interviewer. In addition, respondents may be hesitant to discuss personal, sensitive, or confidential information. In these case, provisions must be made for guaranteed anonymity and confidentiality.

Tests. Tests measure outcomes. A test is useful when you need to measure specific results. Tests can be of people or items. A test can measure a person's knowledge on a subject. A test also can be used to determine if an assembly operates properly.

Work samples. Work samples involve checking or inspecting specific work outputs or work in process. Work samples are most useful when analyzing actual work performance.

Checksheets. Checksheets verify accomplishment of procedures. Checksheets are useful for qualifying or certifying performance exactly as specified.

Data-collection charts

A data-collection chart provides a means to document the information and simplifies data collection. An example of a simple, easy-to-use data-collection chart is shown in Fig. 11.6. This chart shows the discrepancies encountered in a part over a 3-week period. As a discrepancy is encountered, the faulty item is checked. For instance, the board had 3 discrepancies in week one, 1 in week two, and 2 in week three, for a total of 6 discrepancies for the 3-week period. Looking at the bottom of the chart to the component discrepancies in week one, 15 discrepancies in the part were attributed to a faulty component. Again, in week two, 15 discrepancies in the part were attributed to a faulty component, and in week three, 15 more for a total of 45. The totals for all discrepancies for the three weeks are

Board	8
Solder	2
Wiring	12
Connector	15
Component	45
Total all weeks	80

Data-collection charts help in the systematic collection of data. The validity of the data depends on this collection. The customer-driven

DATA COLLECTION CHART

DISCREPANCY	WEEK 1	WEEK 2	WEEK 3	WEEK 4	TOTAL
Board	///	/	//		6
Solder	/	/			2
Wiring	//////	///	////		12
Connector	/////	/////	/////		15
Component	///// ///// /////	///// ///// /////	///// ///// /////		45
TOTAL	31	25	24		80

Figure 11.6 Sample data-collection chart.

team must ensure that the data are unbiased, accurate, properly recorded, and representative of typical conditions.

Data-collection sampling

When collecting data for analysis, it is often impractical to check 100 percent of the items. This would be the entire population of data. A sample of the population may be all that is required. By taking a sample, reliable information can still be collected. A sampling table can help you determine an appropriate sample size. Such tables are usually available from the industrial engineering, quality, or management services personnel in an organization. To reduce the chance for biased results, use a random or systematic method to select samples. A random sample allows each item an equal chance of being selected. A systematic sample selects every fifth, tenth, or twentieth item. This reduces the chance for biased results.

Types of sampling. There are two common types of samples, nonrandom and random. *Nonrandom* sampling is accomplished using judgment. This type of sample usually cannot be verified. In a CDPM environment, nonrandom sampling is recommended only as a predecessor to random sampling. *Random* samples are samples in which each item in the population has a chance of being selected. This type of sampling is most useful in the CDPM environment. Random sampling ensures accurate statistics. Two of the most common forms of random sampling are simple random sampling and stratified sampling.

Simple random sampling. *Simple random sampling* can be accomplished by using a list of random digits or slips. The random-digit method uses a number to represent the items in the population. For instance, imagine that the population consists of 100 items. The items are numbered 1 to 100. Next, the sample must be selected. The sample is selected by the use of a random-number generator or a table of random digits. Suppose that the sample is to consist of 10 items. The random-number generator or table of random digits would select 10 items indiscriminately. With the slip method, each item in the population is also numbered, the numbers are recorded on slips, and the slips are put in a box. The sample is drawn from the slips in the box. These random sampling tools ensure that the sample represents the population so that inferences can later be made about the population from sample data.

Stratified sampling. *Stratified sampling* divides the population into similar groups or strata. There are two methods of stratified sampling. One selects at random a certain number of items from each group or stratum according to the proportion of the group or stratum to the population as a whole. In another method, a certain number of items are selected from each group or stratum and then the group or stratum is given weight according to the proportion of the group or stratum to the population as a whole.

The central limit theorem. The *central limit theorem* states that the mean (average) of the sampling distribution of the mean will equal the population mean (average) regardless of sample size and that as the sample size increases, the sampling distribution of the mean will approach normal, regardless of the shape of the population. The central limit theorem allows the use of sample statistics to make judgments about the population of the statistic. The central limit theorem is an important concept in statistics because it is often impractical or impossible to check the entire population.

Types of data

The two types of data are variable data and attribute data. *Variable data* are data that can be measured. Variable data measure characteristics that have a range of values, i.e., quality characteristics such as thickness, width, and temperature, force, wear, strength, and sensitivity. Variable data are characterized as nominal best, smaller best, and larger best. *Attribute data* are data that can be counted or classified. Attribute data are associated with such characteristics as pass/fail, have/have not, go/no go, grade (A/B/C), and accept/reject.

Data arrangement

Once the data are collected from the population or sample, they need to be arranged in a meaningful way. The arrangement allows observation of such things as the highest and lowest values (range), trends, central tendencies, patterns, the values appearing most often, special causes, common causes, etc. The arrangement of data helps to determine the measures of central tendency and the frequency distribution.

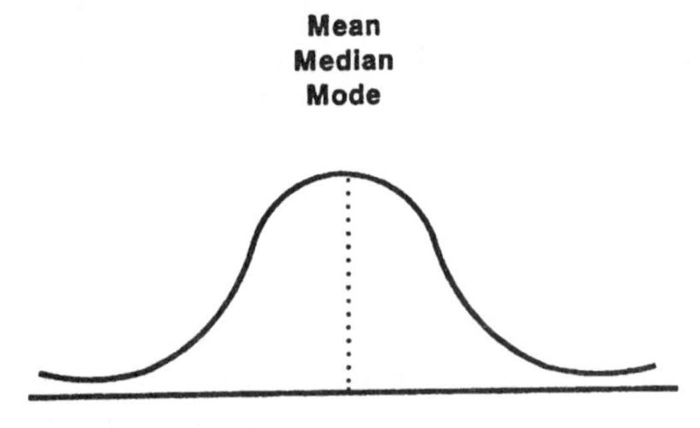

Figure 11.7 Normal distribution curve.

Measures of central tendency. The measures of central tendency are mean, median, and mode. The *mean* is the average of something. The *median* is the exact middle value. The *mode* is the value most often represented in the data. In a normal distribution, the mean, median, and mode are equal. Figure 11.7 shows a normal distribution bell-shaped curve.

Measures of central tendency examples. The sample data are as shown below. Column 1 contains the number of the sample. Column 2 contains the ages of the people in the sample as collected. Column 3 contains sample data arranged lowest to highest:

1.	30	30
2.	33	33
3.	42	35
4.	39	36
5.	50	39
6.	42	42
7.	35	42
8.	47	47
9.	36	49
10.	49	50

The average is calculated by summing all the items and dividing the sum by the total number of items as follows:

$$\text{Mean} = \frac{30 + 33 + 35 + 36 + 39 + 42 + 42 + 47 + 49 + 50}{10}$$

$$= \frac{404}{10} = 40.4$$

The median is the central item in a set of data. Half the items fall above and half of the items fall below the median. In this example, the median is 40.5.

Mode is the value that is most often repeated in a set of data. In this example, the mode is 42.

Data charting

Once sorted, data must be put on a chart. Charts are pictures of the data that highlight the important trends and significant relationships. Charts present the data in a form that can be quickly and easily understood. Charts serve as a powerful communications tool and should be employed liberally to describe performance, support analyses, gain approval, and support and document the improvement process.

When using charts and graphs, label titles and categories for clarity. Keep your charts simple, and report all the facts needed to be fair and accurate.

Many different types of charts or graphs are available and useful in the CDPM process. Some of the most common are the bar chart, pie chart, simple line chart (time plots or trend chart), histogram, and scatter chart. In addition, Pareto charts, control charts, and process-capability charts are helpful for many specific CDPM activities.

Bar chart. A *bar chart* is useful when comparing between and among many events or items. Figure 11.8 shows a bar chart of the information from the checksheet used earlier. The bar chart shows the number of repairs by category arranged from the highest number to the lowest number of repairs.

Pie chart. A *pie chart* shows the relationship between items and the whole. Figure 11.9 shows a pie chart of the information from the checksheet used earlier. The chart in Fig. 11.9 shows each of the repair item's contribution to the total number of repairs for the item. This chart clearly shows that component repair is the most common

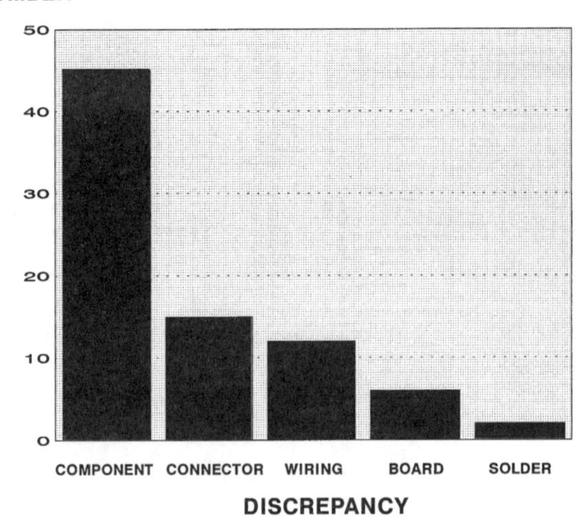

Figure 11.8 Bar chart.

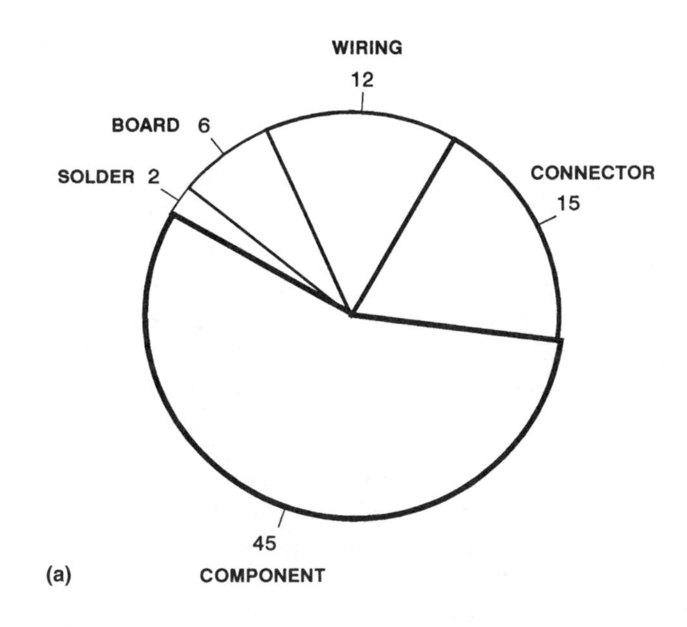

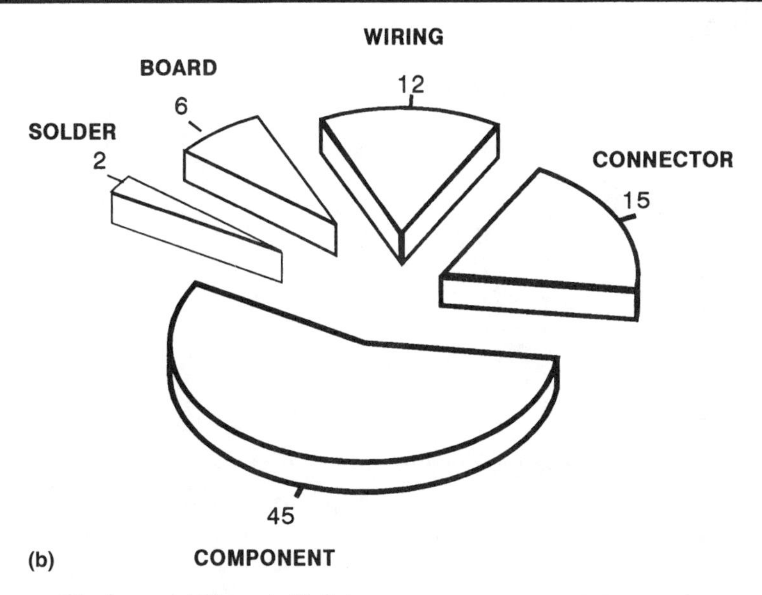

Figure 11.9 Pie charts. (*a*) Normal. (*b*) Cut.

type of repair. In this case, the organization should focus efforts to reduce the number of failures in the components.

Line chart. A *line chart* is used when describing and comparing quantifiable information. A line chart provides insight into statistical trends, particularly over a specified period of time. Figure 11.10 shows a line chart of the information from the checksheet used earlier. This line chart shows the number of repairs by category per week. This is a time-chart form of the line chart. It shows the changes in the discrepancies over the 3-week period.

Scatter chart. A *scatter chart* and its related correlation analysis permit the combination of two factors at once and representation of the relationship that exists between them. A graphic display can help reveal possible relationships and causes of a problem even when links between the two factors are not evident. The pattern or distribution of the data points in a scatter diagram indicates the strength of the relationship between the factors being examined. It also indicates the type of relationship, i.e., positive, negative, curve, or no relationship.

Figure 11.11 shows a series of scatter charts. These charts could be displaying the relationship between the number of items tested and the number of failures. The x axis of the chart is the number tested,

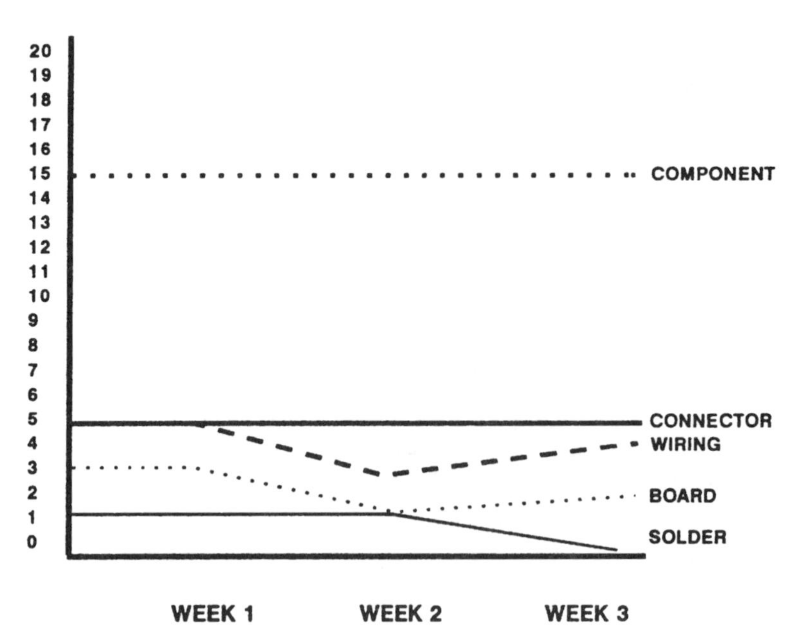

Figure 11.10 Line chart.

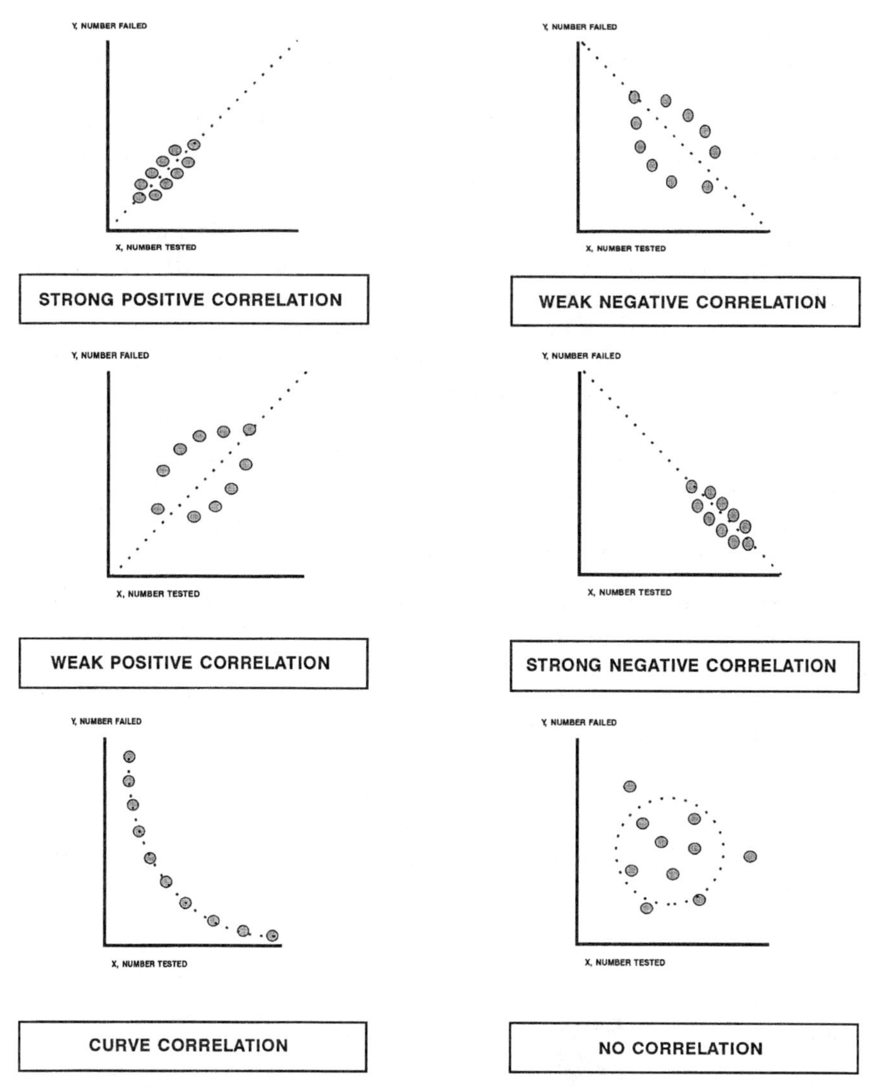

Figure 11.11 Scatter charts. The x axis is the number tested (%), and the y axis is the number failed (%).

and the y axis is the number of failures. First, the information is plotted on the graph. Next, a line is fitted through the scatter diagram. Then the chart is analyzed to determine the relationship. A positive correlation indicates a direct relationship between the factors; i.e., as the number of items tested increases, the number of failures also increases. A negative correlation shows an inverse relationship; i.e., as the number of items tested increases, the number of failures decreases. The relationship could be depicted by a curve. This occurs when the number of repairs changes by some fixed proportion. This is sometimes referred to as the *learning curve.* If no pattern is evident, no correlation exists.

Although a scatter diagram indicates a relationship between two factors, additional correlation analysis is usually required to substantiate the nature of the indicated relationship.

Histogram. A *histogram* is a bar chart that shows frequency of data in column form. The columns may be presented vertically or horizontally. Figure 11.12 is a histogram. This type of data charting is useful in identifying changes in a process. A histogram can provide insight into the performance of a process and appropriate corrective actions by examining its centering, width, and shape. The closer the columns of the histogram are to the center of the chart, the more the process is on target. The wider the spread of the columns from the center, the greater the variation of the process from the target. Any change from a normal bell shape may indicate a problem area. Figure 11.13 shows some examples of various distributions.

Histogram example: Frequency distribution. A histogram is usually a chart of frequency distribution. A *frequency distribution* is a table showing the number of elements in each class of a set of data. It arranges data into classes with the number of observations in each class. Classes are groups of data describing one characteristic of the data. A frequency distribution displays the number of times an observation of the characteristic falls into each class.

In constructing a histogram, the first step is to collect the data. The raw data for the average inventory of work in process of one assembly area over a 15-day period is as follows:

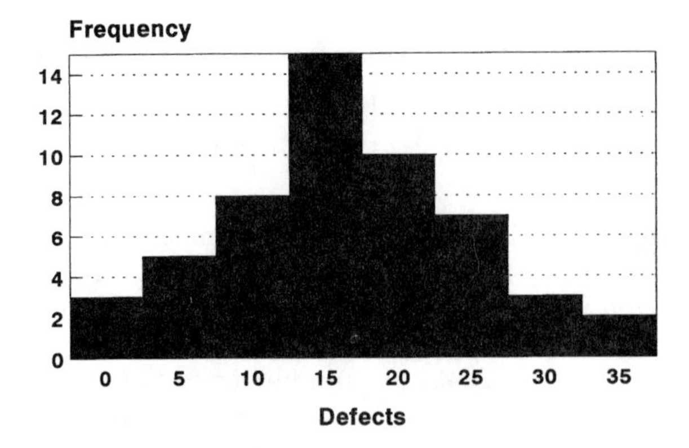

Figure 11.12 Histogram.

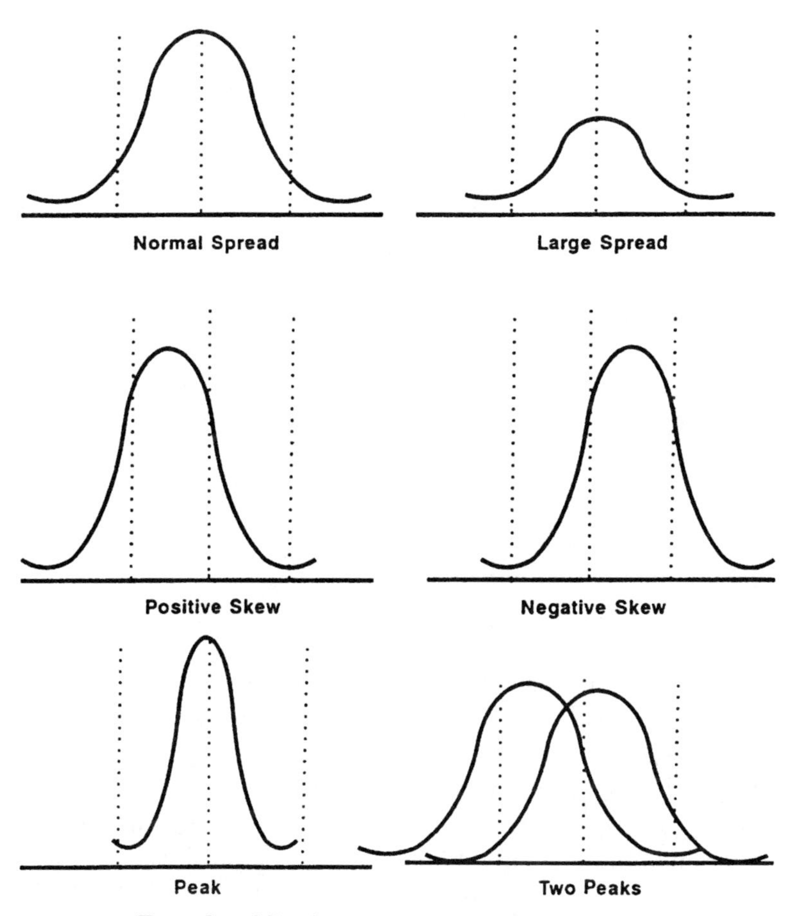

Figure 11.13 Examples of distribution.

1 1 5 2 4 2 3 1 2 2

3 4 3 2 2

Step 2 is to arrange the data from lowest to highest:

1 1 1 2 2 2 3 3 3 3

3 3 4 4 5

In step 3, determine the class intervals, which should be equal. One method for determining class intervals is highest value minus lowest value divided by number of classes. For the example, the formula provides a class interval of 0.8. This gives the following class levels:

1.0 – 1.8

1.9 – 2.7

2.8 – 3.6

3.7 – 4.5

4.6 – 5.3

In step 4, sort the data into classes, and count the number of points.

1.0 – 1.8	3
1.9 – 2.7	3
2.8 – 3.6	6
3.7 – 4.5	2
4.6 – 5.3	1

In step 5, determine the relative frequency and/or the cumulative frequency of the data. Table 11.1 shows relative frequency and cumulative frequency of the example data.

Step 6 is to display the data on a histogram chart.

Control chart. A *control chart* displays the process performance in relation to control limits. Figure 11.14 shows a typical control chart, which displays the data over time and shows the variations in the data. Control charts are designed to illustrate variation.

Control charts are used to show the variation in a number of variables. The control chart allows you to distinguish between measurements that are within the variability of the process and measurements that are outside the normal range and are produced by special causes. For instance, it may be that University College accepts that

TABLE 11.1

	Count	Relative	Cumulative
1.0 – 1.8	3	20	20
1.9 – 2.7	3	20	40
2.8 – 3.6	6	40	80
3.7 – 4.5	2	13	93
4.6 – 5.3	1	7	100
	15	100	100

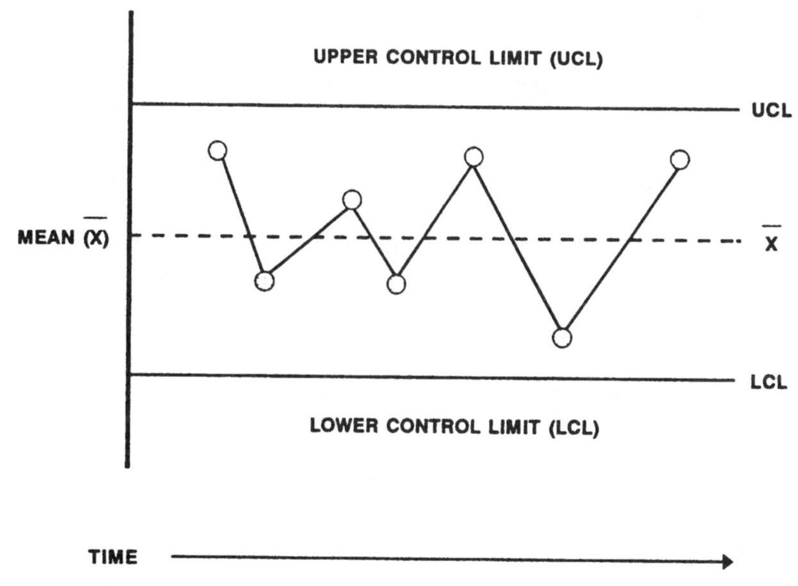

Figure 11.14 Control chart.

90 percent of student complaints are being resolved after the first call. If data suggest that only 70 percent are being resolved after the first call, this is outside the normal range.

The upper and lower parts of the normal range are referred to as the upper and lower *control limits.* These upper and lower control limits must not be confused with *specification limits,* which relate to acceptability of the process output. Control limits describe the natural variation in the process. Points within the limits are generally indicative of normal and expected variation.

Points outside the limits signal that special attention is required because they are beyond the built-in systemic causes of variation in the process. It is necessary to investigate those points outside the control limits. The variations in the data are due to root causes, either common or special causes. Common causes are always present, and the resulting variation in the process output will remain stable within the control limits. Special causes or systemic problems will result in data falling outside the control limits.

These charts help you to understand the inherent capability of your processes, bring your processes under control by eliminating the special causes of variation, reduce tampering with processes that are under statistical control, and monitor the effects of process changes aimed at improvement. Control charts are the fundamental visual display for statistical process control.

Statistical process control is used when a product or service is consistent, meaning that the process is stable or the process outcomes usually fall within prescribed specifications. Statistical process control and control charts are useful when some characteristic of a process output has changed from previously known and predictable levels or the process has changed and the new upper and lower control limits need to be established.

There are many types of control charts. Some of the many types of control charts for variable and attribute data are as follows:

Variable charts

X-bar	Central tendency
R	Range
Sigma	Standard deviation

Attribute charts

p	Proportion nonconforming of total
c	Defects per subgroup for constant sample size
u	Percent nonconformities for varying sample size
np	Number nonconforming compared with total number

Analyzing the data

Once the data have been collected, sorted, and put on charts, they are analyzed to identify the significant findings.

- Ask specific problem identification questions with "what," "when," "where," "who," "how much," "what are the causes," and "what's the impact?"
- Identify underlying causes.
- Clarify expected outcome.

Pareto analysis. One specific type of analysis is a *Pareto chart.* In the late 1800s, Vilfredo Pareto, an Italian economist, found that typically 80 percent of the wealth of a region was concentrated in less than 20 percent of the population. The Pareto principle states that a large percentage of the results are caused by a small percentage of the causes. This is sometimes referred to as the "80/20 rule." Today, this principle is the foundation of problem-solving and management approaches.

Some examples are as follows:

"20 percent of university students account for 80 percent of complaints."

"15 percent of errors produce 85 percent of the scrap."

The exact percentage is not important. In many cases, the percentages may be 10/90, 15/85, 30/70, or even 40/60. The Pareto principle has been proven in numerous situations with various results. The importance of this rule is to focus on the vital few problems that produce the big results. Greater success is probable by concentrating on the vital few problems that bring major results instead of the trivial many that provide minor results.

Figure 11.15*a* shows a Pareto chart. The Pareto chart is simply a bar chart with the data arranged in descending order of importance, generally by magnitude of frequency, cost, time, or some similar parameter. The chart presents the information being examined in the order of priority and focuses attention on the most critical issues. The chart aids the decision-making process because it put issues into an easily understood framework in which relationships and relative contributions are clearly evident.

The Pareto chart is a simple diagram of vertical bars showing first things first. The biggest or most important items are shown on the left, with other items arranged in descending order to the right. The horizontal line indicates what and the vertical shows how much.

The Pareto chart focuses the team on the vital few areas during analysis. It also can be used to measure progress after a solution has been implemented. In addition, it can be used to show results by having before and after Pareto charts. Also, Pareto charts are useful to display consensus during problem solving.

Figure 11.15*b* shows the construction of a Pareto chart. First, data are collected using a data-collection worksheet. In the example, data were collected on five potential underlying causes of engineering changes. The frequency data are listed in column 1 of the Pareto chart worksheet. Second, the percentage of the total for each cause is calculated. The results of this calculation are noted in column 2 of the Pareto chart worksheet. Third, the cumulative percentage is computed. This is shown in column 3 of the Pareto chart worksheet. Fourth, the chart is drawn on graph paper. The horizontal axis represents the causes, and the vertical axis represents the frequency. Fifth, the graph is scaled. Put zero at the bottom and the total at the top, and mark equal intervals in between. Sixth, arrange the causes from highest to lowest. Seventh, construct the bar chart. Start with the cause with the highest number on the left and work right in descend-

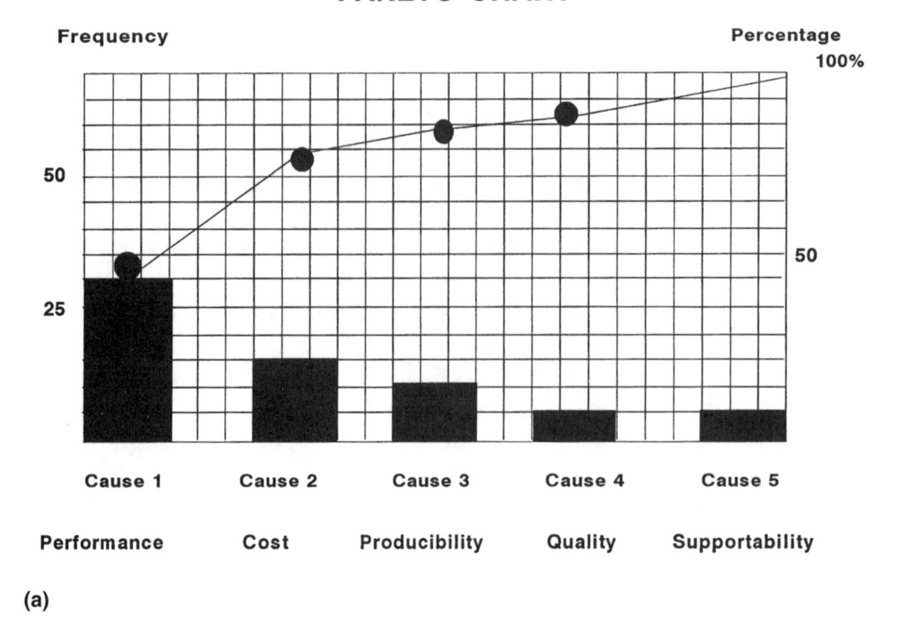

(a)

Pareto Chart Worksheet

	Causes	Frequency (or other measure)	Percentage of total	Cumulative percentage
Cause 1	Performance	30	46	46
Cause 2	Cost	15	23	69
Cause 3	Producibility	10	15	84
Cause 4	Quality	5	8	92
Cause 5	Supportability	5	8	100

(b)

Figure 11.15 Pareto chart.

ing order to the lowest cause. The height of each bar indicates the number of times (frequency) that cause was counted. Eighth, indicate the percentages on the right of the chart. Ninth, put dots to mark cumulative percentages from the Pareto chart worksheet. Tenth, analyze the chart. In this chart, the vital few are performance, cost, and producibility. By focusing on correcting these problems, over 80 percent of the engineering changes can be eliminated.

Variability analysis. As mentioned before, variability exists in everything. The presence of variability is a major obstacle to quality.

Variation is quality's major enemy. For example, if you weigh yourself 10 times, there would be some variation in the measurements. The actual weight is probably near the average of all the measurements. As stated in Chap. 9, variation has common and special causes. By examining the statistical data using statistical process control, deviations from target values can be monitored, controlled, and improved. Variability analysis is an essential tool of CDPM.

Process-capability analysis. Process-capability analysis provides an indication of the performance of a process. It involves measuring process performance in relation to being able to produce the process output within engineering specifications. This is accomplished using process-capability indexes. Two of the most common process-capability indexes are the CP and CPK indexes. The *CP index* gives the ratio of the specification limits to process limits. This shows whether the process is capable of producing within specifications. The objective is a CP that is greater than or equal to 1. The *CPK* index is a measure of the location of the process range within the specification limits. This index provides evidence that the product meets specifications. The objective is a CPK that is greater than or equal to 1. The process is centered when the CPK and CP equal 1. This indicates a capable process.

Force-Field Analysis

Force-field analysis is a technique that helps a group describe the forces at work in a given situation. A force-field analysis chart is shown in Fig. 11.16. The underlying assumption is that every situation results from a balance of forces: restraining forces and driving forces. *Restraining forces* are those things which keep the situation from improving. *Driving forces* are those things which push toward achievement of the goal. Force-field analysis forces the team to examine strengths as well as problems. Sometimes by building on a driving force or strength, a team can bring about the needed improvement.

Force-field analysis steps

1. Define the current status and goal.
2. Identify and prioritize the restraining forces.
3. Identify the driving forces for each restraining force.
4. Identify owners and the level of management best suited to correct the problem.
5. Use CDPM improvement methodology.

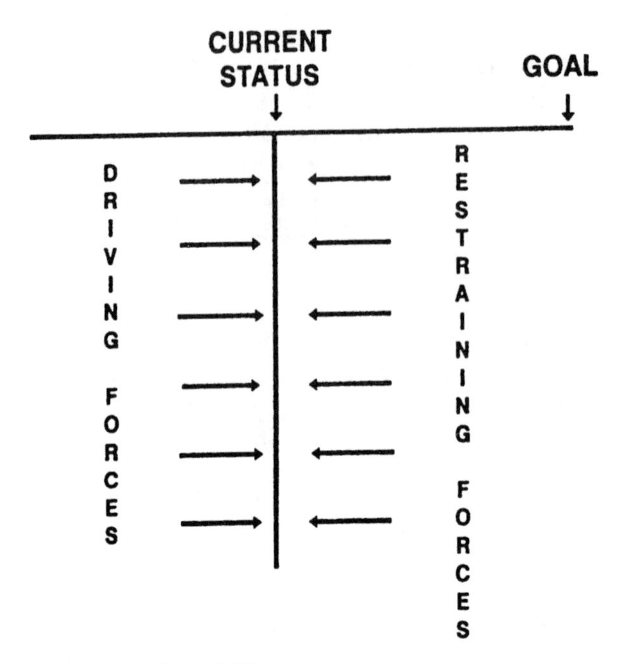

Figure 11.16 Force-field analysis.

Force-field analysis example

In the example shown in Fig. 11.17, the goal is to provide logistics training for engineers. The current status is that no training exists. The restraining forces are management support, funding, courseware, trainers, training equipment, and organizational culture. It is suggested that when performing a force-field analysis the team should select the predominant restraining force first. Then, if needed, other restraining forces can be analyzed one at a time. The restraining force of management is selected to develop the driving forces. The team brainstorms the driving forces. In the example, the team determines that the driving forces that can be used to eliminate or weaken management's nonsupport are facts, communication, and a plan. Next, the team decides which alternatives the team can accomplish. This is shown by *T* for team or *O* for outside the team's control. The team determines that it can provide facts and make a presentation. The other alternative to develop a "doable" plan requires outside assistance. Finally, the team selects an alternative or alternatives to act on.

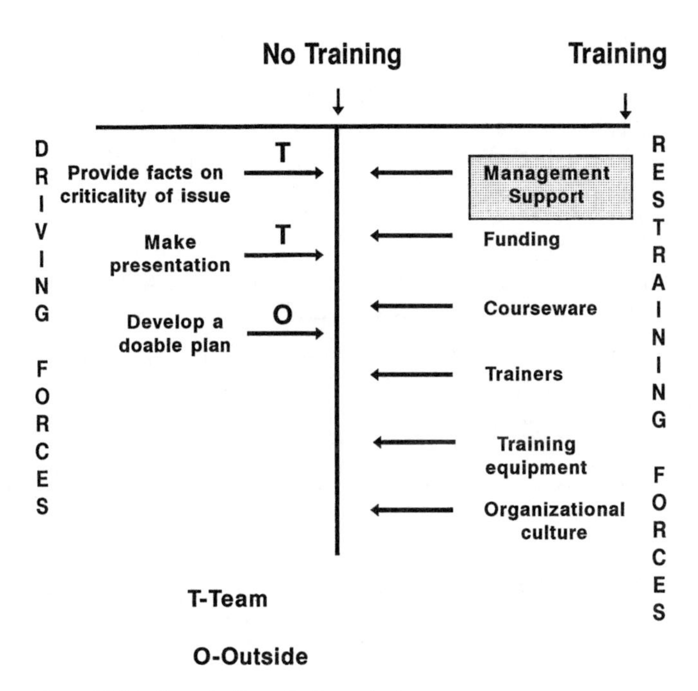

Figure 11.17 Sample force-field analysis.

Main Points

Process analysis is a tool used to improve a process by eliminating non-value-added activities and waits and/or by simplifying the process.

High cost and time areas are primary areas of focus for process analysis.

Analysis of the process for elimination of waits, waste, and simplification is an initial analysis step.

Process analysis and work-flow analysis force critical thinking.

Work-flow analysis targets inefficiencies in work motion.

Elimination of underlying causes is the main focus of cause-and-effect analysis.

The cause-and-effect diagram is sometimes called a "fishbone."

There can be any number of categories in a cause-and-effect diagram.

The underlying or root cause, as determined by cause-and-effect analysis, must always be verified by facts.

Accurate measurements are the key to effective analysis.

Statistics are one method to establish factual data.

The credibility of the data will relate directly to the complexity of the data-collection method. The more complex the data-collection method, the more likely the data will be incorrect.

Always plan the data-collection process.

The central limit theorem allows the use of sample statistics to make judgments about the whole population of the statistic.

There are two types of data: variable data and attitude data.

Use the chart that best shows correct results.

A line chart is used to show trends over time.

Data collection and analysis are used to monitor and improve the process. They should never be used to control or take unfair advantage of others.

Always arrange the data in a meaningful way.

Focus on the vital few items that provide the greater chance for payoff.

Variation is the number one enemy of quality.

Always consider both sides of every issue using force-field analysis.

Chapter

12

CDPM Project Management Tools and Techniques

Focus: This chapter describes how to use specific project management tools and techniques.

Introduction

The customer-driven project management (CDPM) tools and techniques for project management assist the customer-driven project teams in planning, scheduling, monitoring, controlling, and evaluating a project. They also help the customer-driven project teams to select actions that provide the right deliverable at the right time to the customer. Project management tools enhance the performance of customer-driven project, quality-improvement, and process teams at the operating level, which ensures that they produce deliverables efficiently and effectively. The project management tools facilitate the ability of such teams to plan, control, and integrate their work to produce a deliverable within cost, on schedule, and within performance constraints that totally satisfies the customer.

The project management tools and techniques support seven CDPM functions. These functions are performed for the most simple project (5 to 10 tasks) or for the most complex multilevel project (over 1000 tasks). Today, computer software provides practical, easy-to-use tools to aid in the performance of project management functions. The seven CDPM functions are

1. Specify total customer satisfaction.
2. Define the overall deliverable.
3. Establish what needs to be accomplished.
4. Determine who is responsible for performing the work.

5. Develop a schedule to complete the project.
6. Define resources required to complete the project, including people, funds, equipment, and supplies.
7. Manage the project's progress.

Each customer-driven project team performs all these functions for every project. In major projects, the team serves as the key decision maker at major points in the CDPM process. The customer-driven lead team's decisions guide the application of the specific tools and techniques. Figure 12.1 shows a typical CDPM flow with graphic representation of the appropriate project management tool or technique. The input to the project management phase of CDPM comes from the customer-driven lead team and the "voice of the customer." This defines total customer satisfaction and is documented in the contract. Next, a work breakdown structure defines the deliverable. Then the task list determines the specific workload and is used to formulate the project schedule. The project schedule is evaluated for risk before it is finalized. Finally, the project is managed with the assistance of a project management information system.

Below is an outline of key points where the customer-driven lead team targets the use of specific tools and techniques in the CDPM process.

1. Organization and staffing of customer-driven teams constitute the first critical decision point involving a specific project management tool and technique. The customer determines the composition of the customer-driven lead team and the overall project management organizational structure. The project's organizational structure can then be documented on an organizational breakdown structure (OBS).

2. The contract is another critical customer decision point. The contract defines the deliverable specification, contains a statement of work, and lists data requirements. The customer-driven project lead team agrees on the content of the contract.

3. The work breakdown structure (WBS) defines the organization and coding of the deliverable components. Again, the customer-driven lead team determines the format and structure of the WBS. This decision affects the use of the WBS throughout the customer-driven project management process as the WBS changes during the development of the project deliverable.

4. The task list specifies the tasks involved in the project. This is a major step in the customer-driven project management process. Again, the customer-driven lead team approves the list.

5. The project schedule, the plan for carrying out the tasks in sequence on the basis of their interdependency, is another critical

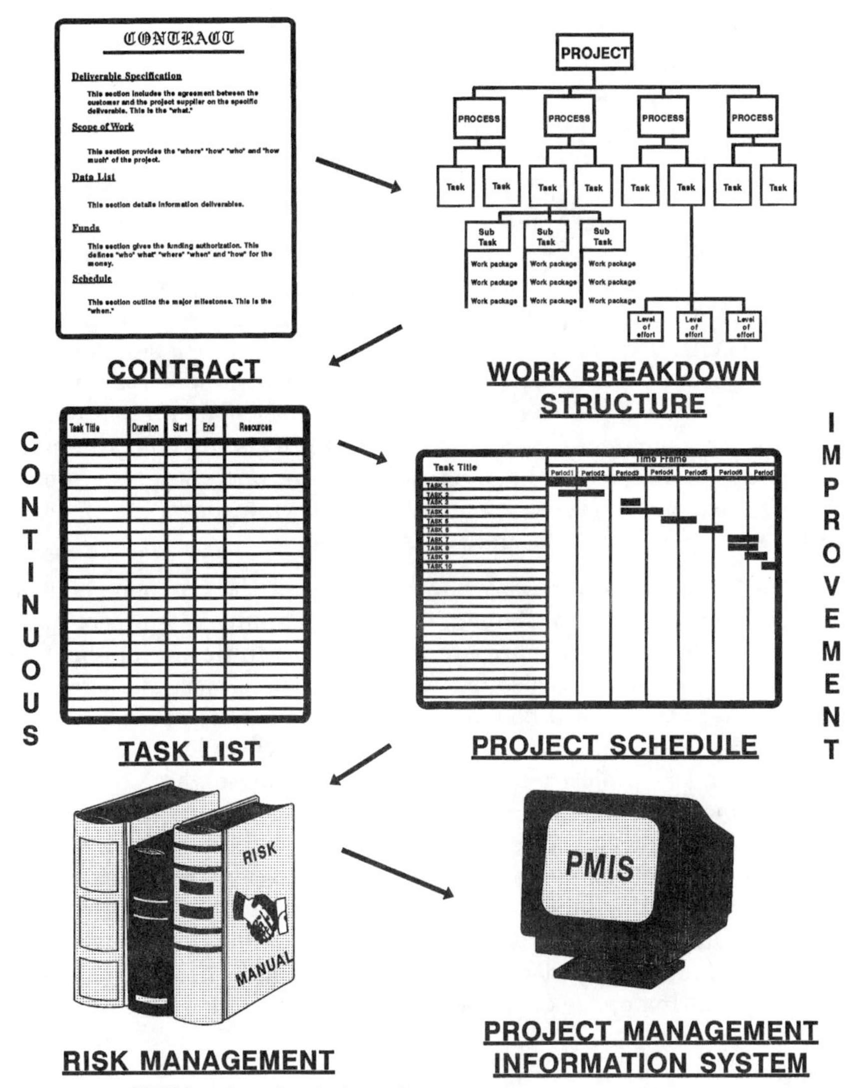

Figure 12.1 CDPM tools and techniques for project management flow.

decision point for the customer-driven lead team. The customer-driven project lead team decides on the approach for project scheduling. The lead team determines the methodology, i.e., critical-path method (CPM), program evaluation and review technique (PERT), or precedence method. The lead team also decides the extent of the use of project scheduling methods. The customer-driven project lead team also determines whether to use manual or automated project scheduling.

6. Risk management is an overall obligation of all customer-driven project teams. Team members must continually assess the risk

involved in each task of the project not only in terms of time and cost but also in terms of the technical feasibility of the task. Again, the customer-driven lead team establishes the system for risk management, and this system influences the use of the other project management tools and techniques.

7. The project management information system (PMIS) provides the project information support for reporting and decision making and captures information from the other project management tools and techniques. The requirements for PMIS must be decided on as part of project planning. During project planning, the customer-driven project lead team decides what information must be available, when it must be available, and in what form it must be available. Once this is known, the lead team decides whether the system will be manual or automated. The automation of PMIS is extremely helpful, especially for project management of a large and complex project. Today's inexpensive project management software allows the effective automation of any project, small or large.

Aside from the organization and staffing of the customer-driven project teams detailed in Chap. 6, the five remaining key project management decision points represent major opportunities for application of project management tools and techniques. The project management tools and techniques will be detailed as follows: (1) contract, (2) work breakdown structure, (3) task list, (4) project schedule, (5) risk management, and (6) project management information system.

The project management tools and techniques are the primary methods used during phase 5, the "take action" phase of the CDPM improvement methodology, to ensure that the project is successful. During phase 7, the implementation phase, project management tools and techniques ensure that the deliverable continually satisfies the customer. Finally, project management tools and techniques provides assistance during phase 8a, closeout of a project.

Contract

In customer-driven project management, the contract is the document that focuses on total customer satisfaction. The customer-driven project lead team agrees on the content of the contract, and the contract describes what the team determines is necessary to perform a successful project. The customer-supplier project steering team approves the contract, and additional approvals depend on the specific contracting regulations of the customer. In CDPM, a system is established to continuously update the contract. The requirements for specific contract parts and their content depend on the particular project. The contract defines the deliverable specification, sets out a state-

CONTRACT

Deliverable Specification

This section includes the agreement between the customer and the project supplier on the specific deliverable. This is the "what."

Scope of Work

This section provides the "where", "how", "who", and "how much" of the project.

Data List

This section details information deliverables.

Funds

This section gives the funding authorization. This defines "who", what", "where", "when", and "how" for the money.

Schedule

This section outline the major milestones. This is the "when."

Figure 12.2 The contract.

ment of work, lists data requirements, and specifies schedule and funding, as shown in Fig. 12.2.

The deliverable specification. The deliverable specification includes the project objectives and expected outcomes. It provides the specific technical and performance information used to develop detailed, lower-level project objectives and the standards for evaluating the performance of the developed system. Specifications are often consolidated into a single document, but they also could be included in separate drawings, references, or regulations.

In general, specifications should address only those characteristics which are important for the deliverable to meet the customer's need. Overspecification reduces the ability of the project team to balance technical performance, cost, and schedules. Conversely, failure to specify a critical requirement can result in costly redesign later in the project's life or a system that fails to perform as needed. Rarely should a specification be immutable. Conditions change, and new information is uncovered during the life of a project. The degree of acceptable change, however, can differ considerably from project to project. In CDPM, the customer-driven lead team develops the initial specifications and modifies existing specifications so that the customer's expectations are met.

Statement of work. The statement of work (SOW) summarizes the work required to complete the project. It is a narrative explanation of how the project will be performed. The SOW should not be a detailed technical document—that is the purpose of the project specification. Neither, in general, should the SOW provide detailed, step-by-step direction for carrying out project tasks. To do so would prevent the customer-driven project teams from applying their knowledge of the technologies involved and the organization's operating policies and procedures to develop a more effective and efficient project.

Data requirements. The data requirements document is a list of all the data items needed for a successful project deliverable. Data requirements are all the project documentation. Data items are reports, studies, plans, engineering drawings, lists, manuals, etc. Data requirements are important because they are project deliverables. The data requirements are a significant contributor to total customer satisfaction. Some typical data requirements are shown in Fig. 12.3. These data requirements must be defined and revised as necessary during the project life cycle.

Work Breakdown Structure

The work breakdown structure (WBS) defines the organization of the deliverable. This is a coded listing of the tasks required to complete a project by levels. Again, the customer-driven lead team formulates the format and structure of the WBS. This tool is critical to the successful planning and completion of the project. As a planning tool, the WBS allows the customer-driven project lead team to organize the project work into levels of activity. Each level of activity is defined further and in even greater detail by process owners. The purpose is to structure each project as a hierarchy of work leading up to the project deliverable and down to each more detailed breakout of the work to be done.

The WBS structure can be arranged in a variety of formats. The format defines the project tasks and their relationship to the total project. A WBS can have as many levels as appropriate for the pro-

Data Items

Project progress reports	Support plan
Engineering drawings	Facilities plan
Data lists	Human resources plan
Aperture cards for drawings	Training plan
Studies	Repair plan
Reports	Supply plan
Qualification test reports	Transportation plan
Mass properties report	Computer resources plan
Stress analysis report	People requirements
Design reviews	Technical manuals
Configuration management plan	Training materials
Master manufacturing schedule	Failure reports
Contract fund status report	Reliability reports
Cost data summary report	Safety assessment reports
Producibility plan	In-progress reviews
Time-phased schedule	Inventory level reports
Action item list	Long lead time items
Safety plan	Life cycle cost data
Quality assurance plan	Material cost data
Test plan	Personnel cost data

Figure 12.3 Sample data items.

ject. One typical six-level project WBS uses the structure shown in Table 12.1.

Figure 12.4 shows the WBS structure outlined in Table 12.1. Level 1 is the top level of the project. It contains the total project package. Level 2 is all the major processes required to complete the project. Level 3 is all the tasks required to perform each of the processes. Level 4 is all the subtasks to accomplish the tasks. All the activities

TABLE 12.1 A Typical Six-Level Project WBS

Level	Item	Number
1	Project	1
2	Process	1.X
3	Task	1.X.X
4	Subtask	1.X.X.X
5	Work package	1.X.X.X.X
6	Level of effort	1.X.X.X.X.X

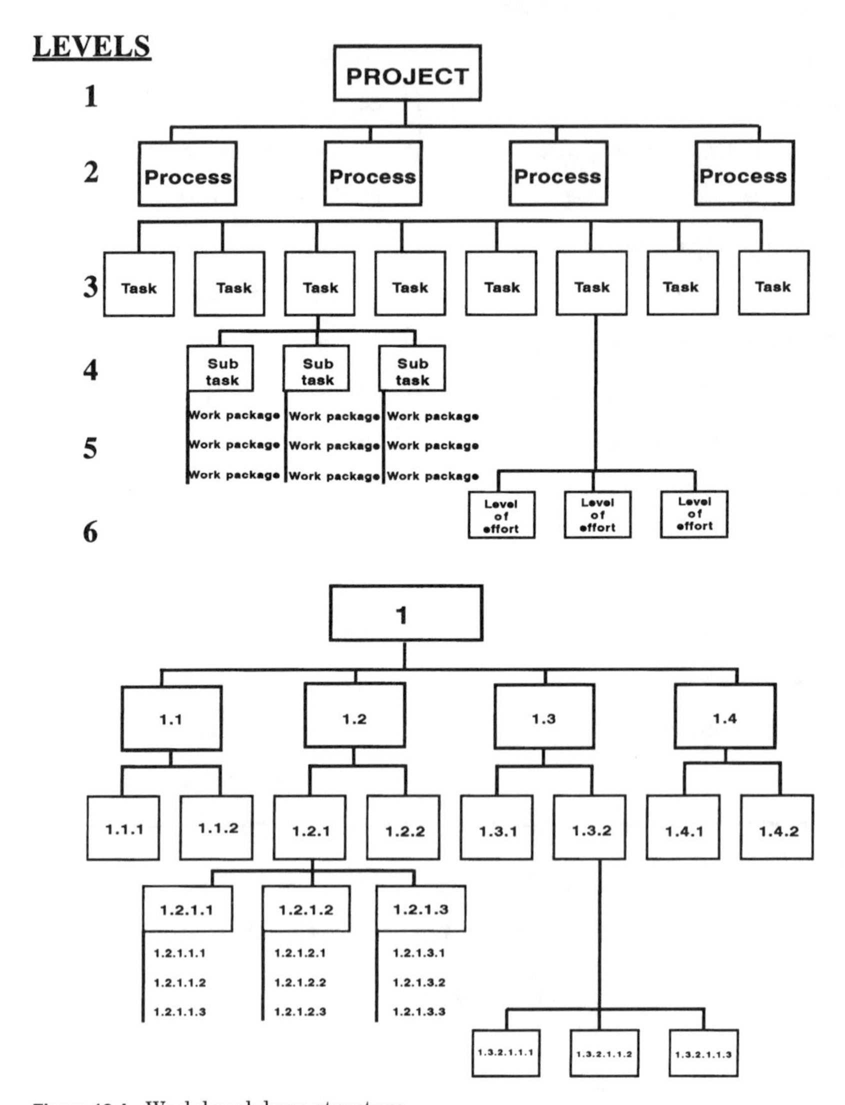

Figure 12.4 Work breakdown structure.

to complete the subtasks are then put into level 5 as work packages or level 6 as level of effort. Work packages include specific activities performed by a customer-driven project team. Level-of-effort work packages are all the activities not specifically connected to a customer-driven project team. Level-of-effort work packages affect performance across many tasks. The level of effort usually involves a functional team, or specific category of team, worker, or equipment type. For instance, the customer-driven project lead team normally is a specific category of team that performs level-of-effort work packages.

Tasks at the first four levels of the WBS provide the summary information necessary for the customer-driven project lead team to perform project management planning, scheduling, monitoring, and management activities. Tasks at levels 5 and 6 are normally used by the customer-driven project teams performing the work.

In addition to the number-based WBS, some WBSs are alphabetically based (Table 12.2) or alphanumeric (Table 12.3).

Figure 12.5 displays an example of a WBS to install a computer system. Any of the coding systems described in the tables could be

TABLE 12.2 Alphabetical WBS Structure

Level	Item	Alphabetical designation
0	Project	0
1	Process 1	A
	Process 2	B
	Process 3	C
2	Task	CA
3	Subtask	CAA
4	Work package	CAAA
5	Level of effort	CAAAA

TABLE 12.3 Alphanumeric WBS Structure

Level	Item	Alphanumeric designation
0	Project	
1	Process 1	A
	Process 2	B
	Process 3	C
2	Task 1	C1
	Task 2	C2
	Task 3	C3
3	Subtask 3-1	C3A
	Subtask 3-2	C3B

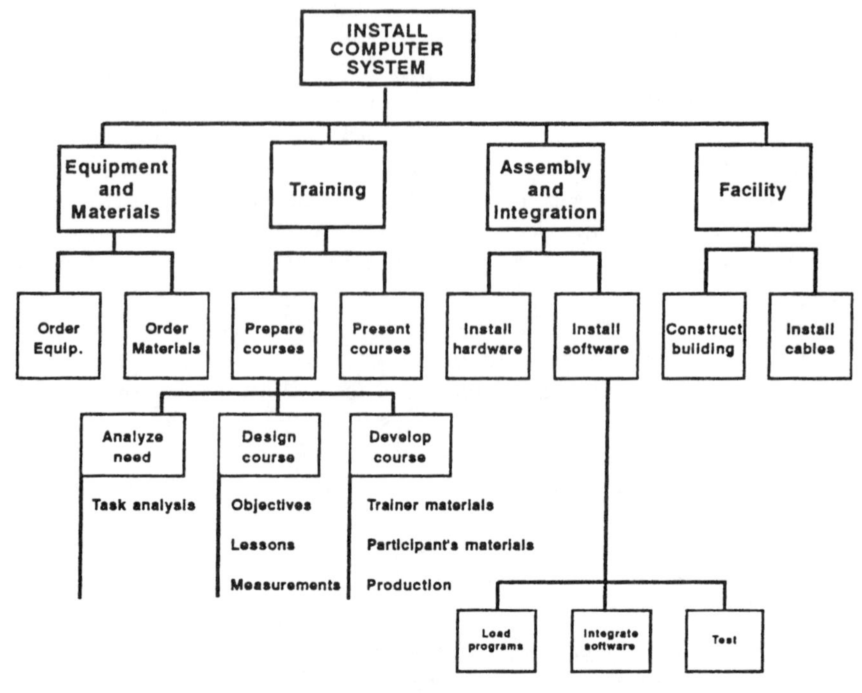

Figure 12.5 Sample work breakdown structure.

used for this WBS. This graphic display of a WBS is called a *tree chart.* It shows project, process, task, and subtask levels of the WBS. The resulting scheme graphically expresses an organizational chart of the work to be performed and serves as a basis for setting up customer-driven project teams.

The WBS has many uses during a project's life. It goes through many iterations as the work becomes more defined, it may change continually from the time the specifications are agreed on with the customer until the project completes. At least seven distinct purposes are served by the WBS.

1. *The WBS defines project team responsibility.* The structured breakdown of the work helps to assign task-level work to customer-driven project teams and gives the customer-driven project lead team a means of developing task descriptions, work outputs, and objectives for each task. Normally, each task or subtask represents a customer-driven project team or team member responsibility, with others in the organization assigned work packages and level-of-effort work.

2. *The WBS sets out organization structure.* The project team is normally structured around, or at least consistent with, the WBS.

This means that each customer-driven project team or team member is identified with one or more task or subtask work element, and reporting to the customer-driven project lead team is accomplished through this system of organization. The WBS serves as the basis for setting up customer-driven project teams. This can be represented by an organizational breakdown structure (OBS). The OBS identifies who is empowered to perform and improve the work.

3. *The WBS allows coordination of objectives.* The necessary coordination of project objectives is supported by the WBS because the interfaces between elements, which are so vital to project success, are made clear in the WBS itself. Coordination of the objectives is also supported by the WBS structure, which shows graphically the work elements as they are to be interrelated across the entire project, not just within the confined arena of one task or another.

4. *The WBS allows control.* The WBS facilitates control by providing a clear basis for monitoring project progress using the structure of the project; thus a specific customer-driven project team or team member is accountable for a task clearly shown on the WBS graphic. Control is exercised through cost, schedule, and technical criteria for each element of the WBS.

The real value of the WBS is that it provides a graphic representation of the entire project with an orderly identification scheme for each level of the project. It provides continuity and a frame of reference throughout the project for interconnections, interfaces between team members, and the critical-path planning process. It describes the whole project, and it is the basis for work authorization, budgeting, and tracking. The WBS is made flexible by the project manager at the end of each project phase; opportunity is provided to update the WBS based on project developments.

5. *The WBS facilitates project scheduling.* If the WBS is turned on its side in a clockwise direction, it begins to resemble the start of a system diagram relating tasks and other elements in terms of sequence. This is the point of departure for developing a task list and project schedule consisting of a critical-path network that identifies all task milestones and task activities and relates them in terms of interdependence.

6. *The WBS facilitates costing.* The cost of each element of the WBS is estimated and controlled by the cost accounting system built into the WBS structure. In this way, the WBS becomes the means by which costs are "rolled up" and captured. Unit costs are derived from this structure in order to document the project cost history.

7. *The WBS facilitates risk analysis.* Each element of the WBS is assessed in terms of the inherent risk involved in completion of that

element. Risk assessment is important in critical-path planning as well as in the process of protecting against unforeseen failures. Risk factors are developed for each element of the WBS through analysis of the work involved by the responsible customer-driven project team or team member.

Task List

The task list includes the development of the tasks involved in the project. Again, this represents a major step in the customer-driven project management process. The task list is developed by the customer-driven project lead team with other customer-driven teams as appropriate and forms the basis for project scheduling.

Figure 12.6 shows the basic task list format, which provides documentation of the following items:

- *Task title.* This item identifies the task. In some cases, it is just a name. In others, it may include a name and an identification code.
- *Task duration.* This item is the total amount of time necessary to complete the task.
- *Task resource(s).* This item includes the human resources, materials, and equipment needed to complete the task.

TASK LIST

Task Title	Duration	Start	End	Resources

Figure 12.6 Task list format.

Task list construction

The task list can be formulated by forward or reverse planning. With forward planning, the customer-driven project lead team sets the start date and then lists the activities from first to last. With reverse planning, the team starts with the end activity and works backward to the first activity. This is sometimes called *backward mapping*. It is usually advantageous to use reverse planning because reverse planning focuses on the project's end result, thus always keeping the project's output objective in view. In reverse planning, the task list is constructed by asking, "What activities are required to complete the project's object?" and then repeatedly asking, "What activities does it take to complete the preceding activity to complete the project's objective?," until all the activities to complete the project are identified.

Task list example

Figure 12.7 shows a typical task list for application of the CDPM improvement methodology. It shows all the steps from beginning to the project closeout. Next, the customer-driven team determines task durations. Then the estimated start and end dates are formulated. Finally, the resources to perform each task are listed.

CDPM IMPROVEMENT METHODOLOGY

Task Name	Duration	Start	End	Resources
CDPM Improvement Methodology	162.00 d	Jan/04/1993	Aug/20/1993	
Define the quality issue	10.00 d	Jan/04/1993	Jan/15/1993	CDPM LEAD TEAM
Understand and define the proce	10.00 d	Jan/19/1993	Feb/01/1993	CDPM LEAD TEAM
Select improvement opportunities	1.00 d	Feb/02/1993	Feb/02/1993	CDPM LEAD TEAM
Analyze improvement opportuniti	20.00 d	Feb/03/1993	Mar/03/1993	CDPM LEAD TEAM
Take action	15.00 d	Mar/04/1993	Mar/24/1993	CDPM LEAD TEAM
Project concept	5.00 d	Mar/04/1993	Mar/10/1993	CDPM PROJECT TEA
Project definition	5.00 d	Mar/08/1993	Mar/12/1993	CDPM PROJECT TEA
Project production	10.00 d	Mar/11/1993	Mar/24/1993	CDPM WORK TEAM
Check results	5.00 d	Mar/25/1993	Mar/31/1993	CDPM LEAD TEAM
Implement improvement	100.00 d	Apr/01/1993	Aug/19/1993	CDPM LEAD TEAM
Project operation	100.00 d	Apr/01/1993	Aug/19/1993	CDPM WORK TEAM
Monitor results	101.00 d	Apr/01/1993	Aug/20/1993	CDPM LEAD TEAM
Continuous improvement	100.00 d	Apr/01/1993	Aug/19/1993	CDPM LEAD TEAM
Project close-out	1.00 d	Aug/20/1993	Aug/20/1993	CDPM LEAD TEAM

Printed: Nov/25/1992
Page 1

Milestone △ Summary
Fixed Delay

Figure 12.7 Sample task list.

Project Schedule

The project schedule involves scheduling, monitoring, and managing the project using network and system diagraming techniques. The project schedule is another critical decision point for the customer-driven project lead team. The project schedule is a team effort of all the teams involved in performing the project.

The project is planned on the basis of the contract, the WBS, and the task list. Using this information, the project schedule organizes the project by describing the sequential relationship of project activities and milestones in a network. The network graphically shows the project plans. It is based on estimates from the customer-driven project teams, who are critical partners in developing the network. Their efforts at communication, cooperation, and developing correct information make the network work, and the network reflects their planning, especially in terms of the necessary interconnections and interdependencies of the project.

Besides the relationships between tasks, customer-driven teams provide their estimates of how long it will take to deliver the work after all events necessary to start it have been completed. A projection is made by comparing schedules with elapsed-time estimates. The estimates are compared with the actual required date of completion. In a project where the estimate exactly meets the target completion date, the project slack will be zero. For such a project, there is no leeway in the schedule for any changes. In a project where the estimate is less than the target completion date, the project slack will be positive. For such a project, there can still be some changes in the schedule without affecting the on-time completion date. If the project estimate is more that the target completion time, the project slack will be negative. A 6-week negative slack means that the activity is estimated to be 6 weeks late. For such a project, some changes must be made to reduce the project estimated time to meet the project completion date. In the planning stage, the customer-driven project teams can do any of the following actions:

- Revise the activity duration times.
- Revise network relationships.
- Revise technical objectives.

Once the time estimates meet the required project completion date, the next step is to determine resource requirements. The customer-driven project teams assign resources for each activity, and resources include people, equipment, and materials. Again, the project is evaluated to determine if it can be completed on time. In this case, the customer-driven project teams can do any of the same actions used for time management. In addition, the team can revise the resource allocations. In this case, the customer-driven project lead team must consider the cost and time constraints of the project.

Once the time and resource estimates meet the project's time, cost, and technical constraints, this establishes the project schedule baseline for monitoring the project's progress. During performance, the project is continually evaluated based on the triple constraint of time, cost, and technical performance, and any variance must be analyzed for possible corrective action.

Once the project is under way, it must be guided and managed continually using the critical path. The critical path represents the longest route between the first event and the last event, and sometimes this critical path changes over the course of the project. In addition, the situation can be complicated by multiple criticality, which requires constant attention by the customer-driven project lead team. Again, the customer-driven project lead team can decide to (1) reestimate activity duration on the critical path, (2) revise network relationships by changing dependencies, (3) revise cost and/or resource objectives, and (4) revise the technical objectives by changing the deliverable specifications.

The project schedule should be seen as support for project planning, scheduling, monitoring, and management. It is a means to an end. It should never be allowed to drive the project so far as to become a dominating factor that replaces good judgment and perspective. There is nothing magic about the critical path itself, but it does provide a valuable frame of reference for monitoring the project and for identifying targets for close attention by the customer-driven project lead team.

The concept of "float" or "slack" is important in project management because it suggests some flexibility in scheduling and allocation of resources. This concept refers to the difference between the duration of an activity and the time available during the entire project to complete it. Activities off the critical path have float; those on the critical path do not. Computer programs provide a format for analysis of float for each activity. In addition, there is often the capability for "leveling" resources where float allows the scheduling of a project activity when the required resources are available at the least cost.

The specific use of project network scheduling is a function of the complexity of the project and the sophistication of the organization employing it. Projects that produce "hard" deliverables are most suited to project network scheduling; because of the certainty of the product, they have less inherent risk. Construction projects are an example of projects producing "hard" deliverables that normally employ network schedules.

For "soft" project outputs, such as computer software development, project network scheduling requires added flexibility. This is so simply because the risk factors are high, the nature of the deliverable is often unclear well into the project, and project success is so much more dependent on "satisfying" than on optimizing.

The network scheduling techniques are the methods described above. The network schedule is not the only way to identify critical tasks or activities. The common bar chart, or Gantt chart, will show which activities precede which other activities, and it will indicate the importance of those activities which in sum will take the longest time. For simple projects, this may be sufficient to plan, schedule, monitor, and manage the project. For most projects, however, a network schedule will provide the information required for successful project management.

Network scheduling

Network scheduling is a project management technique for planning, scheduling, and managing time and estimating, budgeting, and managing resources. Many types of network diagrams can be used for scheduling, including the critical-path method (CPM), program evaluation and review technique (PERT), and the precedence diagraming method (PDM). The concepts are similar in each of these methods, but the differences lie in their orientation to activities or events. Today, the precedence method of network scheduling is used most commonly. Many forms of the precedence method are used, especially in computer software programs. The network scheduling descriptions in this book focus on the precedence method because it uses a combination of CPM and PERT techniques.

Network concepts

Before using network scheduling techniques, customer-driven project management team members must understand some basic network concepts. These concepts include

- Network elements
- Network structure
- Network planning
- Network scheduling
- Network monitoring

Network elements. Every project network has a beginning time, completion time, activity or activities, duration of activities, links, and an end result. This is shown in Fig. 12.8. The *begin time* is the date the

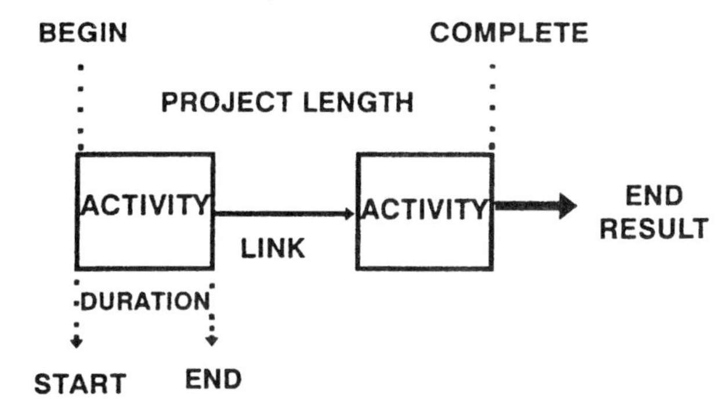

Figure 12.8 Network elements.

project is scheduled to activate. The *complete time* is the date the project is scheduled to finish. The time from the project start to project completion is the *project length.* An *activity* is a task, job, or process requiring accomplishment to perform the project. Each activity has a *duration,* which is the difference between the activity start and activity end times. Each activity also has a *link* to the beginning, end, or another activity. Each project has an *end result,* which is a sum of the outcomes of all the activities in the project.

Network structure. The network structure depends on the software package. Examples of typical network structures are shown on Fig. 12.9. Figure 12.9*a* is from a mainframe-based computer package called Artemis Planning 9000. Figure 12.9*b* is from the Project Scheduler 5 personal computer software. Figure 12.9*c* shows one of several structures from TIMELINE 5 personal computer software.

Network planning. Network planning is based on the WBS and the task list. The customer-driven project lead team determines the actual activities for the network project schedule. Then the team further determines the level of detail for the network project schedule, which could include more or less detail than the WBS and task list. For instance, the team may add events or milestones as activities in the project network schedule. Events or milestones are activities that highlight a particular point in the life of the project. Once the network activities are established, each activity requires an "owner." This establishes accountability for task performance. The "owner" is a customer-driven team or an individual team member. In customer-driven project management, the "owner's" boundaries of empowerment, including authority, responsibility, and resources, are identified during this planning stage.

Activity duration. Each activity has a time duration. This is the time from start to finish of an activity, job, task, or process and involves

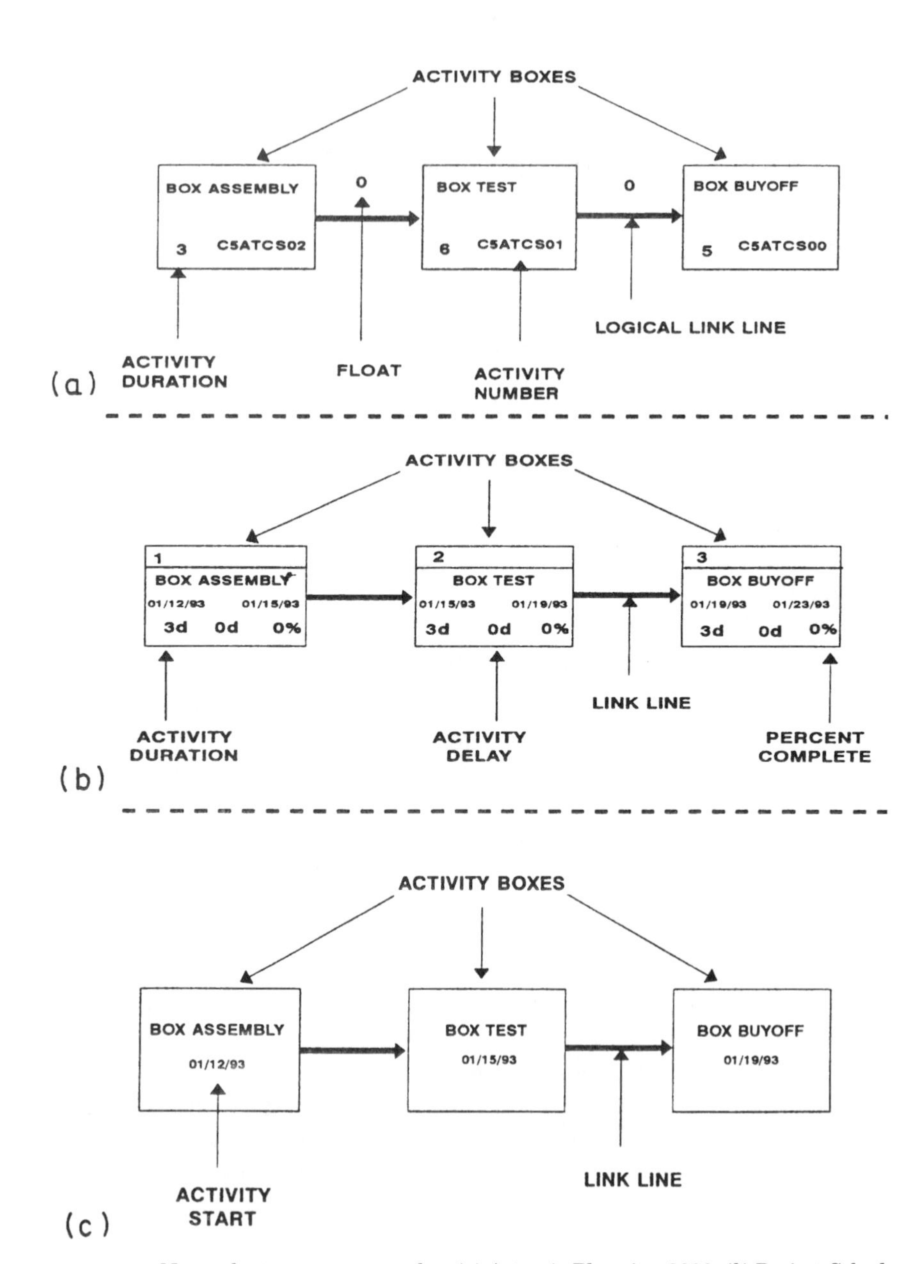

Figure 12.9 Network structures examples. (*a*) Artemis Planning 9000. (*b*) Project Scheduler 5. (*c*) TIMELINE 5.

estimation of the actual number of labor hours using certain skills it will take to perform the task. The initial estimates are based on

1. Normal work flow (usual work week and shifts)
2. All resources available as planned

It is important to determine an accurate initial project schedule. Because of the uncertainty of this original estimate, a range of estimates is normally made, including an optimistic estimate (T_O), a most likely estimate (T_M), and a pessimistic estimate (T_P). The optimistic estimate is the time that the activity can be completed if everything goes exceptionally well. The most likely estimate is the realistic estimate of the time the activity might take. The pessimistic time estimate is the longest time an activity would require under the most adverse conditions, barring acts of God. From these three estimates, an expected time duration (T_E) can be calculated from the formula

$$T_E = T_O + 4T_M + T_P/6$$

For instance, using an optimistic estimate of 5 days, a most likely estimate of 15 days, and a pessimistic estimate of 30 days, calculation of T_E would be

$$5 + (4 \times 15) + 30/6 = 5 + 60 + 30/6$$
$$= 95/6$$
$$= 15.83 \text{ days}$$

In many cases, it may not be necessary to use the detailed time-estimating method described above. It may only be necessary to use one time estimate. In either case, the customer-driven project lead team, with representation from the appropriate customer-driven project or process teams, works to negotiate the original project schedule. This adds credibility and commitment to the project schedule. In addition, activities may be an event or milestone that has no duration. Again, the specific event or milestone must be established by the customer-driven project lead team in conjunction with appropriate customer-driven teams.

Activities characteristics. Figure 12.10 provides a template to diagram activities characteristics, which include the following:

- An activity can be in sequence with another activity. Activities *A*, *B*, and *C* in the figure are in sequence.
- Activities can be in parallel with other activities. Activities *B* and *C* are in parallel.

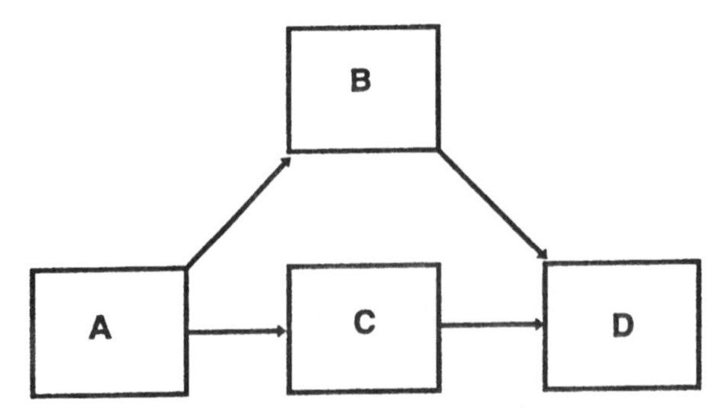

Figure 12.10 Activities characteristics template.

- An activity can be dependent on another activity. Activities *B* and *C* are dependent on activity *A*. Activity *D* is dependent on activities *B* and *C*.
- An activity can be independent of any other activity. Activity *B* is independent of activity *C*, and vice versa.
- More than one activity can start or end at the same time. Activities *B* and *C* can start or end at the same time.

Network relationships or constraints. The network activities have relationships to other activities. Activities have a successor or predecessor relationship with other activities. The activity that immediately follows another activity is the successor of the activity it follows. An activity that is immediately before an another activity is the predecessor of the other activity.

In addition to the predecessor and successor relationship, a relationship can involve the dependency of one activity on another activity. These types of relationships are sometimes called *constraints*. For example, Fig. 12.11*a* shows a finish-to-start relationship between activities *A* and *B*. Activity *B* cannot start until the finish of activity *A*. This finish-to-start relationship is the most frequently used relationship in project scheduling. However, finish-to-finish, start-to-start, and start-to-finish relationships also can exist. Figure 12.11*b* shows a finish-to-finish relationship. In this relationship, activity *B* cannot finish before the finish of activity *A*. Figure 12.11*c* shows a start-to-start relationship. Activity *B* cannot start before the start of activity *A*. Figure 12.11*d* shows a start-to-finish relationship. Activity *B* cannot finish until the start of activity *A*.

Time constraints. Besides dependency constraints, an activity can have additional time constraints that refer to specified time dependencies, i.e., the time constraints between predecessor and successor activities. Such time constraints can be in actual time units or ASAP

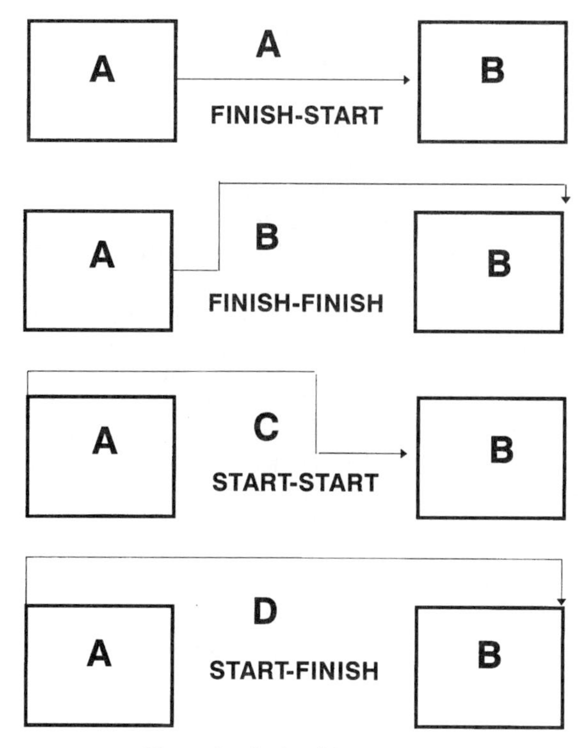

Figure 12.11 Network relationships or constraints.

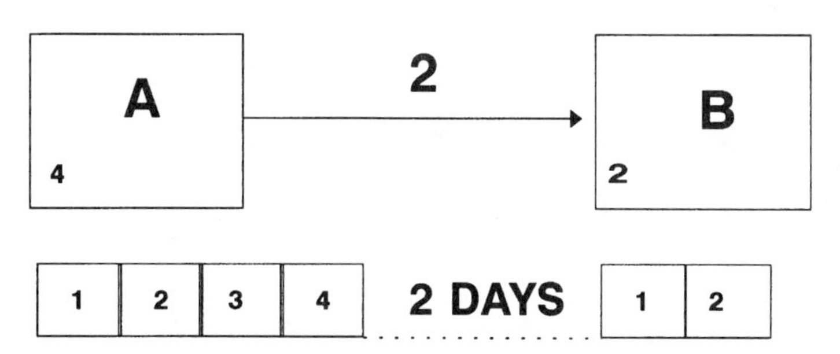

Figure 12.12 Network time constraints example.

or ALAP. Actual time unit constraints are displayed as shown in Fig. 12.12. In this example, there is an actual time delay between the finish of activity *A* and the start of activity *B* of 2 days.

There is also a time constraint for the delay of an activity. This would delay the start of the activity. The ASAP time constraint means "as soon as possible." This option allows the successor activity to start as soon as possible after the predecessor activity. The ALAP time constraint means "as late as possible." This option allows the successor activity to start as late as possible after the predecessor activity. Other time constraints include starting an activity on a specific time, ending on a specific time, or starting and ending on specific times.

Activity hammock. In addition to the other relationships, the customer-driven project team may desire to link a group of activities for reporting purposes. Typically, this is accomplished by linking the activities in a "hammock." Figure 12.13 shows graphically a hammock activity. In this example, hammock activity *D* is used to measure time between the start of activity *A* and the finish of activity *C*.

Network scheduling. Once the network schedule relationships and durations are determined, the network schedule is created. Scheduling involves turning the network plan into a specific timetable of activities with their relationships to meet the project's objectives. Scheduling does the following:

- *Starts the project.* The schedule notifies all the customer-driven teams when to start the project.
- *Communicates project objectives.* The schedule provides an outline of all the project objectives to focus all customer-driven project teams' efforts.
- *Helps establish a disciplined approach.* The schedule advocates a systematic approach to planning, analysis, and action.
- *Establishes "ownership."* The schedule defines accountability for specific tasks to accomplish the project's objectives.

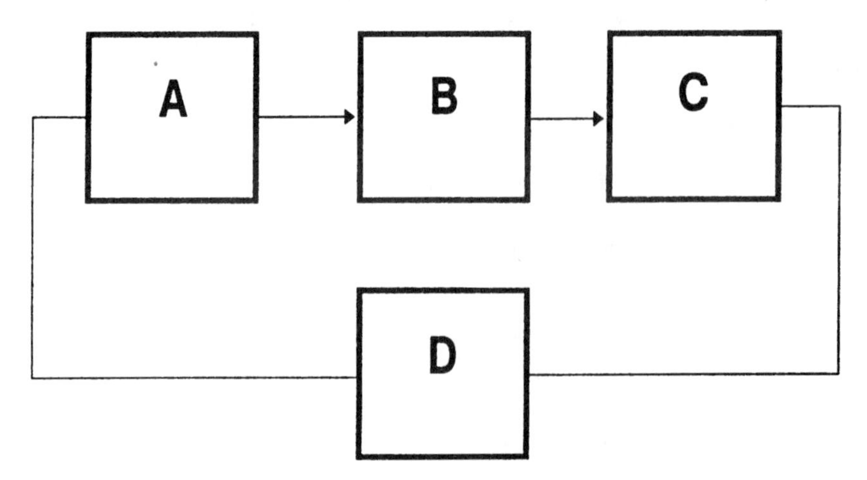

Figure 12.13 Hammock activity.

- *Determines resources.* The schedule allocates specific resources to the project's tasks.
- *Uses graphic techniques.* The schedule provides a diagram of the flow of activities and their relationship from beginning to end.
- *Limits potential solutions.* The schedule allows "what if" analysis that focuses on most probable actions.
- *Ensures visibility of the project's progress.* The schedule continuously evaluates the project's actual versus scheduled accomplishment. This highlights the status of the project at any time.

Network scheduling information is usually shown on a Gantt chart and/or a PERT chart for communication and evaluation. These graphic scheduling techniques allow customer-driven project teams to picture the project, which assists team members in understanding the flow and relationships of the project tasks. A Gantt chart provides an overview of the project, while a PERT chart displays the relationship information.

Gantt chart. The Gantt chart displays the project schedule in bar-chart form. Figure 12.14 is a Gantt chart.

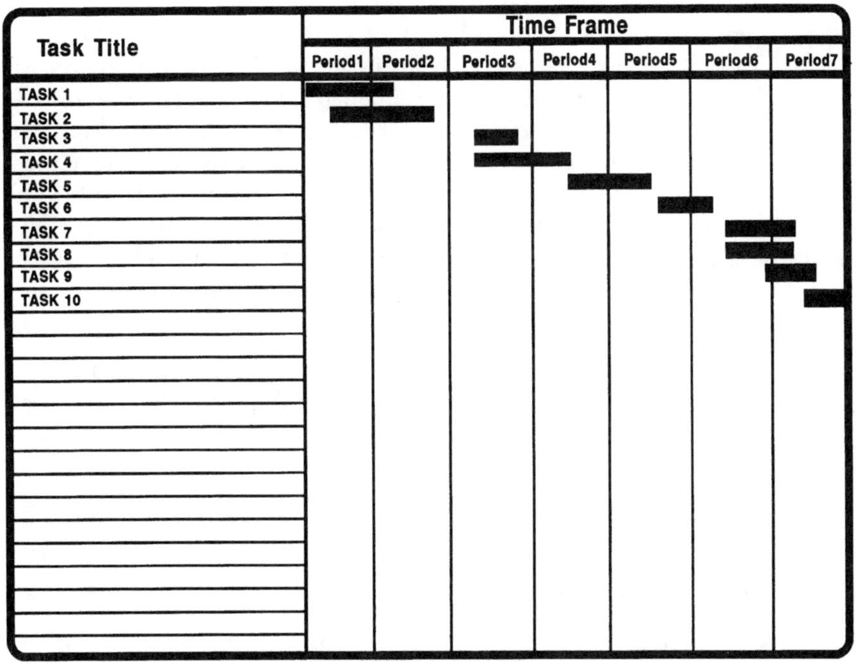

Figure 12.14 Gantt chart.

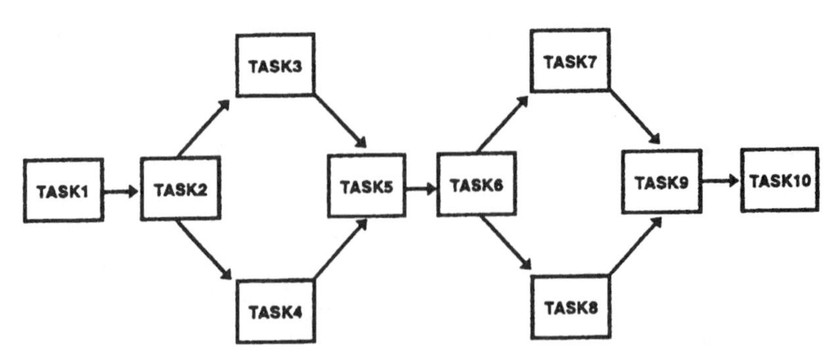

Figure 12.15 PERT chart.

PERT chart. The PERT chart is usually a graphic display of the project schedule with interdependencies. Figure 12.15 is a PERT display. This PERT display contains the same information displayed on the Gantt chart in Fig. 12.14.

Multiple projects. In organizations where more than one project is being performed, the organization needs to evaluate the project schedules in relation to all the other project schedules. Frequently, multiple projects share the same resources. This sharing of resources affects the availability of resources to meet each project's requirements. The customer-supplier project steering team must determine the priorities of multiple projects. The customer-driven project lead team manages resources based on these established priorities. The customer-driven project steering team must determine a process for resolving resource conflicts. In addition, a project may rely on an activity or resource from another project. This relationship must be identified to the appropriate customer-driven project lead teams. In all cases of multiple projects, the priorities, relationships, and constraints must be managed to ensure successful completion of all projects.

Resource estimating and budgeting. Resource estimating and budgeting involve the determination and allocation of the specific people, equipment, and materials for the accomplishment of project. Such activity includes

1. Developing a resource list of all available resources to support the project
2. Allocating the specific resources required to perform to each activity

Network monitoring. Network monitoring starts with planning of the project and continues with conversion of the plan into a project schedule. Once the project schedule is established by the customer-driven project lead team and approved by the customer-supplier project steering team, the work can start. As the work is performed, the network project schedule becomes a primary tool for communicating and correcting the project's progress. The network project schedule provides continuous critical-path analysis and variance analysis of schedule versus actual accomplishment. This assists with the evaluation of the project's progress. In addition, the network project schedule improves decision making through the use of "what if" analysis and presentations or reports.

Network monitoring consists of continually assessing the time, cost, quality, and performance of the project.

Time monitoring. Time monitoring involves critical-path analysis, which focuses on managing the critical-path times by attending to the time durations and relationships of activities. Figure 12.16 shows a critical path graphically. The critical path in this figure is the path with activities *A, B, C,* and *F,* and this path takes 10 units. The path with activities *A, D, E,* and *F* takes 7 units, and the path with activities *A, B, E,* and *F* takes 9 units.

The project's critical path or paths are evaluated for a negative indication. Critical path or paths with a negative indicator finish beyond the required project completion date. Such critical paths require further analysis and possible action.

Once the critical path is determined, it can be analyzed for changing interdependencies, durations, or methodology. Next, the customer-driven team can consider float or slack adjustments, which provide flexibility in project network scheduling. Leaving unscheduled time in a path or between activities can give some leeway in project network scheduling. In some applications, float and slack are the same. In others, there may be different types of float or slack. To simplify the explanation, *total float* applies to path differences and *slack* applies to activity differences.

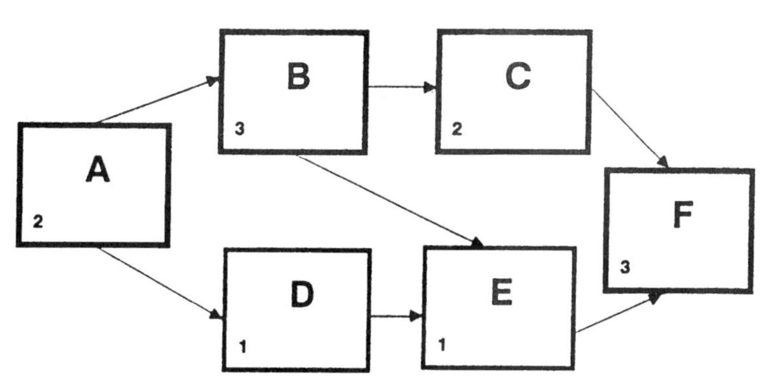

Figure 12.16 Critical-path chart.

Total float. *Total float* is the maximum amount of time a path can slip before it affects the project's end date. Total float is the time difference between the latest allowable date and the expected date. It can be positive, negative, or zero. When the total float is zero, the path and the project completion date are the same. When the total float is positive, the path is expected to be completed before the project's end date. When the total float is negative, the path is expected to move the project's end date beyond the expected completion date. When looking at total float, the more negative the path, the more critical is the path. Total float is important to focus the customer-driven project lead team on paths warranting the most attention.

Slack. Although the major emphasis is managing activities along the critical path, activities not on the critical path also require attention. In some instances, activities not on the critical path require adjustment. Unlike activities on the critical path, activities not on the critical path have slack. *Slack* is the amount of time an activity can be late without affecting the project's completion. The slack is the difference between the earliest start time and the latest start time. Figure 12.17 shows the concept of slack between activities. An activity has an early start (ES) date of 1 and a latest start (LS) date of 3. The early finish (EF) date is day 10, and the latest finish (LF) date is day 12. Thus the slack is 3 days.

Cost monitoring. Cost monitoring consists of resource and earned-value management activities.

Resource monitoring. Once the project is under way, it consumes resources and requires more resources for completion. This requires the management of resource requirements with resource availability. In some cases, the project has a resource overload, with more require-

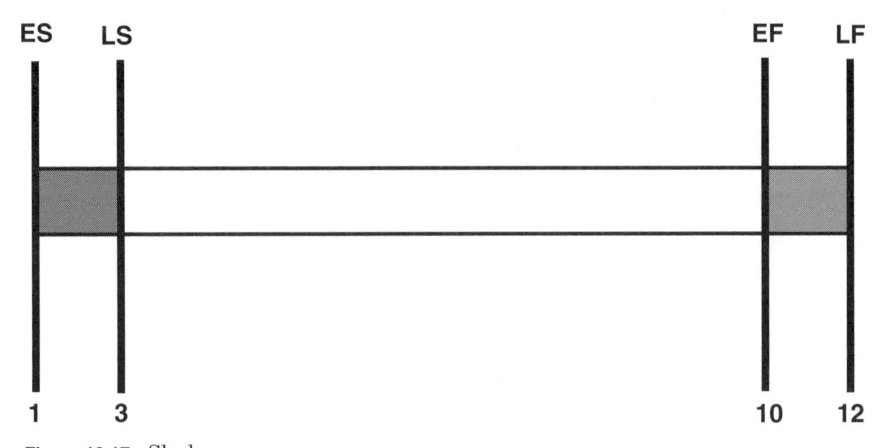

Figure 12.17 Slack.

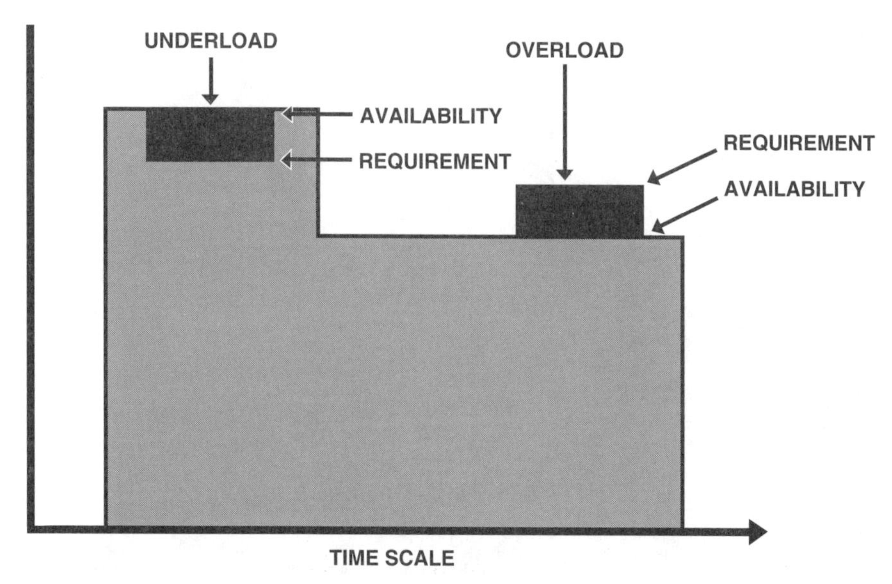

Figure 12.18 Resource underload and overload conditions.

ments than resources. At other times, the project has a resource underload, with fewer requirements than resources. Figure 12.18 displays the resource underload and overload conditions graphically.

In many cases, it is desirable to level resources to distribute them more evenly. This is usually done either time limited or resource limited. Time-limited leveling evens the resources within a fixed end date. Resource-limited leveling smooths the resources, allowing the end date to slip as far as necessary. Figure 12.19 shows resource leveling with limited resources.

Earned-value monitoring. The assessment of earned value is accomplished by variance analysis, a process of defining any schedule, technical performance, or cost deviation from the baseline plans of the project. Earned value is a concept that goes beyond classic budgeting and cost accounting; before a cost is recorded as incurred, this process ensures that the costed work is performed consistently with specifications and customer needs, that it is judged by the task leader and the customer as having met requirements.

The mechanics of earned-value monitoring through variance analysis are fairly straightforward and are supported by available project management software. Variance analysis measures against both budget and schedule requirements. At any monitoring point in the project, usually at the end of a distinct phase or stage, the customer-driven project manager asks each task or activity leader to provide data on the following:

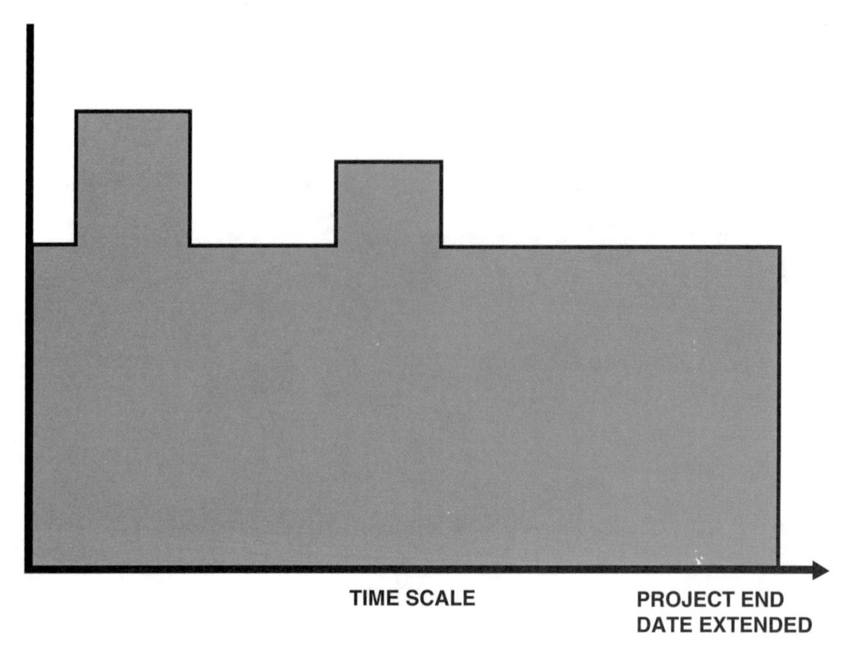

Figure 12.19 Resource leveling (resources limited).

- The actual cost of the work performed to that date
- An estimate of what percentage of the work performed to that date is on schedule
- The percentage of the work in that task which meets performance specifications and requirements
- What the cost should have been according to the original budget for the work, counting only costs of work that meets specifications

In this way, the team can calculate cost variance by comparing budgeted cost of work performed (BCWP) with actual cost of work performed (ACWP). *Schedule variance*—which is earned value—is the difference between BCWP and budgeted cost of work scheduled (BCWS). *Cost variance*—which is the traditional budget report—is the difference between BCWS and ACWP, a straight analysis of costs incurred against original budgeted costs. Earned-value and variance analyses offer an interesting opportunity to apply statistical process control SPC tools to the project management process itself. Run charts can be used to plot average values of schedule variance and cost variance to identify variance that falls outside the upper or lower control limits, suggesting special causes and possible corrective action.

Quality monitoring. This is qualitative monitoring, but it also may involve technical measurement and quality control. The assessment of quality is separable from earned-value analysis because in CDPM, quality cannot be traded off with cost and schedule. If a product or service does not meet the customer's expectations, it makes little difference in the long run that it was produced efficiently. Quality monitoring involves continuously testing the initial outputs of the team, especially the work breakdown structure, and the intermediate products of the process, such as prototype products and service models, against customer expectations. Therefore, it is critical to have placed the customer in the driver's seat through the customer-driven project team structure; the customer literally ensures quality performance continuously throughout the project, not at points at which he or she is "allowed" entry.

Project performance monitoring. At least three kinds of continuous performance monitoring are required in a customer-driven project: (1) the customer's evaluation of whether the project processes and products are achieving total customer satisfaction (e.g., quality monitoring to ensure that the deliverables are the right products and/or services), (2) the customer's evaluation of whether basic cost, schedule, and technical performance specifications are being met (e.g., earned-value analysis to ensure that the project has been managed efficiently), and (3) the project team facilitator's evaluation of team members' performance as team members and team members' evaluations of the project facilitator's effectiveness in building the capacity of the team as a whole.

The evaluation of project progress is a complex process because in many projects the nature of the work and of the deliverable changes during the course of the project because of expectations changing as the team learns and develops new concepts about the customer's needs and expectations. Therefore, it is rarely a question of matching performance against original plans and specifications. In CDPM, it is a matter on continuously monitoring the project's performance against current customer satisfaction metrics.

Presentations or reports. Frequently, the information from the project network schedules is used to communicate, present problems, and recommend corrective action. This is accomplished through presentations and/or reports. Presentations provide status or request action. Project network scheduling provides support for status reporting and recommendations.

Reports are used to communicate various aspects of the project to a wide variety of people. The customer-driven project lead team with appropriate input from other customer-driven teams needs to formulate external and internal reporting requirements. In addition to the regular reports, each customer-driven project team needs the capability to produce or receive reports with the information necessary for successful project performance.

Decision making. The information from the project network schedule is used for decision making. In CDPM, critical decision making should be done by the team having the ownership.

Simple Gantt project scheduling. For most simple CDPM projects, a simple Gantt project schedule is sufficient for project planning, scheduling, and management. For example, suppose a customer-driven quality-improvement team required the implementation of an improved process. The implementation of the improved process is the project. The Gantt chart to implement the process is shown in Fig. 12.20.

Risk Management

Risk management is an overall obligation of all customer-driven project teams. Customer-driven teams must assess the risk in each task of a project continually in terms of schedule, cost, technical aspects, supportability, and programmatics—the five risk facets. This is the task risk. In addition, relationship risk factors involving customers, suppliers, competitors, and internal organizational impacts should be

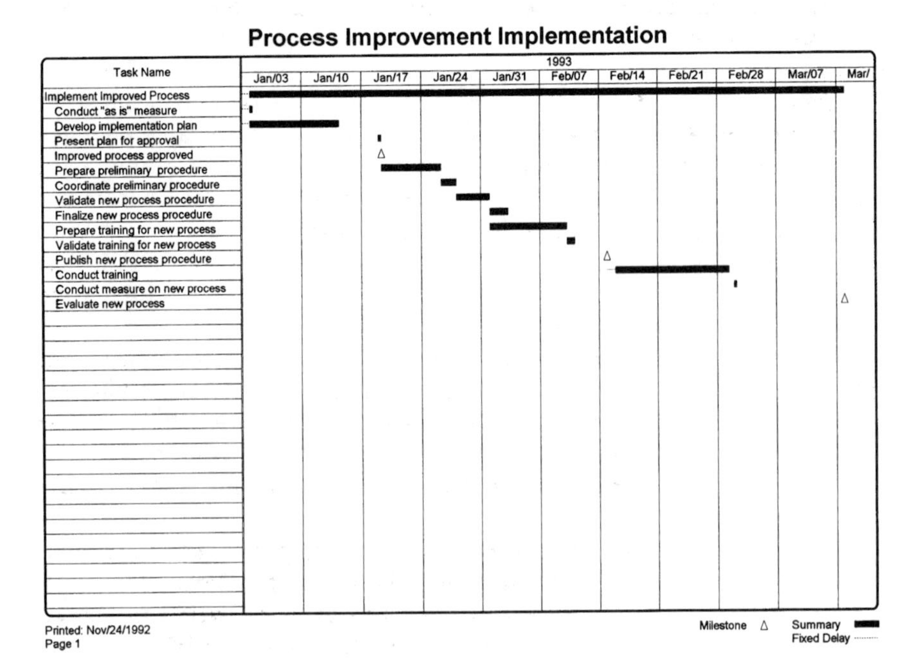

Figure 12.20 Sample Gantt chart for process improvement.

assessed. Again, the customer-driven lead team establishes the system for risk management, and the risk-management system influences the use of the other project management tools and techniques.

No project is without *risk,* which is the probability that a given process, task, subtask, work package, or level of effort cannot be accomplished as planned. Risk is not a question of time; it is a question of feasibility. It pays in the development of the WBS to assign risk factors to each element, separate from the assignments of schedules and milestones for the work. Risk factors can be assigned based on task facets and relationship factors. Later, the elements with high risk factors are given close attention by the customer-driven team, whether or not it is on the critical path itself.

Risk cannot be avoided—it is part of everything we do. In many cases, simply not taking risks leads to more serious risks. Therefore, the issue is not to take or avoid risk but to take calculated risks.

Calculated risk taking

Calculated risk taking is taking risks where possible consequences are well-considered and evaluated with the potential rewards being greater than the acceptable, affordable losses. Calculated risk taking is systematic, reasonable, informed risk taking.

Most definitions of risk highlight the exposure to loss; risk also involves the potential for reward. There are two possible outcomes of risk:

1. Real reward or loss which leaves the decision maker better or worse off than before the decision, that is, closer or further from the objective.
2. Opportunity reward or loss which is an outcome that is more or less favorable than it could have been, that is, maybe another approach would have yielded a better or worse result.

Why take risk? When the customer-driven team faces risk it can be frightening to the team. But, usually, when the risk is the greatest, the potential for reward is also the greatest. Many of us focus on the exposure to loss; it is no wonder we decide to take the path of least resistance, choosing risk avoidance to fulfill a perception of safety and security. Many people choose to act to avoid failure as opposed to seeking success. Some people live in the past and others in constant hope for the future. In the present they simply live by the motto "don't rock the boat." In the workplace, traditional management practices foster fear—leading to unwillingness to take risk. This attitude leads to people's idea of teamwork as agreeing to everything. After persisting for a long time, comfort-seeking actions become the habit.

When faced with a risky situation, people experience stress, and sometimes even physical symptoms, from risk-neutral or risk-avoidance behaviors.

Risk taking has many rewards. It offers the opportunity for organizational success and personal satisfaction. It creates breakthroughs, innovation, and invention. Risk-seeking behaviors lead to success.

Risk-taking environment. A risk-taking environment is critical to customer-driven team success. In a calculated risk-taking environment, each team member is willing to take the risks necessary to ensure the success of the organization. There are many barriers to creating and maintaining an environment conducive to calculated risk taking. Some barriers with suggestions for courses of action follow:

- *Short-term versus long-term view.* The short-term perspective seeks to achieve immediate needs. For example, this is the production level I need for today. The long-term perspective deals with ensuring that downstream the program will be successful. In this perspective, the question is "What can I do today to ensure the entire program is successful?" For example, positive actions might include introducing concurrent product development and producibility early into the design process. The forward-looking customer-driven team looks even further, to the impact on the growth of the organization in the future. This might involve redesigning the whole organization. All these perspectives must be understood. Calculated risk involves the determination of all the short- and long-term implications of a risk situation early enough to take appropriate action. *Possible alternative: Determine the implications for the short and long term for each risk situation, and take calculated risks.*
- *Ignoring the subjective or putting too much emphasis on the objective elements of risk.* From a traditional perspective, three basic types of risk are normally encountered on the job. These involve the so-called triple constraint: performance, cost, and schedule risk. In addition, many projects have risks associated with supportability and programmatic facets. Most risks in the military electronics business have consequences for one or more of these constraints. While some assumptions may be required, it is a straightforward process to develop a connection between all risks taken, the possible outcomes, and these risk facets. These types of risk can be measured objectively by using statistical methods. However, besides the task risk associated with the triple constraint and five risk facets, there is usually another type of risk involving relationships. This could be the values of stakeholders, personal preferences of customers, competitive climate, connection with suppliers, culture of

the organization, and so forth. These types of risks are more subjective, and are, therefore, sometimes ignored in calculated risk taking. *Possible alternative: Clarifying specific criteria for both objective and subjective elements of risk.*

- *Lack of information, control, and/or time.* All situations that involve risk have one or more of the above issues. There is no risk in a situation where you have all the information, complete control, and unlimited time. Therefore, the decision maker can improve chances of success by getting more of the right information, becoming more empowered, or gaining time. *Possible alternative: Determine the optimum amount of information, time, and control to achieve desired outcome. Use technology as appropriate.*
- *Fear of failure or success.* Fear is a major obstacle to risk taking. It comes from external and internal sources. Some external sources of fear include unrealistic schedules or budgets, intense competition, demanding customers, and personal problems. Internal sources of fear come from our internal need to perform. Also, the source of fear can be real or perceived. Although external and real sources of fear are easier to identify, internal and perceived sources are just as inhibiting. Fear results in risk avoidance, low performance, incorrect or incomplete information, perfectionism, procrastination, and dysfunctional decision making. *Possible alternative: Use a team-based systematic risk management process.*
- *Lack of interest in creating a risk-taking environment.* In any risk-taking situation, the organizational environment always directly impacts the tendency to take a risk. The organizational environment must allow failure and mistakes, if breakthroughs are expected within the organization. A risk-taking environment results from constant application of the TQM principles. A risk-taking environment starts with a foundation of trust, builds through a focus on a common purpose, increases as people are more empowered and as they develop additional capabilities, mounts as they use consensus decision making and take advantage of one another's differences, and finally is sustained through the application of a systematic calculated risk-taking process. *Possible alternative: Create and maintain an environment in which risk is welcome.*

Risk-taking principles. Risk taking is essential to making any organization more competitive, dynamic, and vital. Risk taking improves competitiveness through innovation and ingenuity. Risk taking requires a systematic process applied with discipline. When using the risk-taking process, the customer-driven team should consider the following key principles:

*R*equires proactive behaviors

*I*nvolves creative forward thinking looking for breakthroughs

*S*tarts with the recognition that risk is unavoidable

*K*eeps focused on mission

*T*akes calculated risk

*A*ccepts failure as a possibility

*K*eeps risk manageable

*I*nvolves all stakeholders

*N*urtures constructive relationships

*G*ains support

Risk-taking process. There is no "best"technique for risk taking. This section provides a calculated risk-taking "process." The customer-driven team must determine the appropriate content of the process for each particular situation. This content depends on the event and the outcome—the probability of an event occurring and the significance of the outcome. The risk-taking process is as follows:

*R*ecognize the risk

*I*nvestigate the risk issue

*S*eek actions to manage the risk

*K*eep track of progress toward achieving a plan

Steps to calculated risk taking. The calculated-risk-taking process uses a systematic, disciplined four-phase approach to making decisions in high-risk situations. The four-phase approach involves a ten-step process as outlined below.

Recognize the risk.

1. Identify the risk situation.
2. Review mission and objectives as they relate to the risk situation.

Investigate the risk issue.

3. Define the risk issue(s) evaluation criteria.
4. Analyze the risk issue(s)

Seek actions to manage risk.

5. Identify alternatives.
6. Evaluate alternatives.
7. Select a course of action.
8. Gain support from stakeholders.
9. Develop a plan of action.

Keep track of progress toward achieving the plan.

10. Track progress against plan and adjust as necessary.

I. Recognize the risk. Risk is present in some form and degree in most activities. It is critical to success to recognize and manage risk. The first phase of the calculated-risk-taking process—recognizing the risk—involves two steps.

1. *Identify the risk situation.* The identification of risk can come from many sources. Some potential sources of risk identification include output from technical, environmental, and relationship assessments:

Results of understanding the process

Analysis of performance measurements

Evaluation of plans

Views of people

Indication of a problem

Once a potential risk situation is identified, the next activity involves determining if you in fact have a risk situation. A simple decision tree will help you with this activity.

First, state the risk situation. Determine if the situation is certain or uncertain (risky). If certain, there is no risk. Stop the process. If there is any uncertainty, continue to next item.

Second, estimate the potential reward/loss ratio. Ask: "Are there possible significant rewards and losses in the risk situation?" "Significant" varies by the situation and must be determined by the decision maker(s). This determination can be highly objective using statistical methods, totally subjective, or combine objective and subjective methods. Normally, subjective criteria are used during this step. If there are no possible significant rewards or losses, stop the process. If there is possibility of "significant" reward or losses, go to step 2.

2. *Review mission/objectives as they relate to risk.* Once you recognize a potential risk situation, you need to revisit the organization's and team's mission and objectives to understand the really important elements in regard to the risk situation.

Again, there is an evaluation of the acceptability of the risk situation. Ask: "Is the risk situation acceptable or unacceptable in regard to achieving mission/objectives?" Here, too, acceptable and unacceptable will vary by the situation. The decision maker(s) determine(s) the criteria. Again, this can be highly objective using statistical methods, totally subjective, or combine objective and subjective methods. Normally, subjective criteria are used during this step. If the risk situation is acceptable, stop the process. If the risk situation is unacceptable, continue to step 3, in process II.

II. Investigate the risk issue. The second phase of the calculated-risk-taking process—investigate to find the risk issue—involves two steps.

3. *Define the risk issue's or issues' criteria.* Once you determine that there is a risk worth investigating, develop a consistent scheme for rating risk. Make it quantitative with qualitative backup. The risk issue must be described and documented sufficiently to provide some criteria for prioritization of risks for the analysis step. For instance, in a program there could be a risk rating for cost, schedule, performance, and/or some other measurable factor. Again, this can be highly objective using statistical methods, totally subjective, or combine objective and subjective methods. Normally, subjective criteria are used during this step. Heavy mathematical treatment is not necessary at this step. For example, one of the following methods could be used:

A rating scheme provides a framework for eliminating some of the ambiguity associated with many people looking at a risk situation. The rating system should be as simple as possible—for example, high, medium, and low.

Expert interviews provide information from technical experts to set up a gauge for analyzing the risk issues.

Analogy comparisons provide criteria from similar past programs.

4. *Analyze the risk issue.* Risk analysis involves an examination of the change in consequences caused by changes in risk input variables. Sensitivity and what-if analysis are examples of activities that should take place during analysis of the risk situation. During this step use an analysis tool designed to meet your specific objectives.

The results of analysis could be examined depending on your objectives in terms of the following:

Cost/schedule/performance/customer satisfaction

System/subsystem

Funding profiles

Criticality

Consistency with analogous systems

What-if scenarios

Task risk facets

Relationship factors

During this step, the "as is" reward/loss ratio is quantified. In addition, goals are set to specify outcomes. The "as is" reward/loss ratio is evaluated in regard to meeting goals.

III. Seek actions to manage risk. The third phase of the calculated risk-taking process—seek actions to manage risk—involves the next five steps.

5. *Identify alternatives.* For each risk-taking situation, the team must decide a risk strategy to drive actions. Calculated-risk-taking strategy includes several options:

Avoid risk

Assume risk

Control risk

Transfer risk

Share risk

Study risk

Alternative actions are generated based on the calculated-risk-taking strategy.

During this step first look at the process causing the risk situation to determine if there is something obvious that can be done with the process to make the risk acceptable. Ask: "Can the process be changed to make the risk situation acceptable?" "Are there any non-value-added tasks?" "Can the process be simplified?" "Can activities be combined?" "Is there another method available?" "Can people be trained?" "Would written procedures help?" "Will the alternative action produce your objectives?"

In addition to the obvious alternatives, the team should explore "breakthrough" alternatives. For example, challenge the old "rules." Ask: "Does the process need to be reengineered?" "Can information technology provide a better process?" "Are as few people as possible involved in the process?" "Does the process eliminate as many non-value-added activities as possible?"

6. *Evaluate alternatives.* During this step evaluate alternatives based on the selection criteria developed during step 3 in process II.

Ask: "Which of the alternatives will get you closest to your objectives?"

7. *Select a course of action* In this step, the team decides on the "best" course of action that does the following:

Is consistent with mission/objectives.

Meets risk criteria and strategy.

Uses all relevant, available information.

It is important to use consensus decision making to make this decision.

8. *Gain support from stakeholders.* In order to implement the selected course of action, it is critical to get the support of key stakeholders. The stakeholders can help ensure all consequences of a decision are weighed properly. This may involve preparing and conducting a presentation.

9. *Develop a plan of action.* One essential step in ensuring that the risk is managed properly is the development of a plan of action.

IV. Keep track of progress. The last step is the ongoing process of continuous evaluation of progress. The step may lead to a return to key action 1 and the looping through the calculated risk process again and again.

10. *Track progress against plan and adjust.* During this step, the progress and success are measured against the plan to ensure the risk is being managed properly.

Example of risk management: DoD 4245.7M

DoD 4245.7M, "Transition from Development to Production," provides an example of risk management within the Department of Defense. This document provides templates for reducing risk in the Department of Defense for the development and production of systems. The DoD manual uses templates directed at the identification and establishment of critical engineering processes and their control methods. The critical engineering processes are shown in Fig. 12.21. For each of the critical engineering processes, a critical path template is provided that addresses the following areas for each critical engineering process:

- Area of risk
- Outline for reducing risk
- Timeline

For example, the total quality management (TQM) template states the following:

Area of risk. TQM is an organized process of continuous improvement by private defense contractors and DoD activities aimed at developing, producing, and deploying superior materials. The primary threat to reaching and sustaining this superiority is failure to manage with a purpose of constantly increasing intrinsic quality, economic value, and military worth of defense systems and equipments. The armed forces and defense industrial entities may not attain a lasting competitive military posture and long-term competitive business stature without a total commitment to quality at the highest levels. TQM is applicable to all functions concerned with the acquisition of defense materials, supplies, facilities, and services. Being satisfied with suboptimal short-term goals and objectives has adverse affects on cost, schedule, and force effectiveness. A short-term approach leads to deterioration in the efficacy of specific products, the firms that produce them, and the industrial base overall. Major risk also is entailed with the inability to grasp and respond to the overriding importance attached to quality by the "customer" or user activities.

Outline for reducing risk

- The organization has a "corporate level" policy statement attaching highest priority to the principles of TQM. This policy statement defines TQM in terms relevant to the individual enterprise or activity and its products or outputs.
- The corporate policy statement is supported by a TQM implementation plan that sets enduring and long-range objectives, lists, criteria for applying TQM to new and ongoing programs, provides direction and guidance, and assigns responsibilities. Every employee at each level plays a functional role in implementing the plan.
- All personnel are given training in TQM principles, practices, tools, and techniques. Importance is placed on self-initiated TQM effort.
- TQM effort begun in the conceptual phase of the acquisition cycle is vitally concerned with establishing a rapport between the producer and the user or customer and a recognition of the latter's stated performance requirements, mission profiles, system characteristics, and environmental factors. These statements are translated into measurable design, manufacturing, and support parameters that are verified during demonstration and validation. Early TQM activity is outlined in the design reference mission profile template and design requirements template. The trade studies template is used to identify potential characteristics that would accelerate design maturity while making the design more compatible with and less sensitive to variations in manufacturing and operational conditions.
- Design-phase TQM activity is described in the design process template. Key features enumerated include design integration of life-cycle factors concerned with production, operation, and support; availability of needed manufacturing technology; proof of manufacturing process; formation of design and design review teams with various functional area representation; and use of producibility engineering and planning to arrive at and transition a producible design to the shop floor without degradation in quality and perfor-

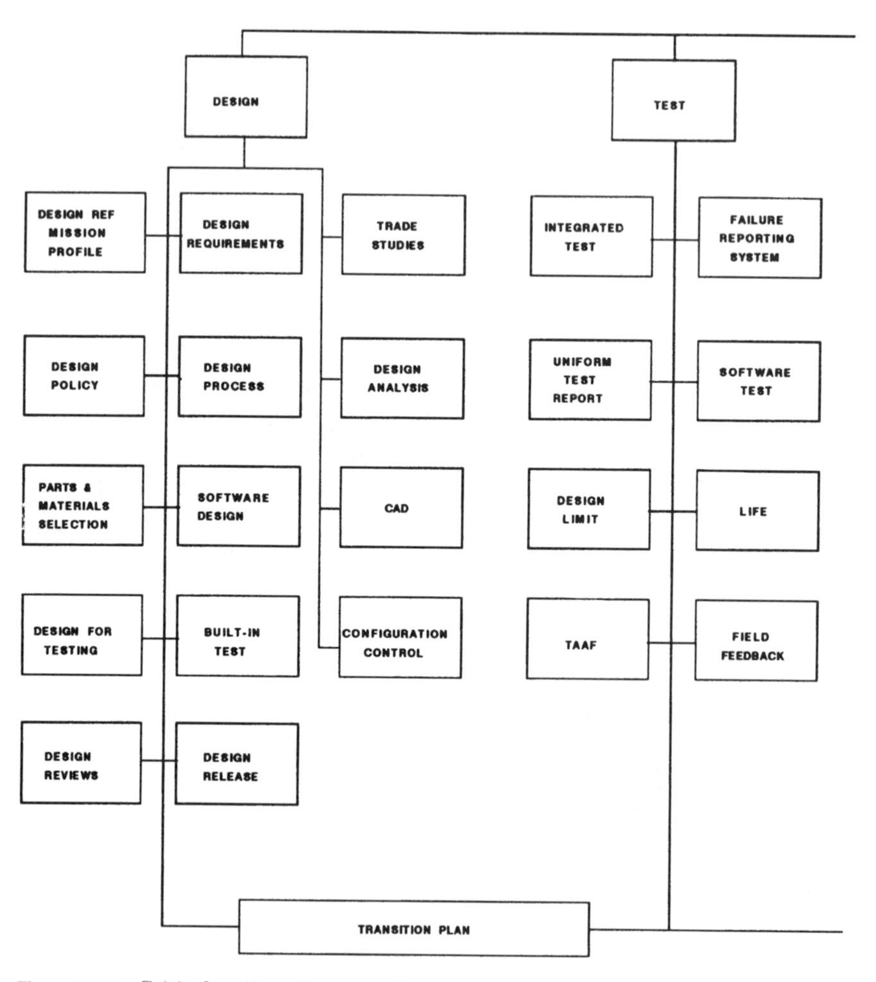

Figure 12.21 Critical engineering processes.

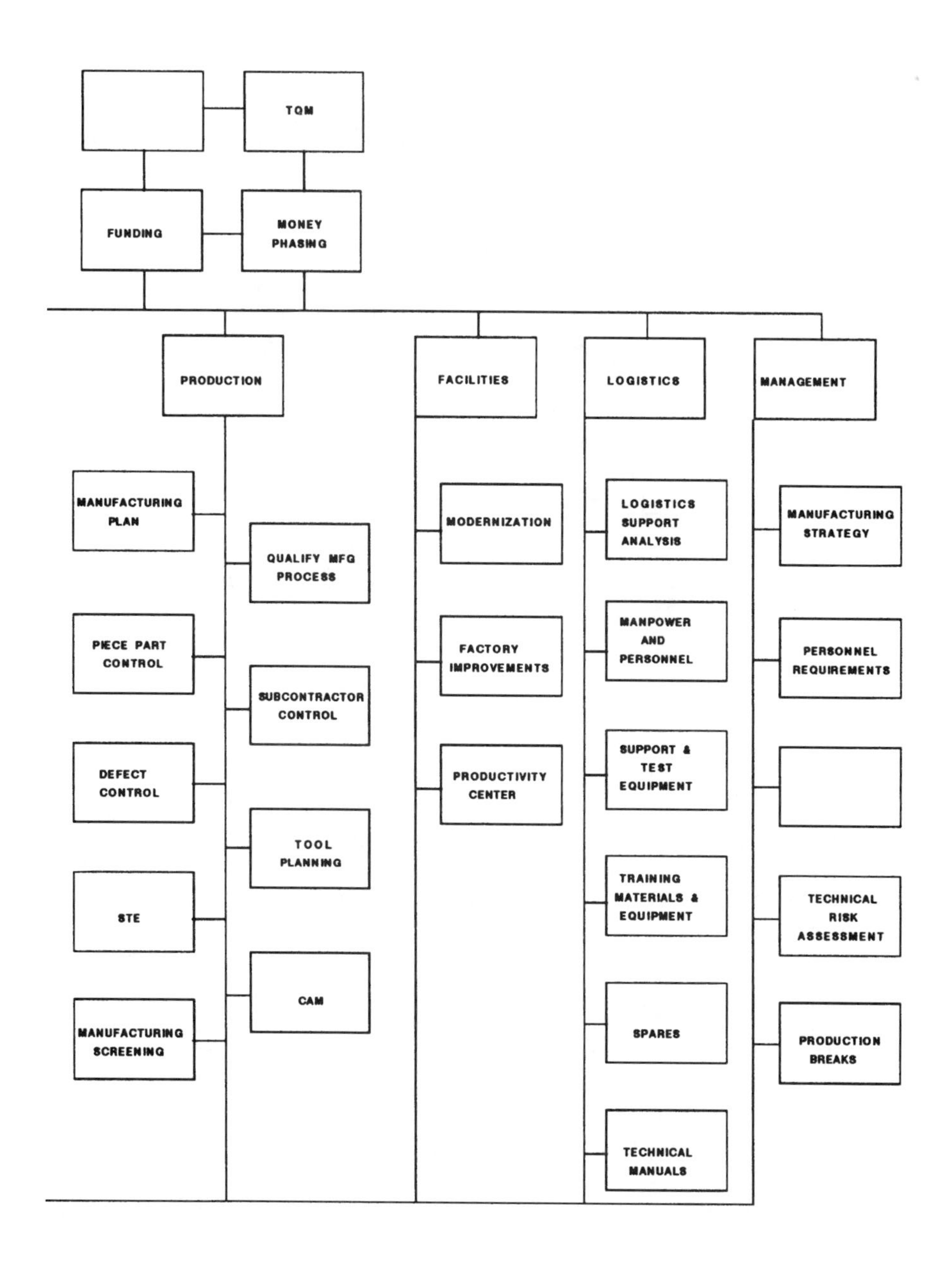
TQM
FUNDING
MONEY PHASING
PRODUCTION
FACILITIES
LOGISTICS
MANAGEMENT
MANUFACTURING PLAN
QUALIFY MFG PROCESS
PIECE PART CONTROL
SUBCONTRACTOR CONTROL
DEFECT CONTROL
TOOL PLANNING
STE
CAM
MANUFACTURING SCREENING
MODERNIZATION
FACTORY IMPROVEMENTS
PRODUCTIVITY CENTER
LOGISTICS SUPPORT ANALYSIS
MANPOWER AND PERSONNEL
SUPPORT & TEST EQUIPMENT
TRAINING MATERIALS & EQUIPMENT
SPARES
TECHNICAL MANUALS
MANUFACTURING STRATEGY
PERSONNEL REQUIREMENTS
TECHNICAL RISK ASSESSMENT
PRODUCTION BREAKS

mance. The design analysis template and design reviews template provide guidance in identifying and reducing the risk entailed in controlling critical design characteristics. Both hardware and software are emphasized (reference the software design template and software test template). A high-quality design includes features to enhance conducting necessary test and inspection functions (reference the design for testing template).

- An integrated test plan of contractor development, qualification, and production acceptance testing and a test and evaluation master plan (TEMP) covering government-related testing are essential to TQM. The plans detail sufficient testing to prove conclusively the design, its operational suitability, and its potential for required growth and future utility. Test planning also makes efficient use of test articles, test facilities, and other resources. Failure reporting, field feedback, and problem disposition are vital mechanisms to obtaining a quality product.
- Manufacturing planning bears the same relationship to production success as test planning bears to a successful test program (reference the manufacturing plan template). The overall acquisition strategy includes a manufacturing strategy and a transition plan covering all production-related activities. Equal care and emphasis are placed on proof of manufacture as well as on proving the design itself. The quality manufacturing template highlights production planning, tooling, manufacturing methods, facilities, equipment, and personnel. Extreme importance is attached to subcontractor and vendor selection and qualification, including flow down in the use of TQM principles. Special test equipment, computer-aided manufacturing, and other advanced equipments and statistical-based methods are used to assess quality and control the manufacturing process.

Timeline. TQM is used throughout the product life cycle.

TQM-oriented defense contractors and government activities concentrate on designing and building quality into their products at the outset. Successful activities are not content with the status quo or acceptable level of quality approach. Such activities respond to problems affecting product quality by changing the design and/or the process, not by increasing inspection levels. Reduction in variability of the detail design and manufacturing process is a central concept of TQM and is beneficial to lower cost as well as higher quality. Defect prevention is viewed as key to defect control. Astute TQM activities are constantly on the alert to identify and exploit new and proven managerial, engineering, and manufacturing disciplines and associated techniques.

Area of risk, DoD 4245.7-M. This part of DoD 4245.7-M stresses the following total quality management principles:

- Continuous improvement
- Total commitment to quality
- Involvement of many functions
- Long-term improvement effort
- Customer focus

Outline for reducing risk, DoD 4245.7-M. This part of DoD 4245.7-M incorporates the following TQM principles:

- Includes a policy statement (vision/mission)
- Pursues a TQM environment
- Stresses a TQM implementation plan
- Fosters "ownership"
- Advocates training
- Includes quality as an element of design
- Encourages measurements
- Includes everything and everyone
- Nurtures supplier and customer relationships
- Encourages cooperation and teamwork

Timeline, DoD 4245.7-M. TQM is used throughout the product life cycle. This part of DoD 4245.7-M highlights the need for

- Quality as an element of design
- Continuous improvement geared toward reduction in variability
- Focus on prevention of defects
- Use of fundamental management techniques, existing improvement efforts, and technical tools under a disciplined approach allowing creativity and innovation

Project Management Information System

A project management information system (PMIS) is required for some projects. The PMIS is the basis for project decisions and therefore represents an important part of project CDPM planning and control. The PMIS should be designed and planned in detail. At this time, the customer-driven project lead team decides what information must be available to provide total customer satisfaction, when it must be available, and what form it must take.

While a PMIS is often computerized, it need not be; therefore, this discussion is generic to CDPM. The PMIS requires the dedication of adequate project planning resources and time to ensure that four basic purposes are met, regardless of project size and complexity:

1. To provide useful data and interpretive information on the project as a whole to key managers and executives in the project and customer organizations. This requires a front-end analysis of communication needs and data requirements.
2. To provide a stable central reference to CDPM teams on the essential nature of the project, its objectives, statement of work and definitions, project team responsibilities, scheduling, and budgeting and to identify the criteria for monitoring and evaluating cost, schedule, and performance progress. The PMIS structures communication on the project.
3. To support critical decisions that must be made at the time they need to be made by CDPM team.
4. To document the progress of the project so that an information trail on the project is available for future reference.

The development and use of a PMIS as described in this section involve two basic activities: (1) the identification of project information needs, based on project functions and decisions, and (2) the development of PMIS "modules," or groupings of data (databases in computer terms), that serve as information sources for reporting on key functions during the project.

Project information needs

Perhaps the most important step in the detailed planning phase of a project is to identify the information needs of the project early in the detailed planning so that there is consistency and continuity in reporting and decision making. This first step requires the identification of project management functions and the basic activities involved so that a foundation can be laid for development of the PMIS itself. To accomplish this, the following characteristics and planning tools of the project must be specified and completed: project objective(s), work breakdown structure, work assignments in the project team, project schedule, Gantt chart and critical-path plan, budgeted cost of work scheduled, project monitoring and evaluation criteria, and reporting formats. These are the building blocks of the PMIS. This package of information is then integrated and described in a narrative text, and this text serves as the basic project information source during the life of the project. Many computer programs for project management facilitate identification of these information needs in various forms.

Project information modules

The second step in constructing the PMIS is to develop modules, or groupings, of information to meet specific reporting and decision needs. Database files, whether automated or manual, are divided into the following three modules:

1. *Planning and Scheduling Module.* This module provides a systematic method for depicting the qualitative goals, time-sensitive relationships, and interdependencies within the project. Formats and frequencies for reporting will include work breakdown structure, Gantt charts, and project networks. Provision is made for variance analysis.
2. *Project Accounting Module.* This module provides for cost-control monitoring, and it is structured from the work breakdown structure.
3. *Narrative Reporting Module.* This module provides for regular narrative reporting at every level of the project. It concentrates on interpreting information gathered on the project in terms of its meaning for the project as a whole.

This modular approach to project information is an essential core activity of the project. Many modules in addition to those given above may be developed to meet special needs such as labor resources, inventory control, document storage, safety, and team personnel performance evaluations. Finally, it is important to understand that the PMIS is developed through a dynamic process of interaction and communication within the project organization as well as with the customer. The PMIS serves, in effect, as the basis for establishing a common purpose and language for the project, as well as providing a common forum for team interplay.

Main Points

Project management tools and techniques assist customer-driven project teams in planning, scheduling, monitoring, controlling, and evaluating a project.

Project management techniques and tools common to all customer-driven project teams include the contract, the work breakdown structure, the task list, the project schedule, risk management, and the project management information system.

The contract defines the deliverable specification, statement of work, and data requirements.

Customer expectations can be determined from phase 1 quality function deployment. In addition, phase 1 QFD provides the basis for the work breakdown structure.

The deliverable specification provides the specific technical and performance information used to develop detailed, lower-level project objectives and the standards for evaluating the performance of the developed system.

The statement of work (SOW) summarizes the work required to complete the project.

The data requirements document is a list of all the data items needed for a successful project deliverable.

The work breakdown structure (WBS) defines the organization and coding of the deliverable components.

As a planning tool, the WBS allows the customer-driven project lead team to organize the project work into levels of activity.

There are seven distinct purposes of the WBS:

1. Defines project team responsibility
2. Outlines project's organizational structure
3. Supports coordination of objectives
4. Facilitates project control
5. Starts project scheduling
6. Structures cost estimating and budgeting
7. Organizes the assessment of risk

The task list includes development of the tasks involved in the project.

A task list is important simply because it provides a basic documentation of

- Task title and identification code
- Task order, predecessor(s) and successor(s)
- Task duration
- Task resources

The project schedule is the plan for carrying out the tasks in sequence on the basis of their interdependency.

The project schedule involves scheduling, monitoring, and managing the project using network and system diagraming techniques.

Communication, cooperation, and correct information are keys to project scheduling.

In project scheduling, the customer-driven project teams can do any of the following actions:

- Revise the activity duration times
- Revise network relationships
- Revise cost and/or resource objectives
- Revise technical objectives

The critical path represents the longest route between the first event and the last event.

Float or slack refers to the difference between the duration of an activity and the time available during the entire project to complete it.

Activities off the critical path have float; those on the critical path do not.

Network scheduling is a project management technique for planning, scheduling, and managing time and estimating, budgeting, and managing resources.

The types of network scheduling techniques include the critical-path method (CPM), the program evaluation and review technique (PERT), and the precedence diagraming method (PDM).

Every project network has a start time, end time, activity or activities, links, and an end result.

Activities can be performed in sequence or parallel. They can be independent of or dependent on another activity. More than one activity can start or end at the same time.

Network activities have relationships to other activities. Activity relationships include predecessors and successors.

Activity constraints include

- Finish to start
- Finish to finish
- Start to start
- Start to finish

Time constraints can be an actual time, as soon as possible (ASAP), or as late as possible (ALAP).

Activities can be linked for reporting purposes in a hammock activity.

Scheduling involves turning the network plan into a specific timetable of activities with their relationships to meet the project's objectives.

Scheduling does the following:

- *S*tarts the project
- *C*ommunicates project objectives
- *H*as a disciplined approach
- *E*stablishes "ownership"
- *D*etermines resources
- *U*ses graphic representation
- *L*imits potential solutions
- *E*nsures visibility of the project's progress

The Gantt chart displays the project schedule in bar-chart form.

The PERT chart is usually the graphic display of the project schedule with interdependencies.

Critical-path analysis focuses on managing the critical-path times by attending to the time durations and relationships of activities.

Total float is the maximum amount of time a path can slip before it affects the project's end date.

Positive total float means the project will finish early. Zero total float means the project will complete on time. Negative total float means the project will be late.

Quality monitoring is the qualitative assessment of technical performance.

Earned-value monitoring involves variance analysis of cost, schedule, and technical performance.

Reports are used to communicate various aspects of the project to a wide variety of people.

Risk management is the continual assessment of the risk in each task of the project not only in terms of time and cost but also in terms of the technical feasibility.

No project is without risk, which is the probability that a given process, task, subtask, work package, or level of effort cannot be accomplished as planned.

The project management information system (PMIS) provides the project information support for reporting and decision making.

PMIS has the following basic purposes:

- Provides useful data and interpretive information on the project
- Provides a stable central reference source
- Supports critical decision making
- Documents the project's progress

Chapter

13

CDPM System Development/Improvement Tools and Techniques

Focus: This chapter provides an awareness of some of the major advanced CDPM tools and techniques for system development / improvement. The actual "how to" of these tools and techniques is beyond the scope of this book because of the complexity of most of these tools. It is recommended that you consult detailed references before deciding to use any of the tools and techniques in this chapter.

Introduction

Frequently a system or process must be developed or completely redesigned to make an improvement. System improvement focuses on the development or redesign of systems. The system can be as complicated as an F-16 jet fighter or a car or as simple as a flight control surface or a car door. The system improvement tools and techniques can be used for any aspect of the process—a system, subsystem, or part. In fact, some of the tools, like statistical process control (SPC) and quality function deployment (QFD), have been used successfully for the continuous improvement of entire organizational systems.

Since the performance of a product is critical to customer satisfaction, this chapter focuses on the system improvement of a product. The product is any output to a customer, including a system, subsystem, or part. A major impact on product performance is the product design, process design, and production processes.

The tools and techniques described in this chapter have specific application in the product design, process design, and production processes. However, these tools and techniques can be used to improve any system in any organization. They are applicable to improving whole systems, subsystems, or parts.

The system development/improvement tools and techniques are used primarily in phase 5, the "take action" phase of the CDPM improvement methodology, to produce or improve a system. These system development/improvement methodologies focus on a product.

This chapter contains the following tools and techniques: concurrent engineering; quality function deployment; robust design; statistical process control; cost of poor quality; and miscellaneous other methodologies.

System development/improvement

As stated earlier, system development/improvement focuses on improving the actual performance of a product through product design, process design, and planning of the production processes. As shown in Fig. 13.1, system improvement starts with the customer. Next, the product and processes are designed. The "voice of the customer" carries through the product design and process design to actual production of the product. Within the product design, process design, and production processes, specific tools are useful for ensuring customer satisfaction. The specific tools are concurrent engineering

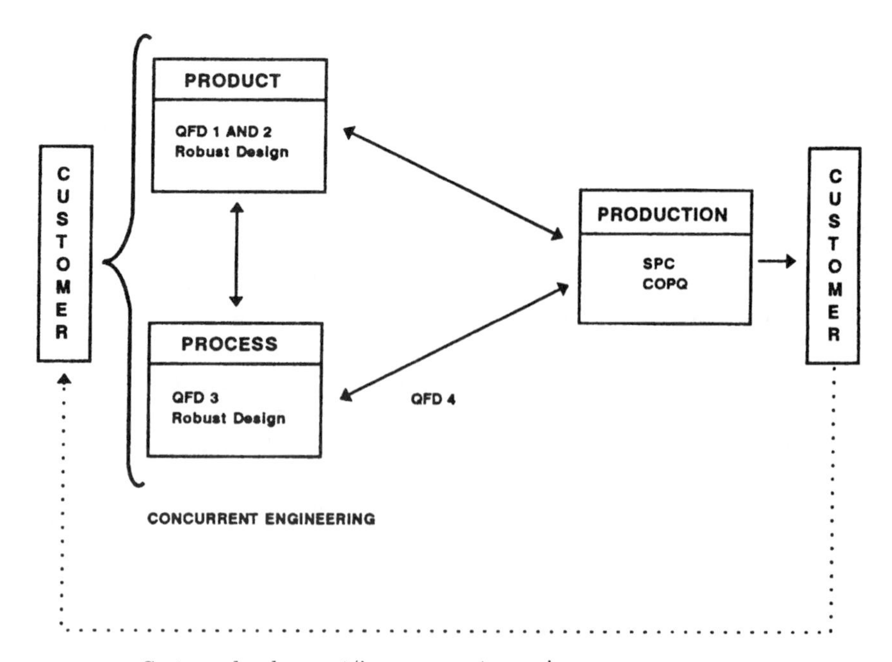

Figure 13.1 Systems development/improvement overview.

(CE), robust design (RD), quality functional deployment (QFD), statistical process control (SPC), and cost of poor quality (COPQ).

Concurrent engineering is useful during the product and process planning and design phases for reducing the time and cost of product development. Quality functional deployment is beneficial for carrying the "voice of the customer" throughout the entire process. Robust design focuses on designing in quality by eliminating loss. Statistical process control is a technique for measuring process behavior during production. Cost of poor quality emphasizes eliminating waste in all the processes.

Concurrent Engineering

Concurrent engineering (CE) is a philosophy and set of guiding principles whereby product design and process design are developed concurrently, i.e., with some product design and process development overlapping. This includes production and support planning. Figure 13.2 shows the difference between sequential engineering and concurrent

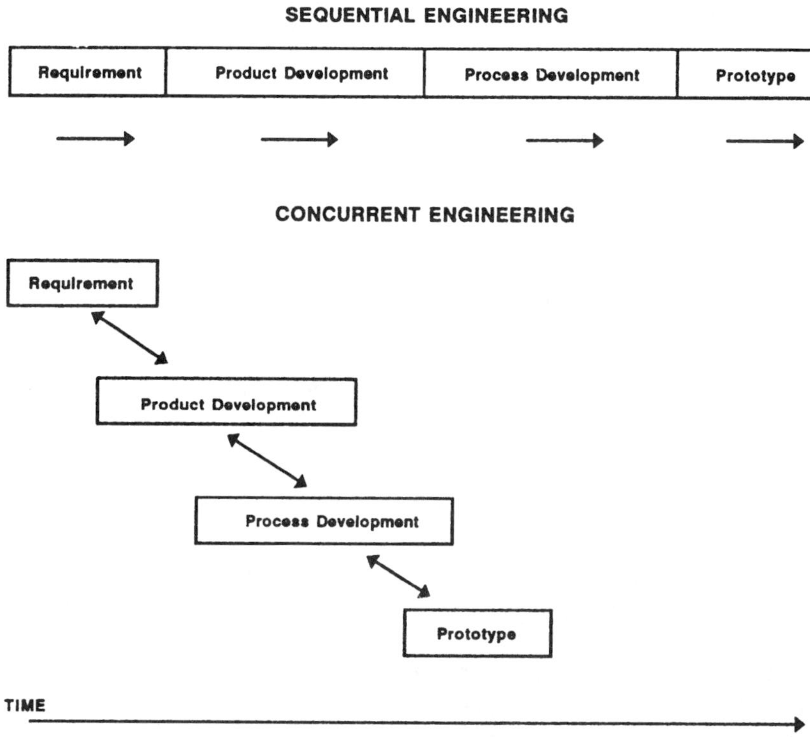

Figure 13.2 Comparison of sequential and concurrent engineering.

engineering. With sequential engineering, the engineering phases are accomplished one after the other. Concurrent engineering overlaps the engineering phases.

Concurrent engineering is a subsystem of customer-driven project management (CDPM) focusing on system and parametric design (refer to robust design phases in this chapter for more detail). Like CDPM and total quality management (TQM), CE requires a management and cultural environment, teams, and an improvement system focusing on customer satisfaction.

The CE philosophy emphasizes a customer focus. It advocates an organizationwide, systematic approach using a disciplined methodology. It stresses never-ending improvements in product, processes, production, and support, and it involves the concurrent, simultaneous, or overlapping accomplishment of the phases of the project. For instance, the concept and design phases are accomplished concurrently. The design and development phases are performed simultaneously. The development and production phases are done with some overlapping activities. In most cases of CE, all the phases contain some overlapping activities. CE requires upper management's active leadership and support to be successful. It accents robust design that decreases loss, and it aims at reducing cost and time while improving quality and productivity. CE uses the latest engineering planning initiatives, including automation, and it forges a new reliance on multifunctional teams using such tools and techniques as quality function deployment, design of experiments, the Taguchi approach, statistical process control, and so on.

A more formal definition from the Institute for Defense Analysis Report R-338 states,

> Concurrent Engineering is a systematic approach to the integrated, concurrent design of products and their related processes, including manufacture and support. This approach is intended to cause the developers, from the outset, to consider all elements of the product life cycle from conception, through disposal, including quality, cost, schedule, and user requirements.

Concurrent engineering steps

1. *Establish a multifunctional team.* Ensure representation from all required disciplines. The team should include representatives from such functions as systems/design engineering, reliability and maintainability engineering, test engineering, manufacturing engineering, production engineering, purchasing, manufacturing test and assembly, logistics engineering, supportability engineering, marketing, finance, and accounting.

2. *Use a systematic, disciplined approach.* Select a specific approach using appropriate tools and techniques.
3. *Determine customer requirements.* Be sure to communicate with customers.
4. *Develop product design, process design, and the planning of production and support processes together.*

Quality Function Deployment

Quality function deployment (QFD) is a disciplined approach for transforming customer requirements, the "voice of the customer," into product development requirements. QFD is a tool for making plans visible and then determining the impact of those plans. QFD involves all activities of everyone at all stages from development through production with a customer focus. QFD phase 1 is outlined in Chap. 8 of this book.

Figure 13.3 shows the four phases of QFD. These phases are product planning, parts deployment, process planning, and production planning. The output from each phase is the input for the next phase. During phase 1, customer requirements are transformed into design requirements. In phase 2, design requirements are converted into a system (part) or concept design. Phase 3 examines candidate processes and selects one. Phase 4 looks at making capable production processes.

Robust Design

Robust design (RD) means designing a product that has minimal quality losses. There are several methodologies associated with robust design. The major ones are traditional design of experiments (DOE) and the Taguchi approach. Traditional design of experiments is an experimental tool used to establish both parametric relationships and a product/process model in the early (applied research) stages of the design process. However, traditional design of experiments can be very costly, particularly when it is desired to examine many parameters and their interactional effects. Traditional DOE examines various causes of performance for their contribution to variation, with a focus on arriving at the most influential causes of variation. Traditional design of experiments may be a useful tool in the preliminary design stage for modeling, parameter determination, research, and establishing a general understanding of product phenomena.

A major approach to RD is the Taguchi approach, which focuses on quality optimization. *Quality optimization* is based on Dr. Taguchi's definition of quality. In his book, *Introduction to Quality Engineering,* Taguchi states, "Quality is the [measure of degree of] loss a product

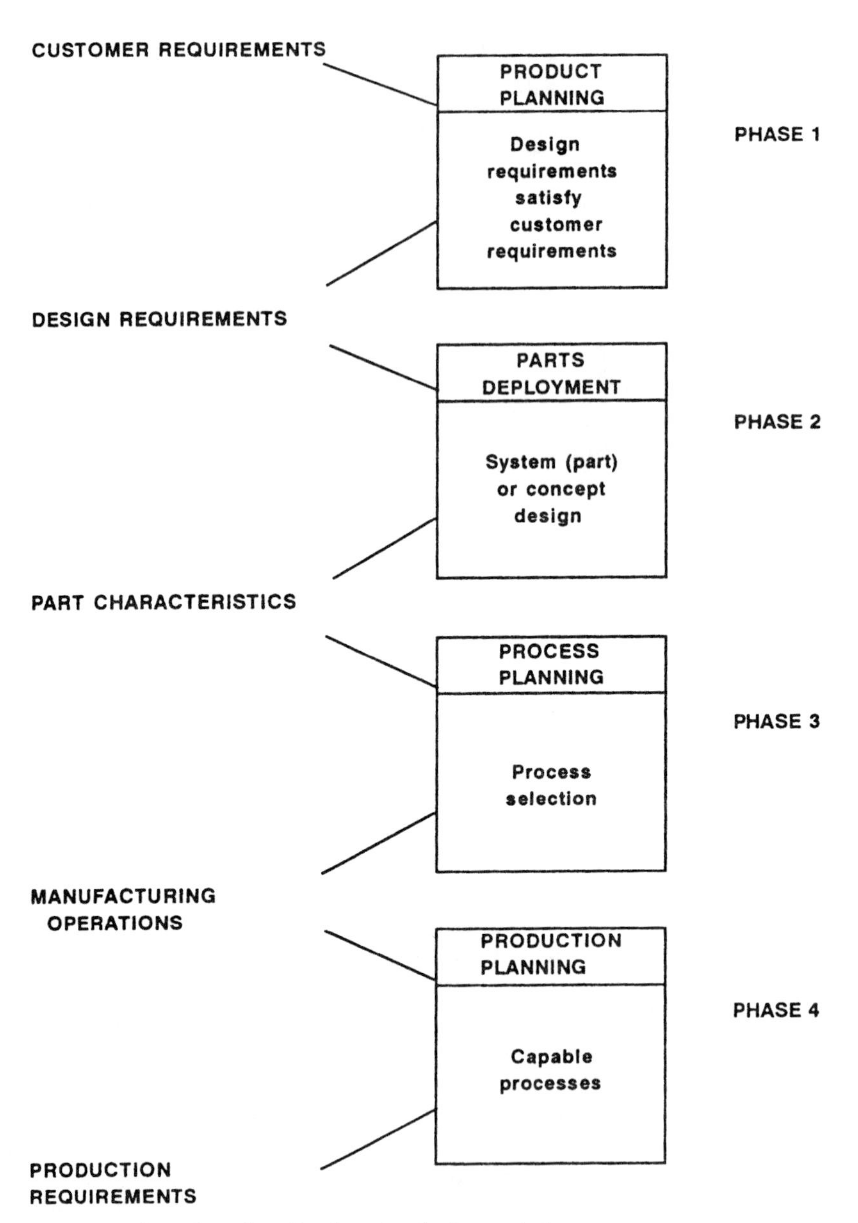

Figure 13.3 Overview of quality function deployment phases.

causes after being shipped, other than any losses caused by its intrinsic functions." Simply put, any failure to satisfy the customer is a loss. Loss is determined by variation of performance from optimum target values. Loss, therefore, in the form of variability from best target values, is the enemy of quality. The goal is to minimize variation by designing a system (product, process, or part) that has the best combination of factors, i.e., a system centering on the optimal target values with minimal variability. By focusing on the bull's-eye, the product, process, or part is insensitive to those normally uncontrollable "noise" factors which contribute to poor product performance and business failures. The Taguchi approach is not simply "just another form of design-of-experiments"; it is a major part of the successful TQM philosophy.

Loss function

The loss function is a key element of the Taguchi approach. The loss function examines the costs associated with any variation from the target value of a quality characteristic. As shown in Fig. 13.4, any variation from the target is a loss. At the target value, there is little or no less contribution to cost. The further away from the target, the higher are the costs. Costs get higher as values of the quality characteristic move from "best" to "better" to "poor" levels.

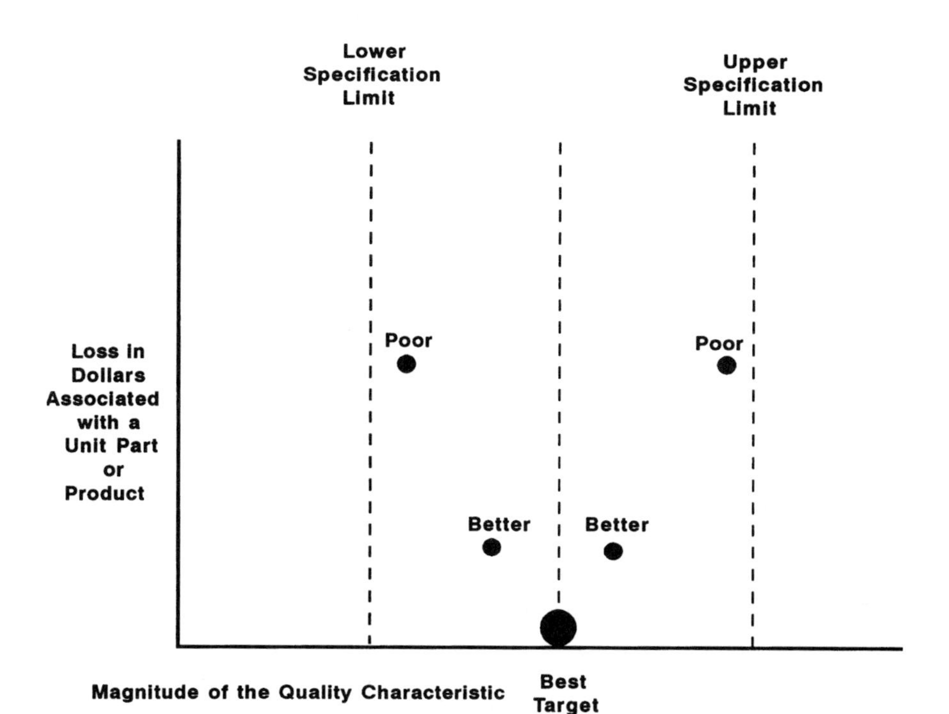

Figure 13.4 Magnitude of the quality characteristic.

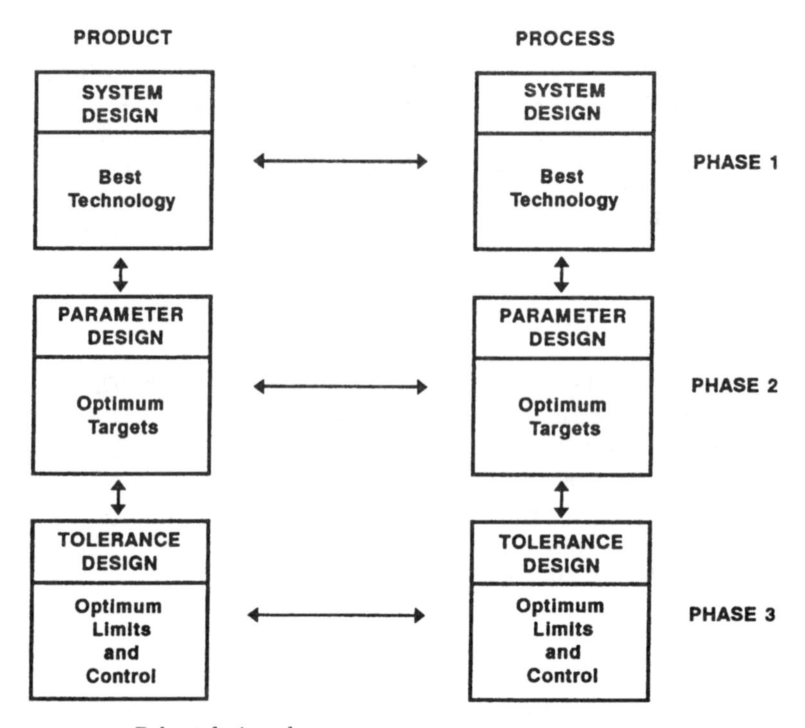

Figure 13.5 Robust design phases.

Robust design phases

In the Taguchi approach, the design of a product or a process is depicted as in Fig. 13.5. Product or process designs have three phases:

Systems (part) or concept design. This stage arrives at the design architecture (size, shape, materials, number of parts, etc.) by looking at the best available technology.

Parameter (or robust) design. This stage focuses on making the product performance (or process output) insensitive to variation by moving toward the best target values of quality characteristics.

Tolerance design. This stage focuses on setting tight tolerances to reduce variation in performance. Because it is the phase most responsible for adding costs, it is essential to reduce the need for setting tight tolerances by successfully producing robust products and processes in the parameter design stage.

Statistical Process Control

Statistical process control (SPC) is a statistical tool for monitoring and controlling a process. SPC monitors the variation in a process with the aim to produce the product at its best target values.

Figure 13.6 shows the major elements of SPC. These elements are a process chart consisting of data plots, upper control limit (UCL), lower control limit (LCL), and the mean for the process. Figure 13.6 also shows the variation in a process. The variation is the result of both common and special/assignable causes. Common causes produce normal variation in an established process. Special/assignable causes are abnormal causes of variation in the process.

Statistical process control steps

1. *Measure the process.* Ensure that data collection is thorough, complete, and accurate.
2. *Bring the process under statistical control.* Eliminate special /assignable causes.
3. *Monitor the process.* Keep the process under statistical control.
4. *Improve the process toward best target value.*

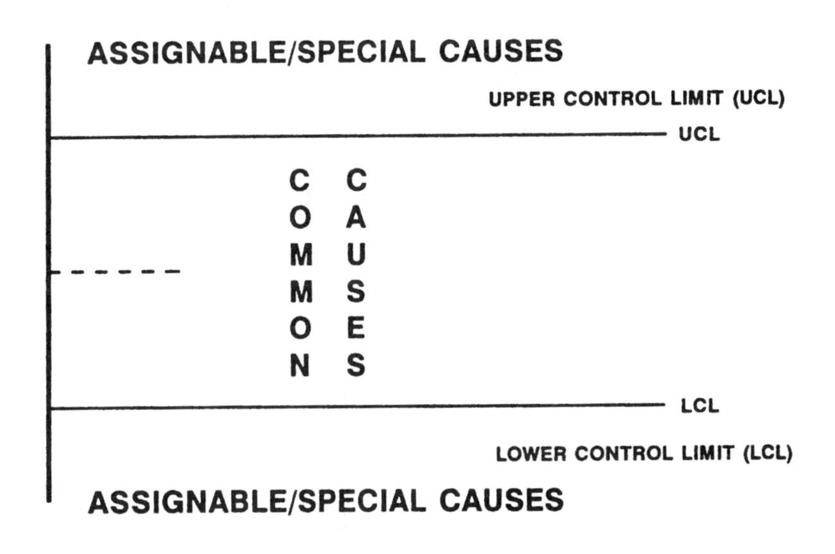

Figure 13.6 Elements of statistical process control.

Cost of Poor Quality

Cost of quality is a system providing managers with cost details often hidden from them. Cost of quality includes both the cost of conformance and the cost of nonconformance to quality requirements. Costs of conformance consists of all the costs associated with maintaining acceptable quality. The cost of nonconformance, or the "cost of poor quality," is the total cost incurred as a result of failure to achieve quality. Historically, organizations have looked at all costs of quality. Today, many excellent organizations are concentrating strictly on the nonconformance costs. This highlights the waste, or losses, due to deviation from best target values. Once these costs are determined, they can be reduced or eliminated through application of the continuous-improvement philosophy.

Typically, the cost of nonconformance includes such items as inspection, warranty, litigation, scrap and rejects, rework, testing, retesting, change orders, errors, lengthy cycle times, inventory, and customer complaints.

Miscellaneous Other Methodologies

There are many methodologies that can be used for system development/improvement. Some of the major methodologies have been included in the tools and techniques chapters in this book. In addition, numerous other methodologies are available to strive for success. Some of the more common systems include just-in-time (JIT); design for manufacturing/producibility, total production maintenance (TPM); mistake-proofing; manufacturing resource planning (MRPII); computer-aided design, computer-aided engineering, and computer-aided manufacturing (CAD/CAE/CAM); computer-integrated manufacturing (CIM); computer systems; information systems (IS); and total integrated logistics (TIL).

Just-in-time

Just-in-time (JIT) is a method of having the right material just in time to be used in an operation. JIT reduces inventory and allows immediate correction of defects. This methodology is used for reducing waste, decreasing costs, and preventing errors.

Design for manufacturing/producibility

Designing for manufacturing/producibility is an approach that focuses design efforts on production processes early in the systems design.

Total production maintenance

Total production maintenance (TPM) is a system for involving the total organization in maintenance activities. TPM involves focusing specifically on equipment maintenance. TPM emphasizes involvement of everyone and everything, continuous improvement, training, optimal life-cycle cost, prevention of defects, and quality design. This methodology is effective for improving all production maintenance activities.

Mistake-proofing

Mistake-proofing, or poka-yoke, is a method for avoiding simple human error at work. The application of mistake-proofing frees the workers from concentrating on simple tasks and allows workers more time for process improvement activities. This is a major measure in the prevention of defects.

Manufacturing resource planning

Manufacturing resource planning (MRPII) is an overall system for planning and controlling a manufacturing company's operations. MRPII is used as a management tool to monitor and control manufacturing operations.

Computer-aided design/ engineering/manufacturing

Computer-aided design, computer-aided engineering, and computer-aided manufacturing (CAD/CAE/CAM) are automated systems for assisting in the design, engineering, and manufacturing processes. CAD/CAE/CAM are used to improve systems and processes, enhance product and process design, reduce time, and eliminate losses.

Computer-integrated manufacturing

Computer-integrated manufacturing (CIM) is the integration of computer-aided design and computer-aided manufacturing (CAD/CAM) for all the design and manufacturing processes. The CIM method improves on CAD/CAM by eliminating redundancy.

Computer systems

Computer systems include a wide range of items such as hardware, software, firmware, robotics, expert systems, and artificial intelligence. Computer systems are a major technological methodology.

Information systems

Information systems (IS) is an automated CDPM system used to focus an organization toward its vision. The IS is used to plan, design, analyze, monitor, and respond to critical strategic information essential to achieving customer satisfaction (internal/external). An information system allows continuous review, analysis, and corrective action.

Total integrated logistics

Total integrated logistics (TIL) is the integration of all the logistics elements involved in the inputs to the organization, all the processes within the organization, and the outputs of the organization to ensure total customer supportability at an optimal life-cycle cost. This method aims at total customer satisfaction by supporting the operations of the organization and the customer. Total integrated logistics can be a major differentiator.

System Development/Improvement Methodologies within the DoD

There are many methodologies used specifically by the Department of Defense (DoD) that have application in the commercial world. Many of the CDPM tools and techniques described in this book can be attributed originally to the DoD or government agencies. The methodologies mentioned in this section are in addition to the tools and techniques described in other parts of this book. Some examples of the more specific DoD TQM methodologies include computer-aided acquisition and logistics support (CALS), in-plant quality evaluation program, R&M 2000, and value engineering. This is not an all-inclusive list. The DoD uses many TQM methodologies in all its agencies to continuously improve its processes focusing on customer satisfaction.

Computer-aided acquisition and logistics support

The computer-aided acquisition and logistics support (CALS) program is a strategy to institute within DoD and industry an integrated "system of systems" to create, transmit, and use technical information in digital form to design, manufacture, and support weapon systems and equipment and apply communication and computer technology to the acquisition and support of major weapon and information systems. CALS focuses on integrating automation between the DoD contractor and the DoD. This is a DoD program to acquire, manage, access, and distribute weapon systems information more efficiently, including all acquisition, design, manufacturing, and logistics information. CALS focuses on an increase in reliability, maintainability, and availability through integration of automation systems. In addition, CALS seeks improvement in the productivity, quality, and timeliness of logistics support while again reducing costs.

In-plant quality evaluation

The in-plant quality evaluation (IQUE) program changes the method by which in-plant government people evaluate contractor controls over product quality. The IQUE program changes some of the traditional methods of evaluation with a TQM approach. The IQUE approach focuses on measuring and continuously improving processes with the aim toward quality (customer satisfaction). It concentrates on the "what" versus the "how." The government provides the "what," and the contractor determines the "how." IQUE implements a cooperative team concept between government and contractors.

R&M 2000

The reliability and maintainability (R&M) 2000 approach is geared to increasing combat capability while reducing costs through R&M practices. It stresses improvements in R&M to increase combat availability and reduce logistics support requirements. The R&M 2000 principles build on the TQM approach. R&M 2000 stresses the need for management involvement (leadership), requirements (vision/mission, involvement of everyone and everything focused on customer satisfaction), preservation (continuous improvement of processes and years of commitment and support), design and growth (training and ownership), and motivation (rewards and recognition).

Value engineering

Value engineering (VE) is an organized effort directed at analyzing the function of systems, equipment, facilities, services, and supplies for the purpose of achieving essential functions at the lowest life-cycle cost consistent with performance, reliability, maintainability, interchangeability, product quality, and safety. This definition comes from DoD Directive 4245.8. This specific DoD system for TQM again stresses the need to improve the quality and productivity of the DoD and DoD contractors while reducing cost.

Main Points

System performance has a great impact on customer satisfaction.

The best opportunity for system performance improvement is during the product design, process design, and production planning processes.

The earlier in the design process the changes occur, the less costly they are.

Concurrent engineering requires a transformation from traditional methods of product and process design.

Concurrent engineering philosophy includes

- Customer focus
- Organizationwide, systematic approach
- Never-ending improvements in product, process, production, and support
- Concurrent design of product and related processes
- Upper management leadership
- Robust design
- Reducing costs and time while improving quality and productivity
- Engineering planning initiatives, including automation
- New reliance on multifunctional teams
- Tools and techniques such as quality function deployment, design of experiments, the Taguchi approach, statistical process control, etc.

Multifunctional teams are the key element of concurrent engineering.

Quality function deployment is the ultimate planning tool for turning customer requirements into products and/or services that satisfy the customer.

All products should be designed with minimal losses.

Variation is the enemy of quality; it must be eliminated during all processes.

Always strive to achieve best target value rather than merely staying within specifications.

Parameter design is an important phase for minimizing losses due to variation from target.

All costs of nonconformance to quality should be identified and minimized.

Not only should processes be under control, but they should also be performing as close as possible to the best target.

Continuously improve all processes, beginning with processes that have the highest losses associated with them.

There are numerous system development and improvement methodologies.

Chapter

14

Specific Applications of CDPM Today and in the Future

Focus: This chapter describes how to apply customer-driven project management to a specific project today. In addition, the chapter explores the many stimuli for the proliferation of CDPM applications in the future.

Introduction

Today there are many applications for customer-driven project management (CDPM). It can be used in any case where the endeavor can be described as a project. A project involves providing a unique deliverable to a customer, usually at a specified time and cost. The project can be simple or complex. The deliverable can be a product, service, or combination of a product and service. The customer can be anyone a deliverable affects, such as an individual or an organization, profit or nonprofit, commercial or government, internal or external. The time to complete the project can be minutes to years. The cost can be anywhere from free to millions of dollars. In other words, almost everything we do for others can be considered for a customer-driven project.

In the future, projects will be the major process for getting things done. Project teams will be formed constantly, perform a project, and be disbanded. CDPM provides a structured, formalized management process for an efficient and effective project by taking full advantage of teamwork using a disciplined approach with appropriate tools and techniques. CDPM provides all the necessary processes to enhance the performance of all types of projects today and in the future.

Preceding chapters in this book described the customer-driven project management philosophy, principles, cycle, methodology, tools, and techniques necessary for the most effective and efficient projects. These chapters provided general guidelines—the basic what and how—on the application of customer-driven project management. This chapter targets the specific application of CDPM. Using CDPM most efficiently and effectively in your organization depends on tailoring CDPM specifically to the requirements of the particular project. This means continuously deciding which actions are necessary to make CDPM work for your specific desired results.

It is important to understand that CDPM is a means to a specified end. It is a process to reach a defined outcome. As shown in Fig. 14.1, CDPM is a process for providing a totally satisfactory deliverable to a customer. CDPM requires each organization to develop its own content within the CDPM process. It is essential to tailor the CDPM content to the specific needs of the project. The content is "owned" by the project organization; it cannot be transplanted, imitated, or cloned. Just as each project is unique, each application of CDPM is distinctive.

In fact, the specific content would not be the same for any project. The content for a NASA-type (complex, long-term) project would be far more elaborate than that for a quality-improvement project (simple, temporary). The NASA-type project would entail the detailed use of most of the customer-driven project management tools and techniques described in preceding chapters of this book. The problem-solving or quality-improvement project would only require the essential elements of CDPM. Each of the projects would require the use of the CDPM philosophy, principles, cycle, and improvement methodology. The major differences would be the implementation of the customer-driven team structure, TQM environment, and project management system. For example, the NASA-type project would require all these CDPM elements in total. The customer-driven project management team structure would include a customer and supplier projects steering team, customer-driven lead team(s), customer-driven project teams, customer-driven quality improvement team(s), and customer-driven process team(s). The TQM environment would require complete application of the VICTORY-C model. The project manage-

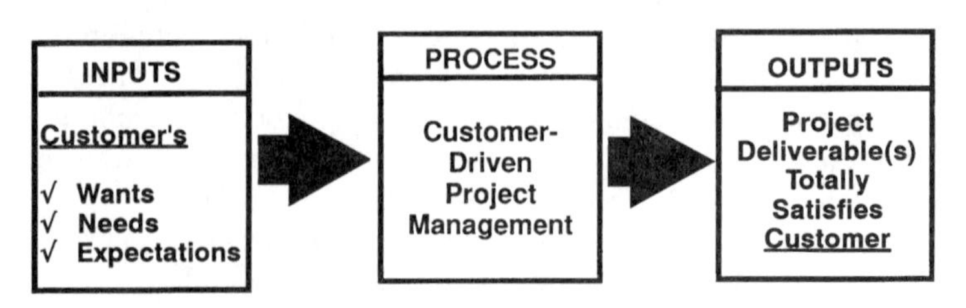

Figure 14.1 The customer-driven project management process.

ment system would automate the project management system of analysis, planning, implementation, and evaluation methods, including the use of all the tools and techniques.

The quality-improvement project would require all the elements, but it would not need the complete detail. The CDPM team structure could consist of just a customer-driven quality-improvement team or a customer-driven lead team and a customer-driven quality-improvement team. The TQM environment would require all elements of VICTORY-C, but they would not need be completely instituted within the total organization(s). For instance, the vision only needs to be a link for the quality-improvement effort. Leadership requires only the top leaders and the people involved in the improvement endeavor. Involvement includes only the people and equipment specifically needed for the project. The organization would use the standard CDPM improvement methodology. The customer-driven quality improvement team would only need the specific training and education just in time to accomplish the task. They would need to "own" the quality-improvement process enough to be able to provide the expected outcomes. In this case, positive recognition by top leaders and reward in the work itself are usually sufficient. The customer-driven quality-improvement team only needs to yearn for the specific outcome long enough to provide the necessary commitment and support to complete the project.

As the practitioner goes through the process, he or she must decide continuously which specific actions to take. Within the CDPM process, the practitioner determines the following:

1. Apply as is.
2. Modify.
3. Does not apply.

CDPM Application Considerations

Each project must have a focus, team, and process. These are basic considerations for all customer-driven projects. They provide the what, why, how much, who, where, and how of the project. The focus, team, and process will vary for each project because of the climate, culture, and customers. Figure 14.2 shows the key application considerations for CDPM.

The specific application of CDPM depends on the external climate, the internal organizational culture, and the external and internal customers. The external climate includes economic pressures, global competition, people issues, and advancing technology, as outlined in Chap. 1.

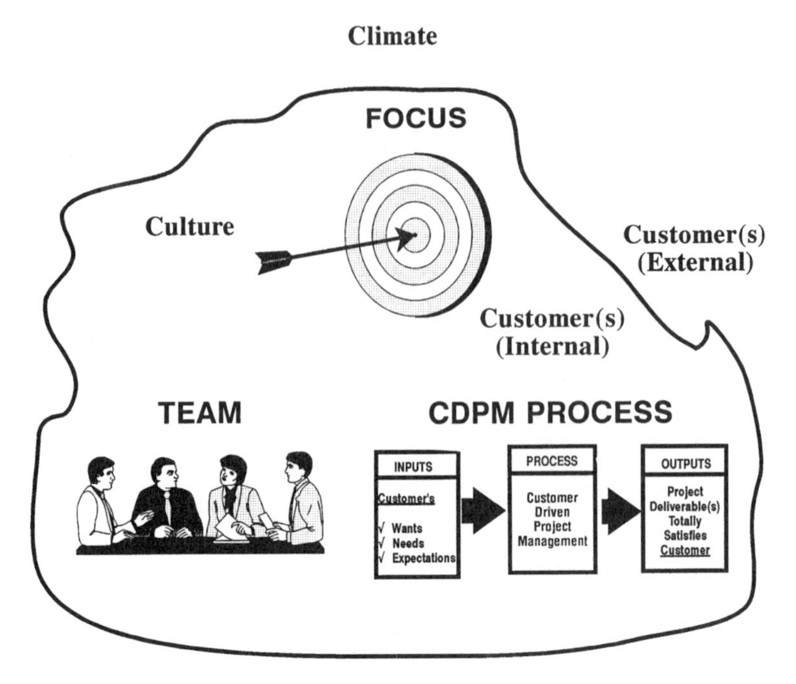

Figure 14.2 Considerations for application of CDPM.

The organizational culture involves consideration of the personalities and abilities of the organizations involved in the project. All the organizations' cultures must be merged to form a successful customer-driven team culture capable of performing the project.

The external and internal customer considerations embody the essence of CDPM. Customers are the drivers of customer-driven project management. The focus of all project efforts is the customer. The customer or the customer's voice guides the project. Therefore, a thorough recognition of all the customers for a project is essential.

The climate, the culture, and customer considerations influence the content of CDPM for a particular project. Specifically, they affect the focus, the team, and the CDPM process.

CDPM Application Process

The application of CDPM requires each practitioner to use the philosophy, principles, cycle, and improvement methodology geared to their particular project. This requires each practitioner to

- Get ready = *analyze*
- Get set = *design*

- Aim = *develop*
- Fire = *implement*
- Re-aim = *evaluate*

This list provides two names for the same processes. For instance, get set and design are both listed on line 2. The first is a common descriptive name; the second is the CDPM application process name. Each of the names can be used interchangeably. The first name is to remind practitioners of the sequence of the application of CDPM. Many times, practitioners try to fire before getting ready, getting set, or aiming. If you do not complete these processes before firing, you never know what you will hit. The second name is to remind the practitioner of the action processes each customer-driven team must continuously apply to make the project successful. The focus is on *action*. Figure 14.3 graphically shows the common descriptive names and CDPM application processes. The CDPM application process involves continuous analysis, design, development, implementation, and evaluation. It is up to each practitioner to provide the specific content within the application for each particular situation and project.

The application of CDPM requires use of the CDPM application processes many times within the CDPM cycle. In addition, the overall CDPM process requires use of the CDPM improvement methodology. Figure 14.4 shows the relationship of the CDPM application processes to the CDPM improvement methodology, providing further detail for the CDPM practitioner. The top-level CDPM process is shown at the top of the figure. The major application processes for use in the CDPM process are shown along the left of the figure. The steps in the CDPM improvement methodology become the subprocesses for the major application processes. These are shown on the right of the figure. At the operational level, the actual actions within each of the subprocesses must be tailored to each situation.

In any project, the CDPM application process and improvement methodology must be used at least once in the overall CDPM process, but the CDPM application process and improvement methodology may in fact be used as many times as necessary within the CDPM project life cycle to make the project successful. For instance, the CDPM application process could be used within step 1 in the CDPM improvement methodology to define the customer or quality issue.

Using the CDPM application process

The complete use of the CDPM application process can take anywhere from a few hours to many years. This section will present sev-

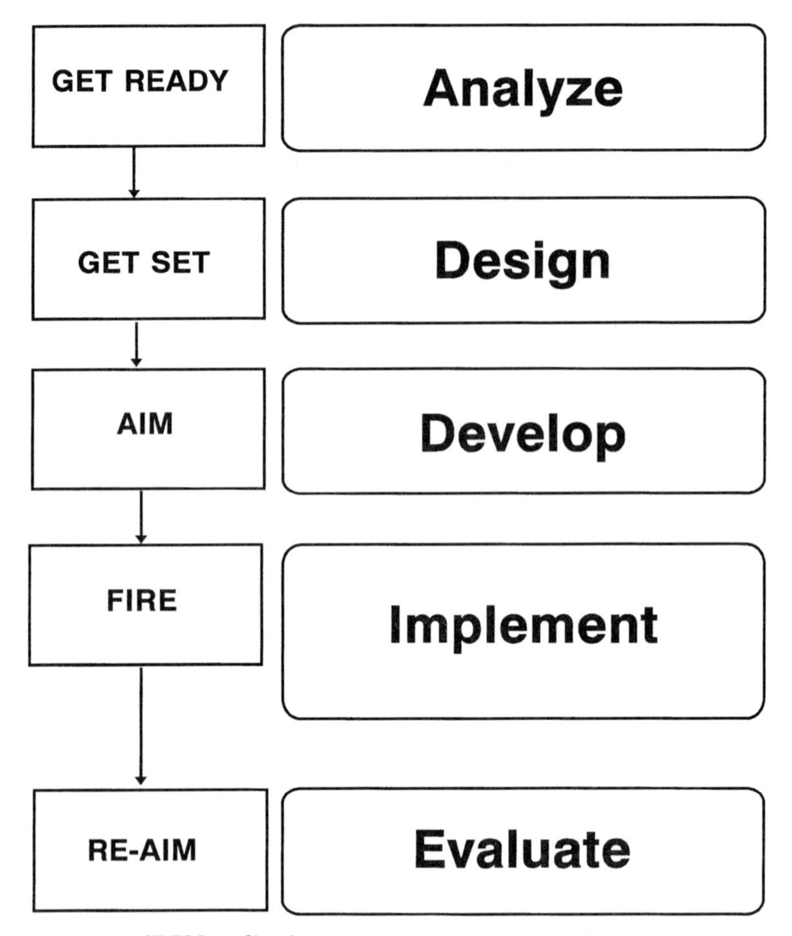

Figure 14.3 CDPM application process.

eral examples of the CDPM application process. In the first example, a simple application of the CDPM process will be explained; here CDPM takes only a few hours. The second example explains the process for implementing CDPM within an organization with a formal structure. Numerous other examples of customer-driven project orientation in current organizations are contained in Tom Peters' book *Liberation Management.* In addition, there are several specific case studies using customer-driven project management in Appendix A of this book.

Improving the lawn's appearance—Example This example is a description of a simple, short-term project. Most of us can relate to lawn care either as a customer or as a supplier. When someone needs their lawn

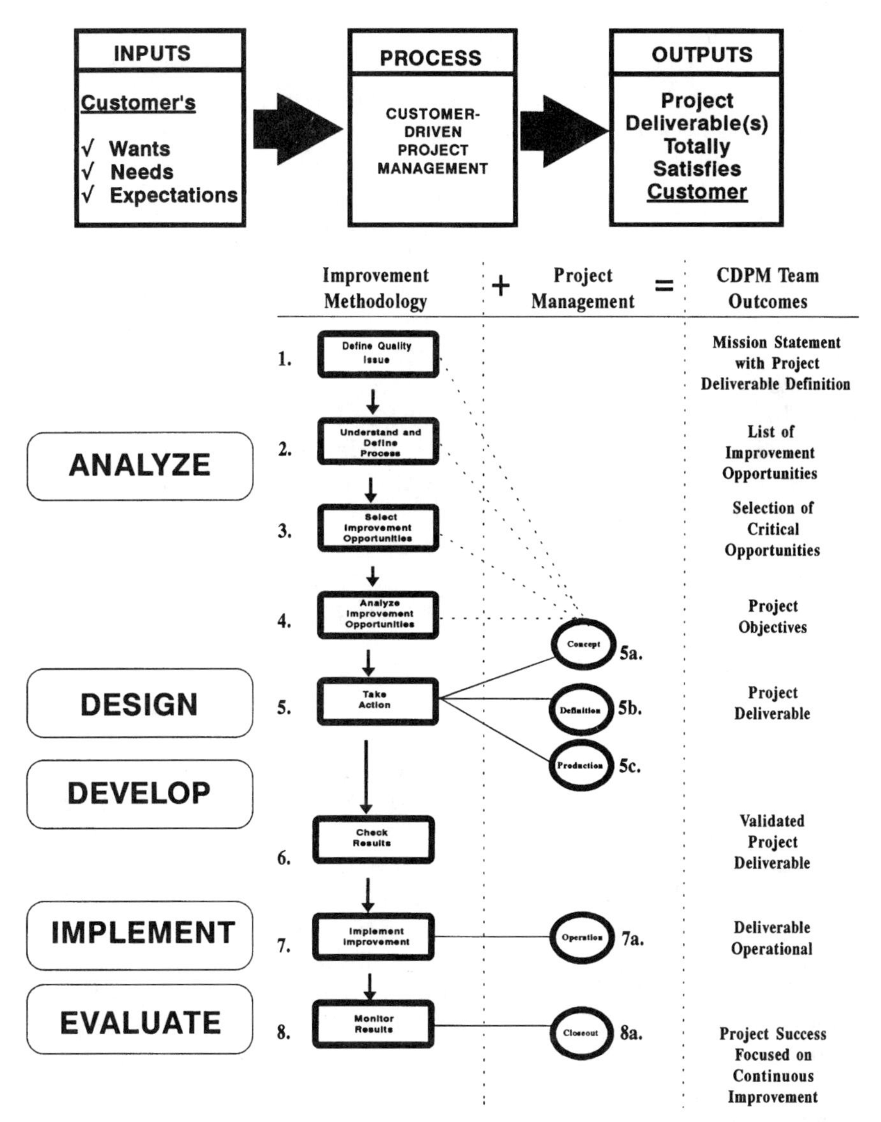

Figure 14.4 Relationship of the CDPM application processes to the CDPM improvement methodology.

cut on Saturday morning, this is a project. The focus of the project is to cut the grass. The schedule is on Saturday morning. The fee is negotiable. The customer driver is the one requesting the lawn to be cut. He or she seeks a supplier. The supplier can be a professional lawn service, a significant other, or any other willing provider of the service. The customer decides the supplier. The customer and the supplier must team to get the lawn cut. The team might consist of just the supplier and customer, or it might consist of other people for specific processes, i.e., lawn edging, sweeping, weeding, etc. Each team member defines his or her specific role. Normally, the customer leads the project, negotiates the requirements, monitors progress, and evaluates results. The supplier gets the lawn cut, determines processes and resources, and performs according to the schedule. Process "owners" perform their specific process, if required.

Next, the CDPM process is used to complete the project. This involves communicating with the customer to determine the customer's wants, needs, and expectations, including the performance, schedule, and price. The performance objective for this project is to improve the appearance of the lawn. The schedule is Saturday morning. The price is no more than $20. The supplier determines his or her own cost. Once the requirements are understood, the CDPM process is used to accomplish the project totally satisfying the customer. The major CDPM processes at the top level consist of the elements in the CDPM cycle. They are analysis, definition, production, operations, and improvement/closeout. When applied, the CDPM cycle becomes the CDPM application processes of analyze, design, develop, implement, and evaluate.

During the analysis stage of the CDPM application process, the customer-driven team defines the customer satisfaction/quality issue, understands and defines the process, and selects and analyzes improvement opportunities. In our example, the mission becomes improving the appearance of the lawn. The team reviews the processes for taking care of the lawn. These processes include cutting, edging, and weeding. Once the processes are defined, the team generates a list of improvement opportunities. A sample of the list could include cutting the grass using a pattern, cutting the grass at a different height, edging using a trench, weeding by hand, and so forth. Using selection techniques, the team determines that the best opportunity for improvement is lawn height. Next, the team analyzes the available information on lawn heights to determine the lawn height that would make the lawn look the best.

Now the team is ready to take action. This is the first step in the definition stage of the CDPM cycle. During this stage, the team designs the project. The team agrees on the concept of lawn height improving the appearance of the lawn. The team decides exactly what needs to be done to cut the lawn to the specified height. It is also extremely important to analyze risk at this point. For instance, the team may decide to cut the grass higher than optimum based on the risk of cutting the grass too short. If the grass is cut too short, the lawn appearance cannot be corrected. If the grass is too high, the lawn appearance can be enhanced by cutting the grass lower.

Next, the team develops a plan of action that includes the contract, task list, and schedule. In this case, the contact is a verbal agreement between the customer and supplier. The task list includes everything the supplier will do to improve the appearance of the lawn. The schedule includes a start and completion time.

Then the team moves to production. Here, the team members start to implement the plan. They develop everything needed to accomplish the project. For instance, the equipment is prepared. In this case, the equipment is the lawn mower. They also cut a section of the lawn as a pilot test. They check the results to ensure that the lawn's appearance will actually be improved.

Once the production is complete, the team is ready to move into operations; team members begin to implement the improvement. They cut the complete lawn at the new height. When the complete lawn is cut, the results are evaluated with the customer.

This evaluation leads to the next stage of further improvement or closeout of the project. If the lawn does not meet their customer's expectations, the lawn could be cut again at a different height. The team would return to the analysis stage to determine a new height. Then it would continue through the same process of taking action, checking results, implementing improvement, and monitoring results. If the lawn meets or exceeds the customer's expectations, the project is successful. This project can be closed out or targeted for continuous improvement in other areas to enhance the lawn's appearance.

Implementing CDPM—Example This example provides another illustration of the use of customer-driven project management. This example involves a more complex, longer-term project than the previous example. This example uses the CDPM process to implement CDPM for use within one organization for one project involving several major functions. The example focuses on the specific processes and content within the processes for one particular application and illustrates one complete cycle of the CDPM application processes within the entire CDPM improvement methodology.

In the example, the organization already has created a TQM environment and performs projects using a project management system. The organization is ready to take the next step, this being customer-driven project management. This would be a typical application in many organizations. This example serves three purposes. First, it describes a detailed service deliverable example. Second, it provides an insight into the efficient and effective use of CDPM within one organization. Third, it highlights many of the main considerations when implementing CDPM in an organization.

As stated in Chap. 4, CDPM in action requires focus, teamwork, and the CDPM process. Figure 14.5 shows the start of a CDPM pro-

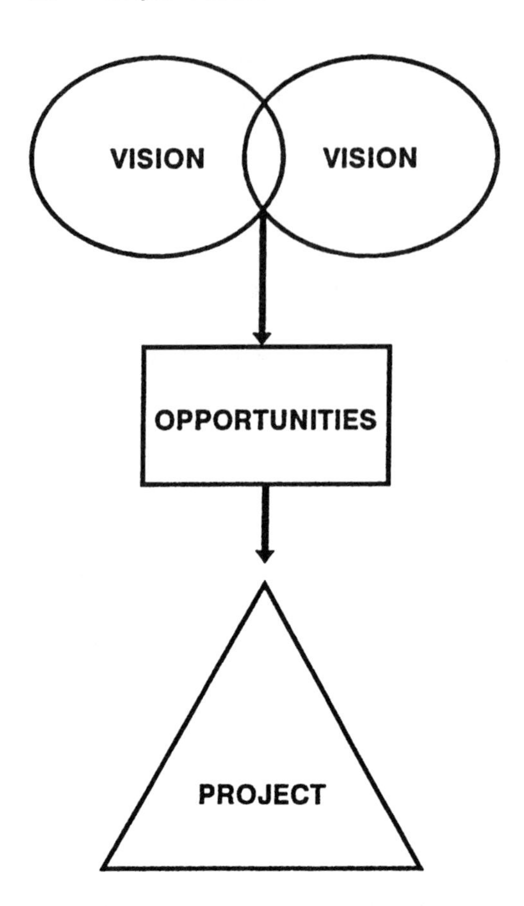

Figure 14.5 Project process start.

ject. The project begins with the desire for some future state or vision. The visions for the two organizations have some area in common, which creates opportunities for joint projects. In our example, the two organizations share one common company vision. In addition, each of the two organizations visualizes its own future state. These two visions have some common area in addition to the one common company vision. These visions lead to the pursuit of opportunities in the common area and make each organization's vision a reality. Through these endeavors, the two organizations discover mutual beneficial opportunities. In our example, one organization requires a certain deliverable from other parts of the organization to make the overall company more successful. This organization is the customer. Another organization within the company "owns" the processes and resources necessary to provide the deliverable. This organization is the supplier. The current deliverable needs to be upgraded to changing external customer expectations, and this progresses into the desire to explore the possibility of using customer-driven project management for this improvement project. In addition, the two organizations' top leadership teams decide they would like to use customer-driven project management for future projects in other parts of the company. In this situation, there is time between the need to implement CDPM and the time to start the product improvement project. This forms the initial focus of the project.

Once a vision for a project is conceived, the two top leaders agree to form with the top executive in the company a customer and supplier project steering team to monitor the implementation of CDPM. This team is led by the top executive who is the customer of the CDPM implementation. The other two top leaders are the nucleus of the team. The initial CDPM project is to implement CDPM. From this team, a customer-driven lead team will be formed to implement CDPM and drive the product improvement project within the organization. The customer and supplier project steering team provides the commitment and support in the form of active participation and initial investment.

Subsequently, the customer and supplier project steering team decides that it does not have enough expertise in CDPM to accomplish the implementation on its own. The team decides to use an outside resource as a coach, facilitator, and trainer of trainers for this initial implementation. This forms the nucleus of the customer and supplier project team. This nucleus team develops an initial vision for the project.

Next, this nucleus customer-supplier project steering team, after some coaching by the outside resource, determines the remaining customer-supplier project steering team members. The team decides to add an internal leader from the organization to become the in-house facilitator. In addition, representatives of the ultimate customer, the major outside vendor, and the union are asked to join the customer and supplier project steering team. They all accept membership.

Once the customer and supplier agree to the initial focus and team for the project, the customer and supplier steering team prepares to start the CDPM application process. The customer and supplier steering team holds its first meeting, the focus of which is to develop the team's mission, teamwork, and the CDPM approach for their situation. It is decided that this meeting will be held off site for 2 days. The agenda for the first team meeting is as follows:

- Provide a CDPM overview.
- Develop a team code of conduct.
- Determine team meeting roles and responsibilities.
- Set the agenda.

- Develop a mission statement.
- Determine each team member's contributions to the team.
- Establish team values.
- Learn about team dynamics.
- Discover how to manage conflict on the team.
- Define how CDPM applies to the situation.
- Review the CDPM implementation flow.
- Decide on the overall content of the CDPM application process.
- Agree on next steps.
- Perform a critique of the meeting.

At the completion of this first team meeting, the customer-supplier team agrees that the next step is to start the CDPM application process. Figure 14.6 shows the CDPM implementation flow for this project. The eight-step CDPM improvement methodology is integrated into this CDPM implementation flow.

The following sections describe the process this customer and supplier steering team used to implement CDPM. The content of the process is the result of the team tailoring the material in this book to the specific needs of the example organization. The process is explained in sequence for simplicity. However, many of the steps within the CDPM application process could be performed simultaneously, concurrently, or overlapping to save time.

Get ready: Analyze. Figure 14.7 shows the content of the CDPM analyze application process for this project, as developed by the customer and supplier project steering team. The CDPM applications process integrates the CDPM improvement methodology with other details as necessary. The following is a description of the process as applied in our example.

Step 1: Define the CDPM implementation issue. The first step in the CDPM improvement methodology is "define the CDPM implementation (quality) issue." This is where the CDPM team finalizes the customer and supplier project steering team's mission.

The team decides to use quality function deployment (QFD) to gain a better understanding of the application of CDPM processes for their specific situation. The team members agree that a complete QFD house will not be necessary for their purposes. They design and use a specially tailored QFD tool with just the elements necessary for their application. The team in our example decides to use assessment results as the "whats" of the tailored QFD "house of quality."

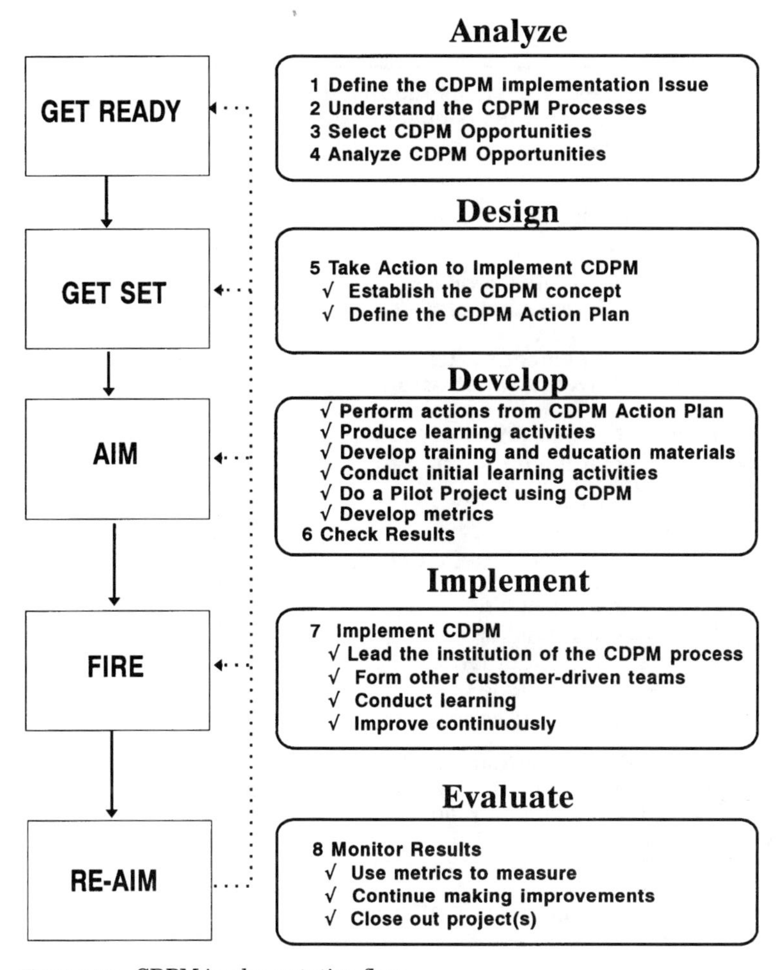

Figure 14.6 CDPM implementation flow.

To complete the necessary QFD "what" element, the customer and supplier project steering team with some coaching realizes that the application of CDPM to each project requires a detailed analysis. This is where the organization assesses the external climate, the organizational culture, the customer(s), and other factors as appropriate. The team asked: How does CDPM

- Fit our needs?
- Fit our organization?
- Fit our people?
- Fit our schedule?

Analyze

1. **Define the CDPM implementation issue**
   - √ Conduct complete assessment
     - ▸ External climate
     - ▸ Organizational culture
     - ▸ Teamwork
     - ▸ Individual competencies
     - ▸ Customer(s)
2. **Understand the CDPM processes**
   - √ Conduct an assessment of:
     - ▸ TQM environment
     - ▸ Project management system
     - ▸ Customer-driven team structure
   - √ Evaluate the organization's readiness for CDPM by determining buy-in to:
     - ▸ CDPM philosophy
     - ▸ CDPM principles
     - ▸ CDPM cycle
     - ▸ CDPM Improvement methodology
3. **Select CDPM opportunities**
   - √ List CDPM opportunities
   - √ Determine selection criteria
   - √ Use selection techniques
   - √ Make decision using consensus
4. **Analyze CDPM opportunities**
   - √ Determine data statistical analysis technique
   - √ Prepare data collection plan
   - √ Collect data
   - √ Chart data
   - √ Analyze the data
   - √ Use results to determine CDPM objectives

Figure 14.7 CDPM "analyze" application process implementation flow.

The first question that the team answered is "How does CDPM fit our needs?" The team's criteria for answering this question are based on the applicability of CDPM to assist the organization to

- Balance technical expertise with customer expectations
- Keep a high interdependence
- Manage information
- Use team problem solving and decision making
- Optimize resources
- Meet world-class quality standards
- Reduce costs

- Meet schedules
- Be flexible and adaptable
- Respond rapidly
- Improve continuously
- Focus on customers

The customer-driven project steering team decides that the customer-driven project management process is exactly what the organization needs to respond rapidly to customers' changing expectations. Again, the team recognized that the customer-driven project management process is required for all applications of CDPM. However, the team needs to adapt the CDPM process to each particular situation.

The second question the team answered is "How does CDPM fit our organization?" The team recognizes that although CDPM can be applied to any organization, large or small, it is particularly useful for most organizations in today's world facing rapid change, rising complexity, and rabid competition, although CDPM has some advantages for organizations in a stable environment with changing customer requirements. For instance, the company has a changing environment with a more demanding customer. Other parts of the organization involved in the project have a stable and unchanging environment with a reliable customer base. The customers in these cases are internal customers. The team recognizes that the CDPM content would differ within the two organizations. The focus, team, and customer-driven project management process would be tailored to the specific situation. In the stable climate, the focus would be on improving the satisfaction of the internal customers. The team would naturally include the process owners, suppliers, and customers. The CDPM process targets projects that solve current problems geared to making the process more efficient. In the changing environment, the focus would be on continuous improvement driven by the customer. The team would consist of a customer leader, a project facilitator, suppliers, a program manager, and process "owners." The CDPM process targets providing deliverables geared to total customer satisfaction.

The third question the team answered is "How does CDPM fit our people?" The team understands that the CDPM process works best in an environment where people are a valuable resource. The team knows that the basic foundations of integrity, ethics, and trust are an essential requirement of all people involved in the CDPM process. With this foundation, the team realizes that its specific application of CDPM can then build the appropriate teamwork and empowerment. With a consensus on these principles, the team agrees that all people involved in the CDPM project must develop the following views:

- Their task is a customer-driven project.
- They are the "owners" of their process.
- They are process- not function-oriented.
- They have a partnership with their suppliers.
- They are focused on the customer.
- They must constantly develop constructive relationships.

The fourth question the team answered is "How does CDPM fit our schedule?" First, the team accepts that CDPM takes time. The schedule has an impact on the use of CDPM. The question the team definitely needs to answer is "Can we afford not to take the time?" The team decides by consensus that the time investment will pay huge dividends in future project and organizational results.

Next, the team examines the many variables in the time factor of implementing CDPM to estimate the actual time it might take in their situation. Team members understood that CDPM takes time to develop the trust to create and maintain the necessary relationships. In addition, they recognize that the time to implement CDPM depends on the specific application. For instance, the largest-scale implementation of a CDPM system across many organizations and many projects could take years to institute. To implement CDPM within an organization for one large-scale project may take the total time of the project. Use of the CDPM process for a small-scale project could be done in a very short time from hours to weeks depending on the project. The team determines that an intelligent approach to its application would increase the velocity of the implementation. The team determines through assessment that there is enough time to analyze, design, develop, implement, and evaluate CDPM on a small scale prior to the start of the new deliverable project. The team assesses all the variables involved in the time to apply CDPM, such as

- Organizational culture
- Previous experience with total quality management
- Previous experience in project management
- Type of project(s)
- Number of organizations involved
- Number of projects
- Complexity of the project
- Relationships with customers

The customer and supplier project team uses some of the sample assessment instruments in Appendix C to help determine the specific situation. The team uses the results from their analysis as the "what" in the QFD matrix. This helps the team gain a deeper understanding of the mission and work involved in achieving success.

Step 2: Understand the CDPM process. Next, the team continues to use the tailored QFD chart to identify the "hows" to perform the "whats." The "hows" are the details of the elements of the CDPM process. Remember from Chapter 1 that the elements of the CDPM process are TQM environment, project management system, and customer-driven team structure.

The team then determines the relationship between the "whats" and the "hows." From this matrix, the team identifies the critical "hows" to implement CDPM for this project. The critical "hows" are the most important processes for the CDPM implementation. The team then thoroughly examines all the identified important processes required to implement CDPM in its specific organization. This is accomplished by dividing the processes among the various customer and supplier project steering team members to get an understanding and definition of each process. Each specific team member "owns" one of the critical processes for CDPM implementation. Each team member then forms a customer-driven project lead team to focus on his or her specific process. These teams usually consist of the customer and supplier project steering team member as the customer leader. In addition, each team adds process "owners," suppliers, and other people as necessary. Each team completes a top-level and top-down process diagram, an input/output analysis, and customer/supplier analysis of its CDPM implementation process. In addition, each team conducts a performance assessment of its process that compares its process with a "world class" benchmark and "actual requirement." In addition, each team completes a brainstorming session on issues, problems, and opportunities associated with implementing its process. Upon completion of this task, each team provides a presentation of results to the complete customer and supplier project steering team. During the presentation, the steering team clarifies assumptions and focuses consensus on each team's findings. From the presentations, the customer and supplier project steering team compiles a list of possible CDPM implementation opportunities.

While the customer-driven lead teams are completing their assignment, the complete customer and supplier project steering team compiles an overall list of issues, problems, and opportunities related to the organization's implementation of CDPM. This list is generated by various brainstorming sessions in the organization and is used by the customer and supplier project steering team to determine the feasibility of buying into CDPM philosophy, principles, cycle, and improvement methodology.

Step 3: Select CDPM opportunities. During this step, the customer and supplier project steering team makes a list of all the opportunities, issues, and problems from step 2. The team then selects the high-priority opportunities and uses rank-order selection techniques to prioritize them. The team decides that the rank order criteria are as follows:

- *Cost:* Could the opportunity be pursued within investment limits?
- *Results:* Would the opportunity produce the desired outcomes?
- *Importance:* Would the opportunity be viewed as critical to the organization?
- *Time:* Could the opportunity be implemented before the new project start date?
- *Effect:* Would the opportunity have the desired effect on the organization?
- *Risk:* Could the opportunity be implemented with reasonable risk?
- *Integration with organization's objectives:* Does the opportunity fit the organization's goals?
- *Authority:* Is the team empowered to pursue the opportunity?

The team reaches consensus on a list of five critical opportunities that have the greatest impact on successful implementation of CDPM for its project. The team decides that these five areas represent the major opportunities for implementing CDPM. They elect to focus on all five of these opportunities, which are

- Leadership
- Systematic thinking
- Tools and techniques
- Teamwork
- Disciplined approach

Step 4: Analyze the CDPM opportunities. During this step, the customer and supplier project steering team again decides to work each opportunity through other customer-driven project lead teams. In our example, the selected opportunities relate directly to the CDPM implementation processes. Therefore, the customer-driven project lead teams with opportunities for analysis just continue with the CDPM implementation project. The objective for each customer-driven lead team during this step is to analyze the CDPM opportunities to determine specific project objectives. Each team uses disciplined analytical tools and techniques to target specific objectives from the selected CDPM implementation opportunities. This involves knowing the current performance of each of the areas of opportunity. In addition, the teams determine the underlying or "root" causes of any issues or problems. From this thorough analysis of the CDPM implementation opportunity, the team determines recommendations for project objectives. Each team reports the results to the complete customer and supplier project steering team. Then the customer and supplier project steering team determines the overall project objectives.

Each team determines the appropriate data statistical analysis techniques for its specific process. For instance, some teams use cause-and-effect analysis to focus on the vital few issues affecting their opportunity with a Pareto chart to verify the analysis, while other teams use different data statistical methods to reach their objective, such as statistical process control, histograms, scatter diagrams, etc.

For clarification of the example, the "leadership" team's analysis process describes one team's actions during this analysis step. This team decides to use metrics to determine leader performance and to evaluate the critical team leader behaviors of individual leaders in the organization. First, the team decides that it is necessary to verify the critical opportunity. Is leadership really an issue? The objective of this metric proves that leadership is in fact a significant opportunity. Second, the metrics need to determine the critical or "root" areas of the leadership opportunity within this particular organization. This information will be used to focus the leadership development on the vital few areas really required to implement CDPM in this specific organization.

The "leadership" team first decides on the following data collection plan. The purpose of data collection is to

1. Provide data to develop metrics to

   Verify the leadership opportunity.

   Determine team leaders' performance in the critical elements of leadership in this specific organization.

2. Required data: Data on overall organization's leadership performance and performance of key leadership elements are necessary.
3. Amount of data: The data will be collected from all affected people in the organization.
4. Data-collection method: The data will be collected by survey.
5. Data-collection "owner": The "leadership" team will conduct the survey over a 30-day period.

To save time, the team combines the data collection for the leadership performance and leadership competencies into one survey. In order to target the vital leadership opportunities, the team performs

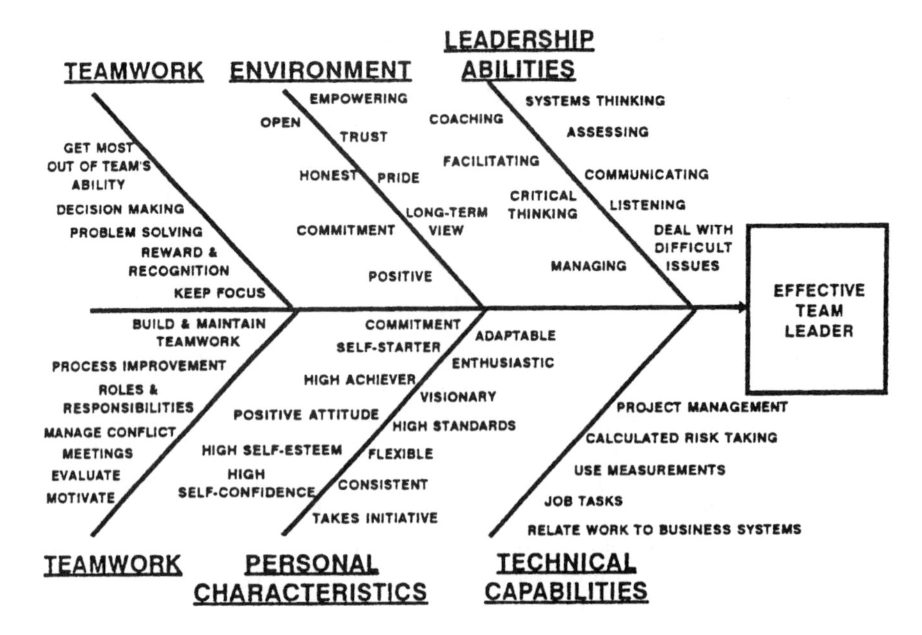

Figure 14.8 Cause-and-effect diagram for effective team leader example.

a cause-and-effect analysis to determine the causes of "effective leadership" in the organization. Figure 14.8 shows the cause-and-effect diagram generated by the "leadership" team. From the cause-and-effect analysis and Appendix B of this book, the team develops a leadership assessment instrument as the tool to gather data for the analysis. Figure 14.9 shows the team's metric description for improving team leaders' effectiveness.

Next, the team develops a team leader report card to survey the team leaders' performance. Figure 14.10 shows the inside of the team leader report card. The report card assesses the overall team leader performance. This simply means an overall rating of the team leader's current performance. Does the leader currently guide the team effectively and efficiently to successful mission accomplishment? In addition, the report card rates the critical behaviors of effective leadership as determined by the "leadership" team. These behaviors are

- Leading by example
- Empowering others
- Acting to optimize resources
- Directing problem-solving and process-improvement activities
- Encouraging teamwork
- Recognizing and rewarding appropriate performance

METRIC TITLE:

IMPROVE LEADERS' EFFECTIVENESS

Operational Definition:

The customer-driven project team will perform assessment and monthly performance reviews using the results of the approved team leader report card.

Measurement Method:

The measurement consists of a team leader effectiveness index compiled as an average from the approved team leader effectiveness report card. The team leader report card will be completed by all team leaders, their immediate reporting official, and their team members. The results of the report card will be compiled into a team leader index. The team leader index indicates the defined key team leader behaviors associated with effectiveness to include: leading by example, empowering others, acting to optimize resources, directing problem solving and process improvement activities, encouraging teamwork, and recognizing and rewarding appropriate performance.

Desired Outcome:

Improve leadership rating in all team leader areas.

Linkage to Organizational Objective:

Meets internal improvement strategy.

Process Owner:

Customer and Supplier Projects Steering Team.

Figure 14.9 Metric description for team leader effectiveness example.

Next, the team collects data from team leaders, their reporting official, and team members. These data are compiled using statistical methods to provide an overall rating and a rating in each of the seven critical areas.

The team puts these data on a chart. Figure 14.11 is the baseline chart, and this chart becomes the basis for metric charts in each of the areas over the long term to monitor progress on leadership issues. The initial analysis of the baseline chart indicates that overall team leader performance requires improvement. Although the organization is performing above average, the organization's leadership performance is below that of the competition. In addition, each of the critical areas needs improvement. Since the organization has embarked on the road to total quality management, its ratings for encouraging

LEADER REPORT CARD

Grade according to the following: "A" excellent, "B" good, "C" satisfactory, "D" less than satisfactory, "F" failing.

	Grade	COMMENTS
Leadership The overall performance of the team leader in guiding the team to successful mission accomplishment.	☐	Leadership
Leading by example The leader walks the talk. Setting the example for other team members and team leaders.	☐	Leading by example
Empowering others Team leader's willingness to share authority, responsibility, and resources with team members.	☐	Empowering others
Acting to optimize resources Team leader uses all resources to their fullest advantage.	☐	Acting to optimize resources
Directing improvement activities The team leader actively leads team problem-solving and process improvement activities.	☐	Directing improvement activities
Encouraging teamwork Team leader creates and maintains teamwork by keeping the team focused, encouraging open communication, and maintaining constructive relationships.	☐	Encouraging teamwork
Recognizing & rewarding appropriate performance Team leader provides personal satisfiers to deserving team members and gives appropriate credit to the team.	☐	Recognizing & rewarding appropriate performance

Figure 14.10 Leader report card.

teamwork, directing improvement activities, and acting to optimize resources are above average and gaining on the competition. The organization is below average on leading by example, empowering others, and recognizing performance. The team agrees that improvement of team leadership needs to be a project objective of the CDPM

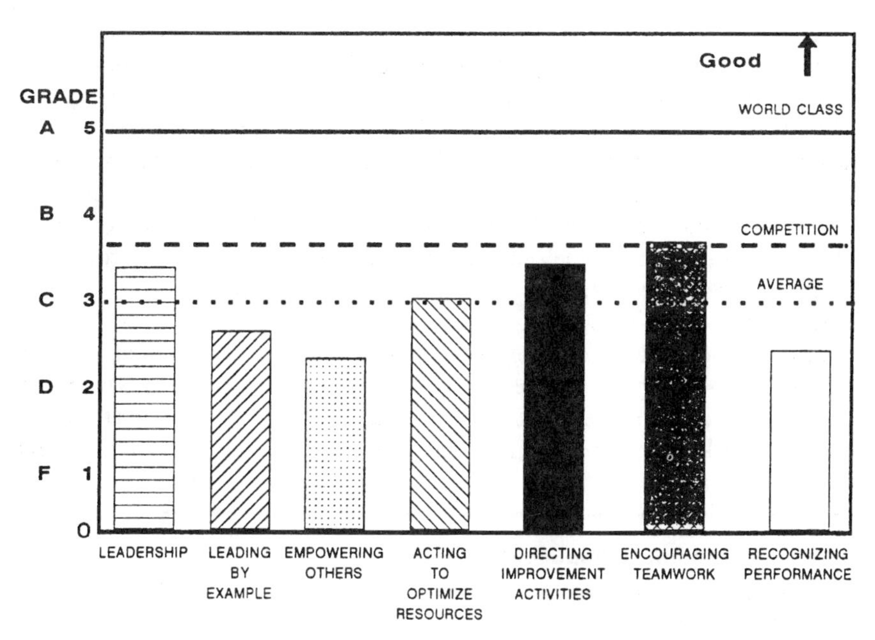

Figure 14.11 Leadership performance chart.

implementation project. This is the recommendation to the customer and supplier project steering team.

The customer and supplier projects steering team reviews the recommendations of all the customer-driven project teams. From these recommendations, the team determines the project objectives.

Get set: Design. The design stage encompasses the concept and definition phases of the CDPM cycle. It also starts the "take action" step in the CDPM improvement methodology. During the design stage, the results of the analyses are used to design the specific CDPM implementation. This is the project deliverable. The customer and supplier project steering team defines the project deliverable in a CDPM implementation action plan, and the CDPM implementation action plan becomes the living document that brings the CDPM implementation project to reality.

Step 5: Take action. Figure 14.12 shows the CDPM design application process implementation flow for this example. In the example, the customer-driven project steering team agrees to devote its full attention to the CDPM implementation project. The team decides to invest

3 days off site to construct the preliminary CDPM implementation action plan. The participants for this meeting are carefully selected to ensure commitment and support. The customer and project steering team designs a matrix to determine meeting participants. The matrix lists each of the project objectives along the left column. Along the top of the matrix is listed the possible customer-driven team members. This list includes the members of all the customer-driven project lead teams participating in the CDPM implementation. The team checks all the people and organizations required for each of the project objectives to implement CDPM, and this list is used to confirm continued membership in the required customer-driven project teams to accomplish the project objectives. In addition, the list provides recommendations for specific participants in the action planning meeting. The team then decides on the specific participants for the meeting, who include the customer and supplier project steering team, the members of the customer-driven project lead team for the pilot CDPM project, other key customer-driven project team members working each

5. Take Action to Implement CDPM

- √ **Establish the CDPM concept**
  - ▸ **Vision**
  - ▸ **Mission**
  - ▸ **Values**
  - ▸ **Strategic link**
- √ **Define the CDPM Action Plan**
  - ▸ **CDPM concept**
  - ▸ **Customer-driven team structure**
  - ▸ **Customer-driven team members contributions**
  - ▸ **CDPM process with tailored content**
  - ▸ **CDPM improvement methodology outline**
  - ▸ **Learning plan**
  - ▸ **Ownership plan**
  - ▸ **Rewards and recognition plan**
  - ▸ **Long-term commitment and support plan**
  - ▸ **Management information system plan**
  - ▸ **Risk-management plan**
  - ▸ **Investment plan**
  - ▸ **Actions = task list**
  - ▸ **Schedule with milestones and network**

Figure 14.12 CDPM "design" application process implementation flow.

of the project objectives, and vital process team leaders (process "owners") as appropriate.

The agenda for this CDPM action plan meeting is as follows:

- Devise CDPM implementation action plan overview.
- Develop CDPM implementation vision.
- Establish mission statement.
- Determine common values for CDPM implementation.
- Align CDPM implementation with a strategic focus.
- Define the CDPM implementation action plan.
- Formalize the CDPM concept.
- Design the customer-driven team structure.
- Discover team member contributions.
- Outline the CDPM implementation process.
- Develop a learning plan outline.
- Establish "ownership" of CDPM implementation processes.
- Determine a rewards and recognition system.
- Get commitment and support (investment).
- Set management information system requirements.
- Outline risk-management procedures.
- Determine actions with "owners" and milestones.
- Get agreement on next steps.
- Perform a critique of the meeting.

The results of this meeting are a preliminary CDPM implementation action plan. This plan constitutes the contract, task list, and schedule for the CDPM implementation.

CDPM action plan. The CDPM action plan consists of the following:

- CDPM concept
- Customer-driven team structure
- Customer-driven structure contributions
- CDPM process
- Learning plan
- Ownership considerations
- Rewards and recognition

- Commitment and support (investment)
- Management information system
- Risk analysis system
- Actions
- Schedule

The CDPM action plan includes all the project management tools and techniques, and the project management tools and techniques include the contract, the work breakdown structure, the task list, the network schedule, risk assessment, and the project management information system. Once the CDPM action plan is formalized, it becomes the project contract. The CDPM action plan defines the customer-driven team structure, and this forms the basis for the work breakdown structure. The actions included in the CDPM action plan equate with the task list. The schedule includes at least a Gantt chart. Risk analysis and a project management information system are also essential elements of the CDPM action plan.

Aim: Develop. The development stage equates with the CDPM cycle's production phase. During this stage, the development actions from the CDPM implementation action plan are performed and the results are checked.

Figure 14.13 shows the CDPM development application process implementation flow for this example. It includes

- Creating the customer-driven team structure
- Establishing the CDPM process with methodology
- Initiating the learning plan
- Determining ownership of the processes
- Developing a rewards and recognition system
- Monitoring the schedule and investment

Specifically, the customer-driven project steering team oversees the development of specific customer-driven project lead teams for each of the CDPM implementation project objectives. These teams create other customer-driven project, process, and quality-improvement teams as appropriate and carry the CDPM implementation project until their specific part of the project is complete. In essence, each of these teams has its own customer-driven project to complete that is part of the overall CDPM implementation project.

In the example, the customer and supplier project steering team with appropriate customer-driven project teams in the development application process performs the following actions:

√ Perform actions from CDPM Action Plan
- ▸ Create customer-driven team structure
- ▸ Establish CDPM process
- ▸ Communicate the CDPM improvement methodology
- ▸ Initiate learning plan actions
- ▸ Determine ownership
- ▸ Design a reward and recognition system
- ▸ Establish initial funding
- ▸ Perform project management activities

√ Produce learning activities for:
- ▸ Organization development
- ▸ Team formation and development
- ▸ Individual knowledge and skills

√ Conduct learning activities
- ▸ CDPM awareness
- ▸ CDPM team leader
- ▸ CDPM team
- ▸ Facilitator(s)
- ▸ CDPM process
- ▸ CDPM tools and techniques

√ Do a pilot project
- ▸ Focus on small success

√ Develop metrics
- ▸ Organization
- ▸ Team
- ▸ Individual

6. Check results

√ Compile lessons learned

√ Make improvements

Figure 14.13 CDPM "develop" application process implementation flow.

- Determine learning activities to develop the organization as a whole, the customer-driven teams, and individual team members.
- Develop and procure training and education materials to support the learning activities. The training and education materials combine the efforts of an outside organization with their particular needs. The organization decides to use internal people to deliver the materials. It decides that the team leaders will be the most appropriate presenters.
- Conduct initial learning activities. These are geared to getting off to the right start. These learning activities focus on the fundamental knowledge and skills necessary to work in and perform the CDPM process.

- Do a pilot project. Each initial customer-driven team targets a relatively short-term project of less than 60 days as its initial effort. In addition, this organization decides to officially start the customer-driven project lead team for the new product project. This team's initial mission involves focusing on internal process improvements for 60 days.
- Develop metrics to focus action necessary to improve the organization, customer-driven teams, and individuals.

Step 6: Check results. During this step, the teams evaluate the CDPM implementation project objectives against current performance to measure progress. The teams each assess their results and compile lessons learned. There is a formal review by the customer and supplier project steering team and customer-driven project lead teams. The teams decide either to proceed to implementation or to solve problems, make more improvements, overcome some obstacles, invent a new process, review the analysis, or further understand the customer (quality) issue or processes before making the process operational. For the example, the customer and supplier project steering team decides to implement CDPM totally for the pilot project while moving to start the CDPM application process in other areas of the organization.

Fire: Implement. Now the organization is ready to implement CDPM. This involves instituting the CDPM process, forming other customer-driven teams as appropriate, conducting perpetual learning, and seeking continuous improvements in projects, products, processes, and people. During this stage, the customer and supplier project steering team becomes an overseer of all the CDPM projects by monitoring the critical outcomes. The team also must actively practice CDPM itself.

Step 7: Implement CDPM. Figure 14.14 shows the CDPM implementation application process flow for this example. This process involves the instituting of the CDPM process by

- Constantly communicating the vision
- Leading by example
- Getting everyone and everything involved in the process
- Building trust and relationships
- Targeting specific results through a disciplined CDPM approach
- Creating a learning organization
- Encouraging pride of "ownership"
- Using consistent and fair rewards and recognition based on performance

7. Implement CDPM
- √ **Institutionalize the CDPM process**
  - ▸ **Constantly communicate the vision**
  - ▸ **Everyone leads by example**
  - ▸ **Get everyone and everything involved**
  - ▸ **Build trust and relationships**
  - ▸ **Target specific results using improvement methodology**
  - ▸ **Create a learning organization**
  - ▸ **Maintain pride of ownership**
  - ▸ **Use consistent and fair rewards and recognition**
  - ▸ **Continue support and commitment**
  - ▸ **Maintain investment**
  - ▸ **Perform project management actions**
- √ **Form other customer-driven teams**
  - ▸ **Customer and supplier project steering teams**
  - ▸ **Lead teams**
  - ▸ **Project teams**
  - ▸ **Quality Improvement Teams**
  - ▸ **Process teams**
- √ **Conduct learning**
  - ▸ **Organization**
  - ▸ **Team**
  - ▸ **Individuals**
  - ▸ **Business systems**
  - ▸ **Technical skills**
- √ **Continuously improve**
  - ▸ **Product(s)**
  - ▸ **Process(es)**
  - ▸ **People**
  - ▸ **Project(s)**

Figure 14.14 CDPM "implement" application process implementation flow.

- Ensuring continuous active support by top leadership
- Maintaining the investment
- Forming other customer-driven teams as appropriate

The implementation of CDPM throughout the organization depends on wide-reaching success with a variety of projects over a long period of time. This means that the organization forms many other customer-driven teams for an assortment of different missions.

During the implementation process, the learning within the organization involves going beyond basic knowledge and skills. It includes changing the behavior of the organization by enhancing the organization and individuals' abilities to perform the CDPM process, tools and techniques, business systems, and technical areas.

The organization focuses the CDPM process during implementation on continuously improving

- Product(s)
- Process(es)
- People
- Project(s)

Re-aim: Evaluate. The customer-driven project steering team stays in business over as many cycles as needed to monitor performance, solve problems, and continuously improve the deliverable. This is a long-term function. In our example, this is the implementation of CDPM. The organization decides to quickly make the customer-driven team structure its primary organizational structure for performing projects. This means that the CDPM process must be integrated into the entire organization. The evaluation application process assesses the complete implementation of CDPM in the organization.

Step 8: Monitor results. Figure 14.15 shows the CDPM evaluation implementation process flow for this example. All the customer-driven teams use metrics to measure

8. Monitor results

√ **Use metrics to assess:**
- ▸ **CDPM process**
- ▸ **Project(s)**
- ▸ **Process(es)**
- ▸ **Leadership**
- ▸ **Teamwork**
- ▸ **Customer satisfaction**

√ **Evaluate project results**
- ▸ **Schedule**
- ▸ **Cost**
- ▸ **Performance**

√ **Continue making improvements**
- ▸ **Return to analyze to enhance**
- ▸ **Continue in operations**
- ▸ **Build new project by starting analysis**

√ **Close out**
- ▸ **Gather lessons learned**
- ▸ **Provide appropriate reward and recognition**
- ▸ **Find other projects for the team members**
- ▸ **Maintain relationships for network**

Figure 14.15 CDPM "evaluate" application process implementation flow.

- Their CDPM process
- CDPM project results
- Process performance
- Leadership effectiveness
- Teamwork
- Customer satisfaction (internal and external)

In addition to the preceding metrics, each team continues to use project management tools to monitor schedule, cost, and performance for their particular project or part of the overall project.

From the results of the evaluation process, the customer-driven project steering team has the following options:

1. Keep the team continuously improving the project
2. Provide a new mission for the same team
3. Close out the project and disband the team

The teams that continue to make improvements advance by continuing in operation to enhance the current project, returning to design or development to advance the current project through reengineering, or going to analysis to relook at the entire project for process invention. The teams that continue with a new mission return to the analysis application process. These customer-driven project teams build on their experience. They still use the disciplined CDPM improvement methodology step by step. In many of the teams, the complete process achieves the mission with greater velocity.

Some customer-driven teams eventually reach closeout, while others will be constantly forming. In our example, the organization provides the following guidelines for closeout of a project:

- Gather lessons learned into an open database.
- Provide appropriate rewards and recognition.
- Maintain relationships for the network.
- Find other projects for team members.

Application of CDPM in the Future

This section provides a stimulus for thinking about the future of customer-driven project management. The application of CDPM will be even more appealing for organization and project groups in the future. The future of customer-driven project management will be shaped by many developments, including the structure of our global economy, new participative organizational structures, new quality improvement techniques, evolving concepts of self-directed teamwork, social and demographic patterns, changes in our social fabric, development of international organizations, changing worker diversity, new technologies, and improving information systems. Many of these factors already point to an increasing use of CDPM. For instance, information technology already allows the customer to drive projects as well as to share all necessary information regardless of geographical location of people or source of information. Information systems combining computer networking and telecommunication systems have already brought about fundamental shifts in how organizations perform and manage work.

The way an organization structures itself to perform its work will consistently reflect broader environmental conditions and societal context. Consequently, customer-driven project management with its strong focus on the customer, teamwork, and a disciplined methodology as a strategy becomes more attractive, because it allows the organization to be close to its customers, decentralizes the work, empowers workers, fosters small, flexible groups, and enables the kind of autonomy that many workers will seek in the future organization. These strategies target flexibility, adaptability, and maneuverability, which allow any organization the ability to act rapidly and proactively to vary conditions for future success.

Changes in traditional forces support the increasing use of CDPM

In the past, powerful forces have acted against the kind of reengineering of the project management process that places the customer as the leader of the CDPM project team. Project and product teams have traditionally considered the customer to be an outsider to the process. Seven basic factors led to this way of thinking:

1. The customer was rarely in the same place as the project team. Thus, spatial distance gaps *automatically* meant communication gaps, closed only by expensive transportation and paperwork reporting systems.
2. The customer was typically *assumed to be* uninterested in project or product design, an activity that involved technical expertise and design, and production capability beyond the customer's intellectual grasp (thus rationalizing the need for a project team).
3. The project team often felt "interrupted" in doing *its* work by unknowing customers who intervened in project or product planning and control at the wrong times. Their behavior was seen as inconsistent with the neatly laid out schedule of the project.

4. Since the project team often concluded that the customer was largely unaware of his or her own needs, let alone how to meet them, customer involvement was always seen as a threat to systematic project control, particularly as a project moved beyond design into production.

5. Project leadership was seen as requiring hands-on and personal access to the team. Thus, the project manager was seen as on-site and within a stone's throw if possible from all team members.

6. Project facilitation—the process of "enabling" the project team to perform, building team cohesiveness and trust and aligning work with interests and capability—was seen as the project manager's job, along with his or her "mainline" function of directing the development of the project or product deliverable.

7. Narrow views of project success often stemmed from a supplier's view of the financial performance of the project; if it made money it was successful; if not, it was not. Under CDPM there is *shared investment in project success*; thus, both meeting customer expectations *and* requirements and financial performance of the project are weighed during development and delivery of the project.

For these reasons, the project team was never seen as including the customer, especially in any key leadership role. The boundary around the team was clearly drawn to exclude the customer, who was seen as the "beneficiary" of its work rather than as an integral part of its working processes.

These seven forces are being overcome in many organizations today. In the future, these factors will be less of an influence. The customer does not need to be in the same geographical place as the project team. Today's information and telecommunications systems allow real-time communication almost anyplace on the globe. More and more customers are becoming directly involved in the project. It is common to see the customer as part of the project team. Project leadership will become increasingly more competent, with the use of leadership by everyone as the focus for success. With project leaders able to provide the vision and empower others to perform, CDPM will proliferate. As the new role of project leaders transforms the organization, the need for teamwork between suppliers and customers will enhance project facilitation capabilities. Most organizations are beginning to realize the need for shared ownership of project success.

The future of CDPM depends on blending CDPM into the total organization's business system with the optimum use of information systems.

Integrating CDPM with business systems. The future use of CDPM essentially depends on successful integration with fundamental business systems. In many cases, this will require process improvement, invention, or reengineering to optimize business systems. Totally new business systems are needed to support the future organizational environment, and will result from a fundamental relooking at everything an organization does—from financial systems to human resources programs, marketing activities, logistics support, procurement processes, and all internal policies and procedures. The process might involve inventing new paradigms, reengineering the project management process, a restructuring of the project teams, reexamining communication patterns, reestablishing roles and responsibilities, and rewriting administrative procedures.

In the future, business systems will allow all project phases to be performed more or less at once, with natural sharing of the work load. Communication patterns can focus on the customer. Roles and responsibilities will target maximizing the team contributions to the value of the deliverable. Team members can more effectively fix on changing customer expectations and needs by using first-hand information. Administrative procedures can be streamlined and simplified since more and more information will be shared through information systems. Communication can occur through electronic means. Information is updated instantly, and it is available to everyone simultaneously. Thus, burdensome administrative and contractual requirements stemming from old ways of doing business are eliminated.

Information systems. Information systems allow the customer to drive the project and everyone involved to share necessary information. CDPM is most effective when implemented in concert with an organizational strategy that fosters the optimum use of information systems to radically restructure the customer's relationship to the project team. The CDPM team does not simply "listen" *to* the customer; it is driven directly *by and through* the customer. Information systems enable the customer to communicate with everyone on the customer-driven team. They facilitate all aspects of CDPM, from assessment of customer expectations to evaluation of the customer's satisfaction with the deliverable. Information technology and new telecommunications systems now make it possible to eliminate virtually *all* of the barriers to full integration of the customer into every aspect of, and every decision in, CDPM.

CDPM will be facilitated in the future by the worldwide availability of integrated CDPM software. Computer packages that allow real-time information for continuous use by CDPM teams will improve the use of CDPM. These packages must continually enhance the total synthesis of all the information needs of the CDPM organization. For instance, computers will need to develop scenarios of the future and perform more complex what-if analysis. In addition, the information systems must allow the constant assessment of customer wants, needs, and expectations, monitoring of the performance of all processes, and continuous evaluation of customer satisfaction.

New enabling technologies. As the gaps of distance and time are closed globally through information technology—from portable computers with modems to wide-band data communications, from interactive video-disk and FAX systems to teleconferencing—it becomes easy enough to include any team member, including the customer, in virtually any team decision. Thus, the customer, or process owner, can be as intimately involved in the project team as is desired—and in CDPM organizations the customer can serve as the project leader.

There are four steps that an organization—whether supplier or customer—can take with its project management and information technology systems to assist its project or product teams to become fully customer-driven.

1. Tie the customer to all team members through electronic mail and other communication systems, so that all text and graphics involved in the project can be instantaneously transmitted. For instance, prototype designs can be graphically communicated to anticipate customer requirements and production and assembly difficulties.
2. Rather than putting the customer at the end of the chain for approval at every step, as in traditional teams, put the customer in the driver's seat to direct each stage of the project work by developing a software template for the customer to make virtually all the key strategic design decisions and to "kick off" each project stage with objectives and goals through the project electronic communication system.
3. Give the customer performance review and "outplacement" authority over the project team members, thus placing in the customer all the traditional authority of the project manager to assess the contribution team members make to meeting customer needs and to take them off projects if appropriate.
4. Document each step of the process so that each project experience becomes a learning process for both customer and project team.

Needless to say, a support system for CDPM requires considerable investment in information technology, if such systems do not exist as part of the company's communication and information infrastructure. Such a system must first be envisioned before it can be justified, designed, produced, and acquired. Thus, the corporate visions of both the supplier and customer must include a new paradigm regarding the involvement of customers in project planning and control. "Seeing" the customer as project leader is the first step in finding methods and systems to serve that purpose.

For this reason, we believe that CDPM, and a system of information technology to support it, is a necessary ingredient in any corporate strategy and vision statement. It must be part of the value system of the corporation, an integral part of its culture. It is not the only required element, since so many other factors enter into the success of project and product management firms. But without it, project teams will lack the means to fundamentally reduce time-to-market processes. More importantly, without it, project teams will lack the means to ensure that customer expectations are exceeded, since they will inevitably lose sight of the customer as a critical juncture in the planning and control process.

The future of CDPM

CDPM provides a major contribution to the variety of management approaches required for the future. Organizations of the future will need to apply a variety of approaches to be successful in a global marketplace. There will be no one best approach or panacea. CDPM offers a gradual expanding of capabilities of an organization beyond project management and total quality management. It is flexible and adaptable to many organizational situations. CDPM is a management approach geared to success today with a view toward the future.

Main Points

Today, customer-driven project management can be applied to any project.

CDPM is the means to a specific end.

CDPM requires each organization to develop its own content within the CDPM process.

It is important for each organization to "own" its particular CDPM application.

The specific application of CDPM depends on the external climate, internal organizational culture, and external and internal customers.

The CDPM application process requires each practitioner to

- Analyze—get ready
- Design—get set
- Develop—aim
- Implement—fire
- Evaluate—re-aim

The CDPM application process can be used for any project short or long, simple or complex.

The CDPM application process may be used many times for a project, even many times within itself.

The CDPM process remains the same for each project, but the content within the CDPM process will be different for every project.

The application of CDPM will be even more appealing in the future.

CDPM's allure stems from its facilitation of successful projects.

Emerging technologies will be a major stimulus to the use of CDPM tomorrow.

Appendix

Cases

Focus: This appendix presents short case studies in four areas: case 1: logistics communication; case 2: computer-based training; case 3: higher education; and case 4: health care facility communication systems.

Case 1: Logistics Communication

This case is from Warner Robins Air Logistics Center (1926th Communications-Computer Systems Group), which won the 1991 Federal Quality Institute Quality Improvement Award (case presented compliments of the Federal Quality Institute). The case has been adapted and enhanced to build on the published version to illustrate how the work might have progressed under the eight phases of the CDPM system.

The eight phases

1. *Define the quality issue.* The issue was prompted by complaints from internal customers at the center about the inability of the communications system to meet customer needs. A quality-improvement team was selected to identify the problem and the customers, define the process, and explore available options. Once identified, the user-customer was placed in the role of customer project leader, a communications project manager was appointed team facilitator, and the team was structured to represent all the department functions. The user-customer represented the end user of the communications equipment and systems.

After an initial organization meeting, the team quickly confirmed the limited capability of communications equipment to meet expanding Air Force Base customers. Delays and disconnections in the local area network were becoming a major issue. A statement of the problem was developed and consensus reached that the team had identified a major problem, one that would require analysis of root causes and some "fixes" to current network systems but perhaps also a "reengineering" of the whole system.

2. *Understand and define the process.* The team prepared a flowchart of the network support system under the leadership of the customer project leader. The team identified a series of over 100 steps in the entire process of providing data communications capability from the generation of the need for data through identification of needs and requirements to the point of access and use. After doing so, the team concluded—backed up by the customer—that the issue was the customer's expectations for the performance of the data communications capability and the inability of the system to meet those expectations.

   It was clear in this process that the customer project leader was developing a clearer view of the expectations and needs of her colleagues as well; thus she was speaking not only for herself but also for many other users as well. There was good engagement between the team and customer, facilitated by the project facilitator.

3. *Select improvement opportunities.* Again under the leadership of the customer project leader, the team brainstormed the issues and supplier-customer steps involved in the data communication process to find opportunities for improvement. The team found that a key access point was obstructed because of the current communications equipment and decided to focus on this point and to proceed to root-cause analysis. Eventually, the team selected this obstruction as the key first opportunity.

4. *Analyze the improvement opportunities.* Here the team used a cause-and-effect diagram to categorize underlying causes for the target problem, a point in the system where access was denied. This effort was aimed at ensuring that the team did not focus prematurely on solutions and specifications before it had a good idea of the customer's problems in the target area.

   The team found that the problem was inherent to the software and hardware platform and that improvements would be difficult without major system enhancement. The team also found that it had to reengineer the system to meet expanding needs and that a new process capacity was in order.

5. *Take action.* During this step, the team performed a quality function deployment (QFD) exercise. The QFD "house of quality" pro-

duced a structured set of customer requirements in the areas of data quality and reliability, effective implementation of the new system, and cost. The team's recommendation was to introduce leading-edge laser technology, the proposed specification for a new system linkage in the communication process.

This generated a project planning and control process. The team obtained project management support and training in project management software systems, chose TIMELINE, and proceeded through the concept and definition phase of system development. The team took responsibility for moving the process from analysis to project delivery and employed project management techniques, working with the procurement process and the ultimate system manager as process owner.

Employing project management techniques and tools, a new communications system was produced. The sequence of steps included

a. Develop the quality function deployment "house of quality."

b. Translate to an overall picture of the deliverable through a work breakdown structure (WBS).

c. Identify the key tasks embedded in the WBS and enter them into TIMELINE.

d. Line the tasks up in terms of interdependence and indicate predecessor in task form.

e. Display the resultant critical-path network.

f. The customer project manager authorizes the beginning of work, followed by authorizations to proceed with other tasks.

g. Progress is gauged through earned-value analysis performed by TIMELINE based on estimated costs to complete.

h. Final deliverable is approved by customer.

6. *Check results.* Once procured and implemented through project and procurement management techniques, the new system was checked in practice by the team. It was found to have expanded the capability for connectivity by as many as 15,000 customers and eliminated use of less capable and more costly equipment. The team indicated that this project deliverable saved $3,300,000!
7. *Implement the improvement.* Here the team developed a new administrative procedure manual for users and a training program for new system managers and users.
8. *Monitor results for continuous improvement.* The team continued to monitor the data communication process and eventually found new system enhancements, proceeding through the eight-step process again.

This is a classic case of how total quality management (TQM) in the front end and project management as implementer and deliverer can combine for a substantial increase in customer service. The developmental nature of the project was identified early as the primary focus of the quality-improvement team. Assessment of customer needs and analysis of processes, problems, and root causes all helped the team assure itself that the planning and control of the ultimate deliverable would be a "value-added" one. Performance for the customer and overall costs and benefits were the points of departure for assessing the effectiveness of the results, not just a narrow assessment of project cost and schedule.

This case illustrates the potential of project management tools and techniques to add legitimacy and new meaning to the quality-improvement effort. Use of project management software such as TIMELINE to translate from customer requirements to work breakdown structure gave this team a new project implementation function. It was apparent from the professional outputs of the software that this team was going to implement change, not just recommend it. The project thus took on more significance to all the participants and stake holders in the process, whether suppliers or customers. The quality team had become a project team, distributing effective graphics and tables showing the project schedule and budget to develop the deliverable.

Case 2: Computer-Based Training

This case involves identifying and meeting internal customers' needs for learning and training systems. The issue here is improvement of the process through which automobile mechanics learn the basics of their trade. The project involves a Prince George's County, Maryland, project and discusses the development of EDUCATE, a pilot effort targeted at the vocational high schools in the county. Below is a discussion of how the project moved through the eight steps of the customer-driven project management process. The case represents a good example of how a quality-improvement effort requires good project management tools to succeed.

The eight phases

1. *Define the quality issue.* The issue was prompted by the increasing costs of training automobile mechanics and the limitations of traditional training methods. The team was established to accomplish two objectives: (1) perform a quality-improvement analysis of the current training systems from the customer standpoint and (2) design, produce, and test a new approach to meeting customer

needs. An automobile mechanic was named to lead the team as customer project leader, supported by a project team facilitator.

The issue defined by the team was the process of adult learning, in this case automobile mechanics' course work delivered in a classroom, and the need to fundamentally reduce the costs of education while increasing quality. The issue was, "How can the process of delivering key courses to automobile mechanic students as customers be enhanced while reducing the costs to the provider and the student?" The customer was the automobile mechanic, and the analysis required a full understanding of the automobile mechanic's operational environment.

2. *Understand and define the process.* The process was defined in terms of four phases, curriculum development, course development, instructor staffing, and delivery and discussion of course content, all leading to a learning outcome evaluated through traditional examinations. The process was outlined to have 45 steps from beginning to end, and a work flowchart was prepared. Once completed, the work-flow process was approved by the customer project leader.
3. *Select improvement opportunities.* The process was reviewed by the team to find key quality weaknesses. One step in the process was seen by the team to be critical in restricting the delivery of automobile mechanics course material to students. This was the step in which classes were closed because of student/instructor ratio standards and/or because of the lack of available training space and facilities. Both the customer and supplier were often disappointed and frustrated by this step, which seemed to be driven by reasons not always clearly important to the automobile mechanic's learning process.
4. *Analyze the improvement opportunities.* In the "fish-bone" analysis, many root causes of this problem were determined by the team. The key root cause was the lack of an alternative mode of educational technology and delivery that could open up learning of content through means other than classroom attendance. Behind the root cause was an outdated theory that assumed that the training had to be delivered in a classroom and that a particular ratio of students to instructors had to be maintained to ensure effective training.

   The team clearly felt that "reengineering" was in order; a new delivery system would have to be described and implemented.
5. *Take action.* With the customer a very present force as leader of the team, the team decided to recommend that it take on the job of producing the new educational technology. The team became a project team at this point, identifying needs and design criteria through the QFD approach and developing a work breakdown

structure, a critical-path network for the project, and a budget using TIMELINE software.

Through brainstorming and the use of QFD, the outlines of a new computer-based training program were identified, and the team acquired the budget support to proceed into project design.

Employing project management techniques and tools, a new, distance-based training system based on computer-assisted tools was produced. The sequence of steps included

a. Develop the QFD "house of quality."

b. Translate to an overall picture of the deliverable through a work breakdown structure (WBS).

c. Identify the key tasks embedded in the WBS and enter them into TIMELINE.

d. Line the tasks up in terms of interdependence and indicate predecessor in task form.

e. Display the resultant critical-path network.

f. The customer project leader authorizes the beginning of work, followed by authorizations to proceed with other tasks.

g. Progress is gauged through earned-value analysis performed by TIMELINE based on estimated costs to complete.

h. The final deliverable is approved by customer.

6. *Check results.* The new system was installed and tested after 1 year as a pilot operation. The system increased customer satisfaction in many areas but raised new issues stemming from the lack of direct instructor support. As is often the case, the new system raised new opportunities for quality improvement.
7. *Implement the improvement.* Once these new issues were resolved, using root-cause analysis, the team was empowered to implement the new system. Implementation required training trainers, developing new computer-based materials, and changing some marketing approaches.
8. *Monitor results for continuous improvement.* The team decided to stay in place to monitor the automobile mechanics' training process and will return to the first step of the analysis to identify new opportunities for continuous improvement.

Case 3: Higher Education

While there is much that makes higher education different from a private business, they share a common interest in satisfying customers. In this sense, a good education—and the learning process that it requires—is a consumer product; it is measurable, can be marketed, and has lasting value to the consumer. When the education process is narrowed down to the actual operational procedures and activities involved, such as teaching technique and delivery of appropriate instructional materials, it is clear that these procedures can be pinned down rather easily, just like, for instance, an order-processing procedure. Thus educational institutions in increasing numbers are using total quality tools and techniques to improve their delivery of products and services to customers.

In 1991, the University of Maryland, University College organized a total quality management process. The process is heavily oriented toward building teamwork across units in both administrative and academic operations. Over 30 teams were organized to address key University College processes such as faculty development, phone communication, scheduling classes, student financial aid, course development, and contract management.

This case covers one of these teams, the course-development team, and how it approached the process of putting the customer in charge of the quality-improvement effort in the complex business of designing and producing adult education courses and course materials. The discussion will follow the eight steps.

The eight phases

1. *Define the quality issue.* The quality effort was triggered by a clear indication that the internal customer of course development, the Assistant Dean for Open Learning, was clearly unhappy with the instructional materials he was getting. His needs for high academic quality products were not being addressed in the course-development process. The internal supplier office, the Office of Instructional Development, knew that if the internal customer was not satisfied, then the ultimate customer, the student, would not be satisfied either. Thus, it decided to address the quality issues squarely.

   The office decided to use the quality-improvement approach. The process began by putting the customer in charge of a team of designers and publishers to define the problem, outline the course-development process, and identify opportunities for improvement. Weekly meetings were held with the customer leading the discussion and serving, in effect, as team leader. The issue began to surface. While the customer office was looking for presentable and analytical print-based course guides for its students, many of whom did not attend class regularly and needed to keep up through instructional materials, the upstream supply office was oriented toward new educational technologies, not print-based products. The supplier was marketing new technology with little success, and the customer was looking for high-quality printed course guides written clearly and concisely by leaders in their fields.

It is important to point out that had the customer not set the agenda for the early discussions and served as project leader, it is not likely that the supplying office could have developed an acceptable response to the situation alone. This is the value of customer-driven project management; it closes the gap between the project team and the customer early, when the deliverable is being conceptualized and designed, not later when the issues are confined to narrow command and control decisions.

2. *Understand and define the process.* The team entered the second phase with the objective of defining the current operating process of planning, designing, and producing print-based course guides and proceeded to prepare work flowcharts and step-by-step analyses to understand the system. It involved a complicated interrelationship between project team leader, course author, editor, peer reviewer, word processor, instructional designer, graphics specialist, and program customer (curriculum specialist), along with several internal checks and balances. An analysis of the whole process was produced as the basis for finding opportunities for quality improvement. The bottom line was that there had been no concerted effort to understand the customer's requirements and needs and that the supplier office had been trying to market its products based primarily on what it determined to be the needs in education technology.
3. *Select improvement opportunities.* The team analyzed the process to find ways to listen to the customer, even to put the customer in charge of the process. Clearly, the culture of the organization would have to change to listen more attentively to the customer's needs, and perhaps some adjustments to staffing and office systems would be necessary to emphasize good publishing procedures and the academic integrity of the instructional products. Most important, the supplying office would need a methodology for translating customer needs to project design specifications.

   There were quality breakdowns in the process itself which signaled the lack of attention to the printed product. For instance, desktop publishers and editors were not included in key early steps of the process in which the customer's needs were being articulated and where preventive measures could be taken to reduce defects and problems in the eventual printed product. This conclusion was facilitated by reviewing the work flow documents and brainstorming problem areas.
4. *Analyze the improvement opportunities.* The root cause was found to be the lack of a clear and shared "mental model" of the process

that included the customer and all the project team members in understanding the problem. A more specific problem was the lack of early involvement of the desktop publisher and editor functions in course development so that the customer's concerns for quality printed materials could be addressed early. In addition, hand-offs from designer/author to editing were not effective, and the internal customer needs of the publishing function—finished copy, well designed—were not being met.

There were several iterations of root-cause analysis before a priority was set on the first corrective action to be taken to address the customer need, the early involvement of the publisher and editor function.

5. *Take action.* This is where the team decided to change the course-development process by planning a new deliverable to address the priority root cause, the lack of a project management model—and a shared vision of how that would work—that involved the customer, publisher, project manager, and editor in the process. The team identified customer needs, essentially the need for a model explaining the new system, and then translated these needs using QFD into a work breakdown structure roughly equivalent to the outline for the new manual.

   The procedure was basically as follows:

   a. The customer-driven team develops the QFD "house of quality," structuring the customer needs and requirements.

   b. Translate to an overall picture of the deliverable, through a work breakdown structure (WBS), essentially the outline of a new course-development manual.

   c. Identify the key tasks embedded in the WBS and enter them into TIMELINE.

   d. Line the tasks up in terms of interdependence and indicate predecessors in task form.

   e. Display the resultant critical-path network.

   f. The customer project leader authorizes the beginning of work, followed by authorizations to proceed with other tasks.

   g. Progress is gauged through earned-value analysis performed by TIMELINE based on estimated costs to complete.

   h. Final deliverable is approved by customer.

   Development of the model required the involvement of several course participants, who proceeded to reinvent the process for the first time under the direct leadership of the customer serving as project leader. The team became at this point a customer-driven project management team. Project objectives, a task list, and a project schedule and budget were developed, and the team initiated a brainstorming process with the customer project leader to confirm that it was on the right track, facilitated by a team facilitator.

The team produced a new project management manual, providing for early involvement of the publisher and editing function in course development with the customer, and developed a process of training and education that would ensure more sharing and communication during the process.

6. *Check results.* The team entered the measurement stage with some misgivings about measuring the impacts of this particular improvement on the process. Issues of personal accountability were raised; if staff members were going to be asked to measure how well the system was working, would they be held accountable personally and in the performance-appraisal process for the results? Or could the process be kept on the high ground?

   The team decided to remove this issue by obtaining approval to confirm with every participant that data and measurements of process performance would not be included in performance appraisals. Once this issue was resolved, the team decided on key indicators of process quality:

   - Time elapsed in process by course-development phase
   - Cost of major processes
   - Lost time due to unclear customer requests
   - Performance of support systems (photocopying, mail, word processor equipment)
   - Workload changes

   At the same time, the team developed a hypothesis that there was a direct relationship between the ability to select the right course author and keep the author throughout the process and the ability of the project team to meet the customer's quality, time, and budget objectives.

7. *Implement the improvement.* Once the new system is in place and tests and measurements suggest that the system is performing better on the basis of the chosen indicators, the team will go into implementation, writing procedures, developing training materials, and ensuring that the new procedure is internalized in the office. Whether the improvements can be attributed directly to the corrective actions taken will not be a major focus for the team. The assumption will be that corrective and preventive actions of this sort in a complex process set off new dynamics which themselves will serve to improve quality. The point is to affirm that the process capacity has increased, not necessarily to attribute all the increase to particular actions.

8. *Monitor results for continuous improvement.* The team will stay in place to monitor the process. It will likely return to some fundamental issues, such as the lack of an effective system to include graphics and other media in print-based course development, and move through to corrective action again. In this case, the team will become a project development team, designing, acquiring, and testing computer software systems to meet the needs it brought to the surface in the initial process.

Case 4: Health Care Facility Financial Communication Systems

This case addresses an information system project that evolved out of a quality-improvement analysis in a health care facility. This particular facility, operated by a foundation, serves chemically dependent patients and their families.

The eight phases

1. *Define the quality issue.* A quality team was established by the foundation management to define the problem of providing useful financial information to internal managers to enable them to make cost-effective decisions. Although there were many customer issues in the current system, the symptom addressed was an $800,000 loss on one single contract.

The team was under the leadership of a customer representative, a member of middle management. Under her direction, the team brainstormed and came up with a problem statement that indicated that financial reporting was not meeting the customer's need for timely and accurate information in a format that would allow key decisions to be made when they needed to be made.

2. *Understand and define the process.* The team outlined the process of providing information to the six operational departments of the foundation. The team found in the workflows that there were major disconnects in logical and physical data links, that the reporting system step in the process did not meet the needs of the customer, and that the manual filing procedures were inadequate in the view of customers who used them.

3. *Select improvement opportunities.* Here the team looked for major defects in the system. The team estimated that the absence of physical data links between the six departments was costing an estimated $1540 a week in terms of redundant labor alone. Manual transfer of information in different formats became the major target.

4. *Analyze the improvement opportunities.* Guided by the customer, the team stepped back from the analysis to look at underlying

causes, using a cause-and-effect analysis diagram. After exploring several "people and technology" issues, the team decided that the major root cause of customer dissatisfaction was that the accounting manager was the only source of key data and that when he was unavailable, customers went unserved. The root cause of the problem was that financial data were being considered personal data, not corporate data.

5. *Take action.* Here is where the team took on the job of designing and delivering a new system to "make the data corporate." Interviews were conducted with all customers to confirm needs identified in earlier analysis and brainstorming, and user acceptance tests were designed and verified. A QFD "house of quality" was constructed, and then a work breakdown structure was developed to provide a graphic picture of the deliverable, including hardware and software, user screens, and database structure. A project schedule, Gantt chart, critical path, and resource schedule were developed.

The team then guided the project through definition, production, and operations and testing to produce the new system description and manual. If effect, the team was designing a new mental model of the system, as well as a new operating procedure.

6. *Check results.* The team had developed a user acceptance plan and in this step implemented it. Users had reviewed a prototype of the system, thus this step was relatively easy.

7. *Implement the improvement.* The foundation is now running the new system and using it to make financial and resource decisions. The team continues to monitor and will begin to write manuals and procedures for the process, in addition to a training guide.

8. *Monitor results for continuous improvement.* Here the team will recycle back into the operation of the new system, find new opportunities to enhance it through customer feedback, and perhaps move again into project definition and delivery.

Appendix B

Personal and Professional Impacts of CDPM

Focus: This appendix explores some key personal and professional issues raised by customer-driven project management. The purpose of this discussion is to concentrate on the key building block of CDPM—people. In the United States, quality still comes back to individuals growing and developing in an organization truly concerned with them as people. To be successful in our culture, the concepts of customer-driven project management must be made relevant on an individual basis.

Introduction

The purpose of this discussion is to put some personal and professional meaning on the CDPM process. We start the process of change and empowerment ourselves, through the mental models we use and the personal decisions and choices we make in the daily routine of the workplace. While the employer and the organization can open up opportunities, the individual must take advantage of those opportunities for anything to happen. We must want to be empowered to be empowered, and we must seek out opportunities to grow and develop to be educated. We must draw the boundaries of our own jobs wide enough to allow creative uses of the new talents and competencies we acquire across larger and larger parts of the system.

But how should we "play" in such a world, one which, on the one hand, holds out so many opportunities for growth and empowerment in meeting customer needs but, on the other, seems so determined to place accountability for failure at the level of the individual. What if the organization does not support my empowerment and does not have the infrastructure to help me widen the boundaries of my job?

The purpose of this section is to speak personally to each reader as a potential or current member of a customer-driven project team and to every person in an organization going through the quality awareness process. The discussion is designed to help you deal with the opportunities that CDPM and the total quality movement bring to the workplace and specifically to help you participate effectively in customer-driven project teams.

Three sections are provided to stimulate your thinking about what all this means:

1. Self-development
2. Using performance appraisal to get feedback
3. Empowerment

Self-Development

Each individual member of the CDPM team is responsible for his or her own specific evolution. Nothing will change unless there is awareness and understanding of one's potential in the context of CDPM and the curiosity and drive to listen to the customer. The organization and its leadership can help this happen through providing self-development tools and techniques and through communicating with and empowering each employee, but the organization cannot make it happen alone. It requires the commitment of each working person, regardless of role or level in the organization. In effect, the focus on employee participation and involvement has never been more clear; those who take advantage of it will grow personally and professionally and participate directly in making the organization more competitive both domestically and globally. Those who do not may find themselves regretful later, perhaps in different times with less stress on empowerment and self-development, that they did not take up the challenge when it was available.

Self-development starts with an attitude, one that welcomes feedback from several sources inside and outside the organization and which is able to integrate that feedback into productive learning and growth. Because the whole process of looking at one's self through others' eyes raises questions of competency and comparison, it is not easy for team members and people in general to develop a healthy attitude about accepting feedback, especially if that feedback is negative.

Self-development requires an attitude of openness. This openness evidences itself in many ways in the workplace and indicates a willingness to solicit and learn from others about your performance. Customers, peers, employees, and leaders will provide useful data if they feel the solicitation is genuine and will welcome the opportunity to be candid. In most cases, they have harbored views of you long before you requested feedback; thus to open the conversation is often comforting for those who participate.

Here is the basic area of inquiry as one approaches the question, "How do I engage in a self-development process in the organization so that I can identify the best opportunities for my contribution to serving customer needs and continuous improvement?" Answers to this question can be solicited formally through questionnaires, but the richest feedback does not come from the written word. It comes from the daily interactions one has with customers and suppliers and from the regular feedback that customers give about the "service."

1. Do I do the right thing right the first time? This issue begins with the individual's sense of accountability to completing the job to be done, adding value to the work flow as it comes in and out of personal control, and ensuring that one is not expecting someone else to fix the problem later. This challenge is at the heart of the enlightened approach to work. Ensure that the inspection and appraisal functions are eliminated so that each point in the work flow can add value and not commit costly time and effort to doing work over. In CDPM processes, this means that as projects move from front-end analysis and planning in the first several phases and that as the project deliverable is designed, produced, and installed, each team member assumes that his or her individual component of the work breakdown structure is done right first, according to the needs of the next customer down the line in the project development process.

There is in the literature about bureaucracy and administration the theory of the "tyranny of small decisions." This theory indicates that the progress and productivity of an organization stem not from big investment decisions but from the many small decisions that individuals make as the work progresses. In customer-driven project management, many decisions are made that combine in effect to create the quality of the process itself. It starts with each member of the team and each party to the process feeling responsible and empowered to do the right thing right the first time and to inquire of the next customer down the line if it is not clear what the right thing is.

2. How do I come across in the organization with respect to serving internal customers, doing it right the first time, and continuously improving myself? This problem has to do with "the book" on an individual, those stories and anecdotal images that gather almost inevitably among peers, bosses, and colleagues about individuals. One can seek out "the book" on one's own performance, but candid feedback is often difficult to get if one of the issues in the "book" is one's inability to listen and seriously address opportunities to improve service to one's internal customers.

3. How do I contribute to gaining consensus in the team, and what is my role in moving the group from its orientation and dissatisfaction stages into resolution and production? This criterion has to do with whether one is a team player or an individual performer. Getting to this issue in self-development involves understanding the cultural and social underpinnings of American workers and professionals. We have not been trained to work in teams; rather, we have trained to do our "thing" and to pass it on, to hand off "my work" to the next workstation. Part of the philosophy of CDPM has to do with the ability to rise above this attitude and to enter into a full commitment to working every customer's agenda as we work our own. This means that in the analysis of opportunities, one is willing to contribute to the brainstorming process and to educating the team. This means that in the development of corrective actions and project development one is willing to work within the context of meeting customer's needs down the line, not "handing off" the work or leaving it on the stump hoping someone will pick it up.

4. What professional development opportunities do I seek in improving my grasp of my job? How about the tools and techniques of CDPM? This issue has to do with how willing you are to pursue training and education as the organization opens up possibilities to do so. CDPM opens up new ways to learn and grow with the movement toward customer-driven work, but the energy to follow up on these opportunities must start with the individual, not the training or human resources offices.

5. What is my emotional investment in teamwork and CDPM, and how much effort and energy am I able to bring to the process? This is a difficult issue to articulate, but it is characterized by the "affective" side of the workplace. Attitude drives behavior, and training and education guide the effectiveness of behavior in the workplace. The beginning of individual fulfillment in the improvement process is the appreciation that the broadest and most meaningful benefit of CDPM in the workplace is the emotional uplift it represents to all those who participate in it.

The process of responding to the opportunities that CDPM brings to the organization begins with self-development and work planning—designing one's own career path based on feedback from others in key areas of CDPM. This planning process does not look upward into the "home" organization but rather looks outward into the customer's environment. Traditional upward work planning orients individuals to training and education that equip them for higher and higher levels of responsibility to manage resources and people. In contrast, outward work planning is oriented toward learning internal and external customer needs and processes so that one can fine-tune the ability to work in a team process serving customer needs. Self-development is really "peer" assessment. Work plans for the former include training and education in supervising others and controlling programs; work plans for the latter include training and education in quality improvement, project management, customer service, understanding a market and a client base, and team facilitation, management, and leadership.

In a customer-driven organization, work planning does not start with discussions with the boss on program objectives, goals, and performance appraisal criteria, as in traditional performance management systems. It begins in a peer-assessment process that never ends—and which is driven by feedback from key colleagues and customers in many areas. These areas of personal inquiry include, for instance, your ability to work in teams and your customer orientation. Appendix C provides some self-assessment questions that can be addressed in such a process, either conducted formally through questionnaires or informally through open communication and dialogue. The idea here is to find your strengths and play to them.

Individual CDPM competencies

An individual is "competent" when he or she is able to take certain knowledge and skills and successfully apply them in a given customer-driven organization. Knowledge and skills are demonstrated as behaviors when they are applied. Successful application is the key. Knowledge and skills have little or no impact on the business if they are not applied successfully.

Each individual is personally responsible for developing the right competencies. Competencies help to define a blueprint leading to successful performance. Major customer-driven project management individual competencies follow:

- Ability to work in teams
- Customer orientation
- Ability to see work flows and the big picture
- Ability to do high-quality work
- Effective and efficient use of resources and time
- Ability to communicate
- Interpersonal relationship skills
- Conceptual skills
- Problem-solving and decision-making skills
- Technical abilities and job knowledge

- Personal initiative
- Coaching, mentoring, facilitating, and leading skills

Learning process

An individual can become competent through a variety of ways. Figure A.1 shows how a continuous learning process develops competence. The learning process starts with knowledge as the foundation. Knowledge is gained mainly through self-development and education. Skills are the next element of the learning process. They can be gained by training and experimentation. Next, the person needs to apply the knowledge and skills on the job. Facilitation and on-the-job training help this process. Finally, the learning process involves experience. Coaching and mentoring can be used to improve the learning process at this stage. All this learning takes place in the context of the organizational culture. A person may be competent in one organizational culture and not competent in another.

Development process

Each individual needs to use a development process to become competent within the specific organizational culture. The development process as shown in Fig. A.2 consists of the following:

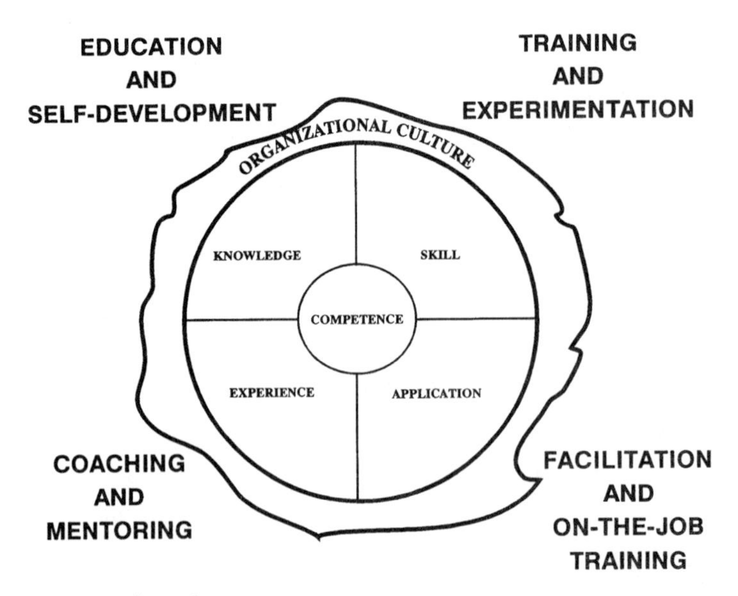

Figure B.1. Learning process.

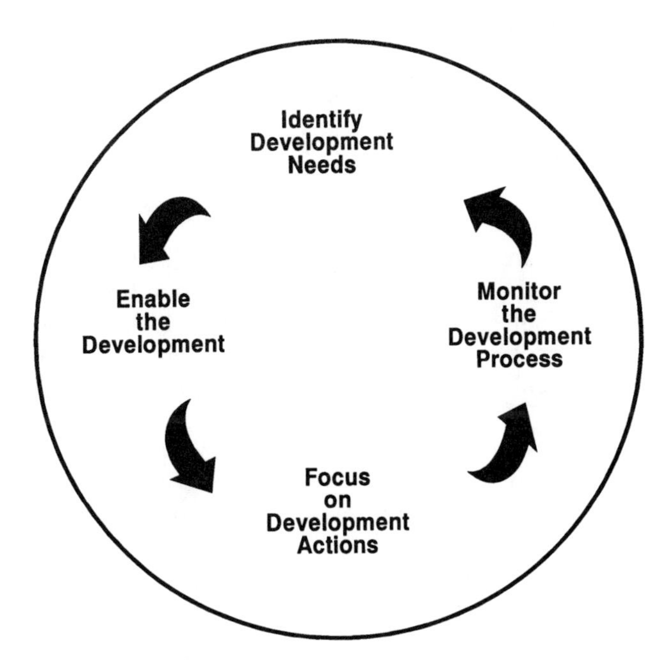

Figure B.2. Development process.

- *Identify development needs.* Use the self-assessment instruments to start the development process. This tool will assist you in the identification of individual development opportunities. It is always advisable to have others also complete assessment instruments to provide you with additional feedback.
- *Enable the development.* Determine high-priority individual development and growth needs. Development needs are short-term needs, and growth needs are longer-term needs. These needs should be included in an individual development action plan. The action plan provides the what (actions), when (milestones), and how (education, training, coaching, mentoring, on-the-job training).
- *Focus on development actions.* The individual takes appropriate action. This helps build competence to improve the individual's effectiveness.
- *Monitor the development process.* Regularly review progress on the individual action plan.

Using Performance Appraisal to Get Feedback

While most discussions of performance appraisal proceed from the manager's viewpoint, this discussion proceeds from the individual's viewpoint. How do you use the appraisal system to continuously improve? The assumption behind CDPM is that one seeks every possible avenue for feedback, even the often "dreaded" performance appraisal system. CDPM assumes that performance appraisal can be a useful way to gain valuable feedback about your ability to work individually but think collectively.

Customer-driven project management requires a high degree of alignment between individual members' goals, team goals, and organizational goals for meeting customer requirements. This alignment can be ensured through self-development as described above, but there is another important step in the customer-driven project management process: the process for gaining feedback on your individual and team performance.

Performance appraisal gives everyone in an organization a unique chance to improve personally and professionally, but there is an important practical reason for focusing on improvement in the appraisal process. It is because many organizations consider a "meets standards" rating to equate with "job proficiency," the competent, high-quality performance they expect from all staff. In other words, you are expected to master your job requirements and to be able to accomplish them simply to meet the standard. And while there are some exceptions to this rule, simply doing a good job does not equate with excellence and a high rating. What is increasingly valued in this process is the willingness to improve the job and the way things are done and to improve personally and professionally at the same time. In general, this means that to achieve high performance ratings, one is expected to continuously improve the way things are done. Thus it pays to focus on improvement.

Here are seven steps that might help you use performance appraisal to achieve *personal continuous improvement.* These seven steps come from a model developed by the Logistics Management Institute.

The process first involves establishing a *vision* for your own improvement effort, then *enabling* that effort, then focusing your *behavior* and your *expectations* to achieve continuous improvement in your job and the processes that you touch in your job, then *helping others* to improve, and finally, *evaluating* your efforts to improve. Here are the steps in more detail.

1. *Envision personal improvement.* Before you can begin to improve, you have to decide that there is a need for improvement and then determine the general emphasis of your improvement effort. You therefore build you own self-awareness of the need to improve and your individual ability to improve. Assessing your relationships within the organization and with your internal customers and your inter-

nal suppliers provides a fundamental understanding of where you are now. From this assessment, you can develop your expectations for your own behavior, and you can begin by creating a personal vision for your improvement. In other words, how do you see yourself improving, and what is your own personal vision for yourself?

2. *Enable personal improvement.* You then make your vision a reality by smoothing the road along which you will travel. This effort starts with educating yourself about which improvements are considered high priority. Seek training for yourself in the skills and principles you, your manager, your customers, and others see as essential to your effort. "Enabling" or "empowering" oneself is a process of learning—learning about your own potential, about the "big picture," about opportunities for training and education, about participating in teams, and about tools such as total quality management which are available to achieve continuous improvement. You also should seek the support of others, not so much from the standpoint of gaining their approval, but more important, by cultivating their help in removing the barriers to your effort. In effect, get yourself ready to improve.

3. *Focus on improvement.* You then focus your improvement effort through establishing goals for that effort and then ensuring that your improvement activities are aligned with those overall goals. Your should develop a clear improvement strategy to guide your efforts and ultimately use that strategy to evaluate the success of those efforts. Making improvement a high personal priority and creating time in your schedule for improvement activities are vital to this effort. They are a clear demonstration to yourself and to others of your commitment to improvement. In other words, plan for improvement.

4. *Improve the job.* Your job may be defined as the collection of processes you "own." You should establish control over your job by defining your processes and understanding how those processes interrelate and relate to others, including your customers and your suppliers. By removing unnecessary complexity from your processes and pursuing small, incremental improvements, you will substantially increase the effectiveness of your own performance in your job, and you will greatly enhance your personal improvement effort. In essence, make the changes in your job that are necessary to make your life easier and more rewarding.

5. *Improve yourself.* You can demonstrate leadership in the overall quality-improvement effort through your commitment to personal improvement. This means that you establish and adhere to a structured, disciplined approach to improvement that clearly defines your goals and requires steady, consistent improvement in your personal performance. You also should facilitate communication between your-

self and others and among others. Remove the barriers you place in your own way, seek the assistance of others to remove the barriers you do not control, and work to eliminate your own fears of change and improvement. This is best done through education and communication with others. Depend on your vision as your guide for improvement, and use that vision to maintain your momentum. Start by looking at yourself.

6. *Help others improve.* Through your improvement effort, you will help your unit as a whole improve. An essential part of your personal improvement effort should be to help others improve themselves and the organization. By training and coaching others, by creating more leaders, by working to create teams and eliminate barriers, and by encouraging others' improvement activities, you will spread your own example and your enthusiasm throughout the organization. Personally, you can make a substantial contribution to the individual improvement efforts of others. Spread the word.

7. *Evaluate your improvement progress.* You then measure your success in your efforts to improve. By measuring your performance against your vision and your plan, and by documenting your improvement efforts so that they may be shared with and used by others, including your supervisor, you will benefit the most from your efforts. Ensure through your evaluation that the improvement effort itself is rewarding and provides further incentive for the continuous improvement effort. Celebrate your success and the success of others.

Empowerment

Empowerment is a personal strategy to improve. We empower ourselves to the extent that we take advantage of the process of ensuring that all employees, and particularly those closest to the customer, have the flexibility and support to meet customer needs and expectations. This process is designed to open up the creative and innovative potential of the organization and to put everyone to work thinking through their jobs in relation to customer needs. But what does it mean to each member of an organization when the organization communicates the message, "Since quality is the job of every employee, you are now empowered to carry it out." It sounds simple, and, of course, it assumes that each member of the organization seeks out and can handle the "handoff" of responsibility, authority, and resources that goes with empowerment. And it assumes that the support is there to make it happen.

To the individual member of a customer-driven organization, the empowerment process can be a vexing problem. This is so because it is not clear that the stated benefits of empowerment, improved quality of work life, professional and personal development, rewards and recognition, new opportunities to assume new jobs and roles, and increased latitude in decision making, are really all so "beneficial" in practice. Let's explore each to see how the individual can assure an appropriate personal response in a CDPM environment.

1. *Improved quality of work life.* Empowerment is intended to allow employees to take more "ownership" of their jobs and thereby to improve the work experience. The assumption is that employees *want* more flexibility and that change in the direction of more flexibility will make employees happier and more satisfied in their work. The underlying issue is whether the individual's quality of work life is indeed determined by having more flexibility. The personal issue here is, "If I accept more flexibility and empowerment in the project team, will I benefit in terms of the quality of my work life and workplace?"

The answer is liable to be, "It depends." It depends on whether empowerment is granted to all my colleagues and team members so that we can truly negotiate relative roles. It does no good and may do harm to the project team if only some of its members are empowered. It also depends on the ability and competence to perform; empowerment brings with it new challenges for creative and critical thinking, conceptual skills, communication skills, and a generally increased self-confidence in the workplace. Finally, will empowerment allow team members to really look at fundamental system and process design problems and correct them, or is the empowerment limited to narrow issues with little long-term consequence for the organization's performance. Thus with empowerment must come education and training and so real dialogue with management before signing on the dotted line.

2. *Professional and personal development.* Do you really *want* to grow and develop in the organization? How do you know when you really want to grow? How can you communicate your interest to the organization? The effectiveness of training and education is determined by the investment of the participants; therefore, there must be legitimate individual commitment before there is effective individual learning and consequent organizational development. Professional and personal development has many angles, but in a customer-driven project environment, professional development typically takes on three basic characteristics:

a. Planning and analytical skills. More and more, teams and individuals are being charged to exercise analytical and critical thinking skills to fully understand the customer's business systems and work flows. This will require training in research design, data collection and presentation, interviewing, and the use of computer-assisted charts and graphs. It will not be enough to be able to collect and analyze data; meaning must be derived and articulated to the customer, placing a high premium on those who can quickly spot variation and nonconformance.

b. Team-building skills. Increasingly, the work force will be called on to exercise more effective teamwork skills and attitudes. This may require training and development in group and meeting management, communication and listening, and facilitation. Rather than taking responsibility for a group's work as the *chairman* always did, the *facilitator* takes responsibility for group development and support. The *team* is now accountable.

c. Presentation skills. Presentation skills include the ability to analyze data, extract meaning and essence, and make oral presentations using briefing materials, computer projection systems, and multimedia systems. Computer conferencing will become common.

3. *Rewards and recognition.* While the emphasis seems now to be on giving recognition, the team must be able to receive recognition and rewards and to model behaviors for others who aspire to more recognition. This means that those who receive rewards have a responsibility to model for younger, more impressionable staff who look to those who are recognized for keys to success.

4. *New job opportunities.* Empowerment brings on new job opportunities because the process widens contacts and relationships in the working environment, thus opening up new career tracks. With this new found flexibility, however, employees need to see their responsibility to accept empowerment with some sense of longer-term commitment to the organization to live out the outcomes of their work. If granted new latitude, one assumes more responsibility for longer-term commitment to the organization, thus cutting expensive attrition.

5. *Increased latitude in decision making.* Remember that empowerment grants more latitude in decision making but also assumes more trust that the decision will be the right one. Thus along with the ability to decide comes the accountability to defend the decision in terms of facts and figures. This adds weight to the argument that empowerment and quality improvement lead to more data-collection and documentation needs and thus to more reliance on the computer in such areas as project management and financial management.

Empowered people

Empowerment is important for customer-driven teams because empowered people:

*E*nergize themselves and others

*M*ake things better and better

*P*romote teamwork

*O*wn their work

*W*ork on "vital" issues

*E*ncourage open and honest communication

*R*ecognize achievements

*E*njoy their work

*D*evote themselves to continuous improvement

Empowerment issues

Ideally, empowerment would not have bounds. Unfortunately, even in customer-driven project management, there are boundaries—traditional boundaries and team-based boundaries. Traditional boundaries are fixed by the organizational structure. For instance, the boundaries are specified by organizational charts, work breakdown structures, or the customer-driven team structure. There is a tendency to view these boundaries as fixing limits of responsibility and authority. In addition, there are team-based boundaries. These boundaries can be viewed in terms of relationships, which can be strong or weak depending on openness and honesty, trust, communication, attitude, motivation, ownership, pride, respect for others, and common purpose. They imply an interdependence. The relationship requires mutual support. These team-based boundaries are permeable and dynamic. The boundaries in a team-based organization are not fixed. They change along with any relationship changes.

In addition to boundaries, there are other limitations on complete empowerment. These include paradigms, trained incapacity, avoided test, or external constraints.

Paradigms

Our paradigm or world view forms a screen for information and ideas. Our paradigm filters our thinking. Only information and ideas appropriate to our paradigm are allowed in. Other information outside of our world view will be filtered out.

Trained incapacity

Our training limits how we can conceive ideas and problems. Our training or particular field of knowledge keeps us from sometimes seeing the "big" picture. No single perspective can give a completely accurate account of what "the issues are" or what "exactly is happening," because all are happening. All perspectives are pieces of the big picture, but our training allows us only to see our view.

Avoided test

Another conceptual limitation to full empowerment common in organizations is our perceived barriers to action. We tend to perceive our ability to act when we observe that others do not act, or only act a certain way. So, we follow the "norm" and don't act either. We tend to follow the behavior of others, thinking that there is some actual barrier to acting differently, when in fact we have never even tried to test that boundary, or limit, or assumption.

External constraints

In addition to the above limitations, there are the boundaries between other individuals, teams, functions, management, external vendors, and customers. The boundaries between the individual, the team, and the external environment require the exchange of information with the environment. The individual and team need to cross external boundaries to provide information, but especially to seek information (i.e., self-assessment, competitive analysis, and so on).

In addition to the above boundaries and limitations, full empowerment is influenced by the following boundary dilemmas:

- *Ambiguous jurisdictions.* Virtually all organizational members belong to more than one group. For instance, each team member belongs to a functional department and at least one program/project team. This causes the person to manage multiple boundaries and memberships. In many cases, the person asks, "Where do I belong? Where do my loyalties lie?" The managing of multiple boundaries may involve:

  The desire to be loyal to old connections and the need to make new connections as well

  The desire to withhold information and other resources, and the requirement for team interdependence

  The desire to hold on to power, and the need to relinquish or share power

- *Mixed motives.* The traditional hierarchical structure tends to promote separation. The separate departments are created in order to be functional parts of the organizational whole. But as organizations grow, the departments become larger, more numerous, and more autonomous. This leads to loyalty to department, often at the expense of the whole organization. This is compounded when the organization is faced with scarce resources, and the departments compete for them. Often departments do not share information in order to acquire more of the scarce resources.

- *Boundary heightening.* Frequently, we tend to heighten the boundaries between us and others; we are also more likely to be loyal to those most like us. This natural phenomenon causes us to be more sensitive and notice differences over similarities. It also makes us most loyal to functional departments. It leads to unproductive defensiveness.

Despite all these issues, empowerment unleashes the full potential of individuals and teams. These issues must be managed to gradually remove boundaries, limitations, and the dilemmas.

The empowerment process

The empowerment process involves the gradual shifting of responsibility, authority, and resources to those people in the organization performing and improving the work. Empowerment involves definition of boundaries, recognition of changes, and monitoring of impacts and constraints on performance. The following is a process for managing the empowerment process:

I. Define the traditional boundaries
    1. Mission
    2. The process
    3. The process beginning and end
    4. Expected results with performance measures (goals/metrics)
    5. Customer(s)
    6. Suppliers
    7. Key responsibilities

II. Identify strategies to manage empowerment
    1. Set the example
        a. Take the initiative to build trust
        b. Reward team as whole, but recognize individuals
        c. Act to develop competence
    2. Take time to develop relationships
        a. Encourage open communication
        b. Get the most out of capabilities and differences
        c. Increase problem-solving and decision-making skills
    3. Encourage the challenging of boundaries
    4. Continually assess the environment for opportunities

III. Develop a plan to manage empowerment
    1. Empowerment Development Action Plan

IV. Support the empowerment

1. Recognize and reward achievements
2. Recognize positive progress
3. Learn from mistakes
4. Celebrate success

Appendix

Assessment

Focus: This appendix provides information to formulate self-assessments in the following areas: Malcolm Baldrige National Quality Award, individual, and team.

Introduction

One of the critical elements of customer-driven project management is the constant assessment of the many factors contributing to success. This appendix and other assessment instruments within this book furnish abundant information to help formulate your own specific assessment instruments.

When you are developing your particular assessment instrument, ensure that the assessment is accurate, valid, reliable, understandable, and usable in your specific organization.

- The assessment is accurate if it provides correct and valuable feedback.
- A valid assessment gives information that is relevant and meaningful to your specific organization.
- A reliable assessment would provide the same results on successive trials.
- The assessment is understandable if it can be comprehended by the person taking the assessment and its results are clear to everyone.
- A usable assessment means it leads to meaningful action within your specific organization.

The format and rating scale of the assessment can take many forms. The following are some examples for your consideration.

Example 1. *Instructions:* Please provide an accurate self-assessment. Circle the appropriate number as it applies to you. The rating scale is as follows:

1 = Never
2 = Seldom
3 = Sometimes
4 = Usually
5 = Always

You tell others how you are feeling.

Never	Seldom	Sometimes	Usually	Always
1	2	3	4	5

If you never tell others how you are feeling, you would circle the 1. If you always tell others how you are feeling, you would circle 5. If you do not fit any of these extremes, you would circle the 2, 3, or 4 depending on your normal behavior.

Example 2. *Instructions:* Please provide an accurate assessment of your team. Check the appropriate box as it applies to your current team. Check the first box, not applicable, if you feel the item does not apply to your team. Check requires *no* improvement if you know the item requires no further development at this time. This is an item the team has already accomplished or the team is performing satisfactory. Check requires improvement if you feel your team requires improvement in this area. This could be an item the team does not currently perform or an item the team leader or the team feels requires improvement. If possible, please have each member of the team complete this team assessment.

Not applicable	Requires *no* improvement	Requires improvement

Example 3. *Instructions:* Please provide an accurate organizational assessment. Circle the appropriate number as it applies to your organization. The rating scale is as follows:

0 = Don't know
1 = Highly inaccurate
2 = Inaccurate
3 = Somewhat accurate
4 = Accurate
5 = Highly accurate

Malcolm Baldrige National Quality Award 1993 Award Criteria

The following are questions to assist you in creating your own assessment tool based on the 1993 Malcolm Baldrige National Quality Award 1993 award criteria. The Malcolm Baldrige National Quality Award criteria provide an excellent source for performing a top-level assessment. Copies of the complete 1993 award criteria can be obtained from

Malcolm Baldrige National Quality Award
National Institute of Standards and Technology
Route 270 and Quince Orchard Road
Administration Building, Room A537
Gaithersburg, MD 20899

In some cases the assessment question is not an exact interpretation of the examination criteria. In these cases, the assessment question is more stringent.

1.0 Leadership

1.1 Senior executive leadership

Are senior executive leaders personally involved and visible in quality-related activities of the company?

Do senior executive leaders actively reinforce a customer focus (internal and external)?

Are senior executive leaders personally involved in creating quality values and setting expectations?

Do senior executive leaders plan and review progress toward quality and operational performance objectives?

Do senior executives recognize employee contributions?

Are senior executive leaders constantly communicating quality values within and outside the company?

Can senior executive leaders recite a brief summary of the company's customer focus (vision) and quality values?

Can senior executive leaders describe how the vision and values serve as a basis for consistent communication within and outside the company?

Do senior executive leaders regularly communicate and reinforce the company's customer focus and quality values with managers, supervisors, team leaders, and other employees?

Do senior executive leaders have an individual development plan to evaluate and improve the effectiveness of their personal leadership and involvement?

1.2 Management for quality

Are the organization's customer focus (vision) and quality values translated into requirements for everyone in the organization?

Can everyone in the organization summarize their principal roles and responsibilities within their units?

Can everyone in the organization describe their roles and responsibilities in fostering cooperation with other units?

Are the organization's customer focus (vision) and quality values communicated and reinforced throughout the organization to all employees by other means than senior executive leadership, i.e., newsletters, bulletin boards, staff meetings, etc.?

Are organization and work unit quality and operational performance plans reviewed consistently from senior executive leaders to work leaders?

Does this include an established type of review, frequency, and content?

Does the organization have a process to assist units that are not performing according to plans?

Does the organization have key methods and key indicators the company uses to evaluate and improve awareness and integration of quality values among all employees?

1.3 Public responsibility and corporate citizenship

Does the organization integrate its public responsibilities into its quality policies and practices?

Does the organization determine or set operational requirements and goals taking into account risks and regulatory and other legal requirements?

Does the organization know how its principal public responsibility areas are addressed within the company's quality policies and/or practices?

Can key people in the organization describe how key operational requirements are communicated throughout the company?

Does the organization know how and how often progress in meeting operational requirements and/or goals is reviewed?

Does the organization look ahead to anticipate public concerns and to assess possible impacts on society that may derive from its products, services, and operations?

Can the key people in the organization describe briefly how this assessment is used in planning?

Does the organization lead as a corporate citizen in its key communities?

Can employees provide a brief summary of the types and extent of leadership and involvement in key communities?

Can employees describe how the company promotes quality awareness and sharing of quality-related information?

Does the organization seek opportunities to enhance its leadership?

Does the organization promote legal and ethical conduct in all that it does?

Does the organization continuously evaluate trends in its key indicators of improvement in addressing public responsibilities and corporate citizenship?

Does the organization consider its public responsibility and corporate citizenship in its responses to any sanctions the company has received under law, regulation, or contract?

2.0 Information and analysis

2.1 Scope and management of quality and performance data and information

Does the organization establish key criteria for selecting data and information for use in quality and operational performance improvement?

Does the organization use any of the following key types of data and information in improving quality and organization?

- Customer satisfaction, external
- Customer satisfaction, internal
- Product, service, deliverable, project performance
- Internal operations and performance
- Internal business processes
- Support services
- Employees
- Supplier performance
- Cost and financial

Does the organization ensure reliability, consistency, and rapid access to data throughout the company?

Does the organization ensure software quality?

Does the organization have key methods and key indicators used to evaluate and improve the scope and management of data and information?

Does the organization use key methods and key indicators to review and update information?

Does the organization have an active process for the following?

- Continuously shortening the cycle from data gathering to access by people needing the information
- Broadening of access to all those requiring data for day-to-day management and improvement
- Alignment of data and information with process-improvement plans and needs

2.2 Competitive comparisons and benchmarking

Does the organization use competitive comparisons and benchmarking information and data to help drive improvement of quality and organization operational performance?

Can employees describe how competitive comparisons and benchmarking information needs are determined?

Does the organization have established criteria for seeking appropriate comparison and benchmarking information from within and outside the company's industry?

Can key people provide a brief summary of current scope, sources, and principal uses of each type of competitive comparison and benchmark information and data?

Does the organization use competitive comparison and/or benchmarking information for analysis of the following?

- Customer satisfaction, external
- Customer satisfaction, internal
- Product, service, deliverable, project performance
- Internal operations and performance
- Internal business processes
- Support services
- Employees
- Supplier performance
- Cost and financial

Can key people in the organization describe how to use competitive and benchmarking information and data to improve understanding of processes, to encourage breakthrough approaches, and to set "stretch" objectives?

Does the organization have an established procedure for evaluating its processes requiring improvement for feasibility of use of competitive comparisons and benchmarking information?

Does the organization have guidelines for selecting and using competitive comparisons and benchmarking information and data to improve planning and organization operations?

2.3 Analysis and uses of company-level data

Does the organization have a process for the analysis and use of company-level data?

Does the organization aggregate customer-related data and results with other key data and analyses?

Does the organization analyze this type of key information and translate it into action?

Does the organization use available information to develop priorities for prompt solutions to customer-related problems?

Does the organization continuously determine key customer-related trends and correlation to support status review, decision making, and longer-term planning?

Does the organization aggregate operational performance data and results with other key data and analyses?

Does the organization analyze this type of key information and translate it into action to support the following?

- Developing priorities for short-term improvements in company operations, including cycle time, productivity, and waste reduction
- Determining key operations-related trends and correlation to support status reviews, decision making, and longer-term planning

Does the organization relate overall improvements in product/service quality and operational performance to changes in overall financial performance?

Can key people in the organization describe how the organization evaluates and improves its analysis as a key management tool?

Can everyone in the organization communicate how analysis supports improved data selection and use?

Can everyone in the organization describe how the organization is pursuing the shortening of the analysis-access cycle?

Can everyone in the organization outline how analysis strengthens the integration of overall data for improved decision making and planning?

3.0 Strategic quality planning

3.1 Strategic quality and company performance planning process

Does the organization have a strategic quality and organization performance planning process?

Does the organization develop strategies, goals, and business plans to address quality and customer satisfaction leadership for the short term and longer term?

Does the organization's business plan consider the following?

- Customer requirements and the expected evolution of these requirements
- Projections of the competitive environment
- Risks: financial, market, and societal
- Company capabilities, including human resource development and research and development, to address key new requirements or technology leadership opportunities
- Supplier capabilities

Can key people in the organization outline how the company develops strategies and plans to address overall operational performance improvement?

Can key people in the organization describe how the following are considered?

- Realigning work processes ("reengineering") to improve operational performance
- Productivity improvement and reduction in waste

Can key people in the organization detail how plans are deployed?

Can key people in the organization describe the method the company uses to deploy overall plan requirements to all work units and to suppliers?

Does the organization ensure alignment of work unit plans and activities with the business plan?

Does the organization have a process to ensure that proper resources are committed to meet the plan requirements?

Does the organization continuously evaluate and improve its planning process?

Does the organization's planning process include improvements in the following?

- Determining company quality and overall operational performance requirements
- Deploying requirements to work units
- Receiving planning input from company work units

3.2 Quality and performance plans

Does the organization have a quality and performance planning system?

Can everyone in the organization summarize for the organization's chosen directions, including planned products and services, markets, or market segments the following?

- Key quality factors and quality requirements to achieve leadership
- Key company operational performance requirements

Does the organization have documentation of principal short-term quality and company operational performance goals and plans?

Do the organization's planning documents include the following?

- A summary of key requirements and key operational performance indicators deployed to work units and suppliers
- A brief description of resources committed for key needs such as capital equipment, facilities, education and training, and personnel

Does the organization have documentation of principal longer-term (3 years or more) quality and company operational performance goals and plans, including key requirements and how they will be addressed?

Does the organization have documentation of 2- to 5-year projection of improvements using the most important indicators of quality and company operational performance?

Does everyone in the organization know how quality and company operational performance compare with competitors and key benchmarks currently and over the 1- to 5-year time period?

Can key people in the organization explain the comparisons, including any estimates or assumptions made regarding the projected quality and operational performance of competitors or changes in benchmarks?

4.0 Human resource development and management

4.1 Human resource planning and management

Does the organization have a human resource and management system?

Does the organization have human resource plans?

Do the human resource plans address the following?

- Development, including education, training, and empowerment
- Mobility, flexibility, and changes in work organization, processes, or schedules
- Reward, recognition, benefits, and compensation
- Recruitment, including possible changes in diversity of the work force

Do the human resource plans distinguish between short-term (1 to 2 years) and longer-term (3 years or more) actions as appropriate?

Can key people in the organization (other than human resources personnel) describe how the organization improves its human resource operations and practices?

Can key people in the organization (other than human resources personnel) outline key goals and methods for processes/practices such as recruitment, hiring, personnel actions, and services to employees?

Does the organization have key performance indicators, including cycle time for focusing on human resources improvements?

Are key performance indicators used to measure improvement in human resources areas?

Does the organization evaluate and use all employee-related data to improve the development and effectiveness of the entire work force?

Is this information used to provide key input to overall company planning and to human resource management and planning?

Does the human resources improvement process address all types of employees?

Are employee satisfaction factors used to reduce adverse indicators such as absenteeism, turnover, grievances, and accidents?

Does the organization have an evaluation methodology to continuously monitor and review key employee satisfaction factors?

4.2 Employee involvement

Does the organization have an active process to encourage employee involvement?

Does the organization promote ongoing employee contributions individually and in groups to quality and operational performance goals and plans?

Can key people in the organization describe the principal mechanisms the company uses to promote ongoing employee contributions?

Does the organization normally provide quick feedback to individual and group contributors?

Does the organization encourage the following?

- Employee empowerment
- Increased responsibility
- Employee "ownership" and pride of workmanship
- Innovation and creativity

Does the organization have a process to increase employee involvement?

Can everyone in the organization briefly summarize the principal roles and goals for all categories of employees?

Does the organization use key methods and key indicators to evaluate and improve the effectiveness, extent, and type of involvement of everyone in the organization?

Does the organization monitor and take action on trends in the most important indicators of the *effectiveness* and *extent* of employee involvement for everyone in the organization?

4.3 Employee education and training

Does the organization have a learning system that considers at a minimum employee education and training?

Does the organization determine the learning needs for everyone in the organization?

Does the organization's learning needs analysis include the types and amounts of quality and related education and training for everyone in the organization?

Does the organization's learning needs analysis take in consideration differing needs of each person?

Does the organization have a documented learning plan for everyone in the organization?

Does the organization's learning plan include the following?

- Linkage to short- and long-term plans, including companywide access to skills in problem solving, waste reduction, and process simplification
- Growth and career opportunities for all employees
- How employees' input is sought and used in the needs determination
- Summary of how quality and related education and training are delivered and reinforced

Does the organization outline methods for education and training delivery for all employees?

Does the organization provide on-the-job application of knowledge and skills?

Does the organization conduct a quality-related orientation for all new employees?

Does the organization have a process to evaluate and improve its quality and related education and training?

Does the organization's learning system include an evaluation process which supports continuous improvement of the training and education process?

Does the organization's learning system evaluation process assess training and education impact as it relates to on-the-job performance improvement that betters outcomes in key quality and operational performance improvement goals and results?

Does the organization's learning system evaluate the growth and progression of all categories and types of employees?

Does the organization's learning system monitor trends in the *effectiveness* and *extent* of quality and related training and education based on key indicators of each?

4.4 Employee performance and recognition

Does the organization have an employee performance and recognition system?

Do the organization's employee performance, recognition, promotion, compensation, reward, and feedback approaches for individuals and groups, including everyone in the organization, support the organization's quality and operational performance goals and plans?

Does the organization's employee performance and recognition system address the following?

- How the approaches ensure that quality is reinforced relative to short-term financial considerations
- How employees contribute to the company's employee performance and recognition approaches

Does the organization use key methods and key indicators to evaluate and improve its employee performance and recognition approaches?

Does the organization's performance and recognition system evaluation include the following?

- Effective participation by everyone in the organization
- Employee satisfaction factors information
- Key indicators of improved quality and operational performance results

Does the organization's performance and recognition system monitor trends in key indicators of the *effectiveness* and *extent* of employee reward and recognition by employee category?

4.5 Employee well-being and satisfaction

Does the organization have a health and safety system?

Does the organization include quality-improvement activities in the health and safety system?

Does the organization ensure that well-being factors such as health, safety, and ergonomics are included in quality-improvement activities?

Does the organization's health and safety system include principal improvement goals, methods, and indicators for each factor relevant and important to the organization's employee work environment?

Does the organization use root-cause analysis to determine how adverse conditions are prevented for accidents and work-related health problems?

Does the organization's health and safety system focus on prevention?

Does the organization's health and safety system include all the special services, facilities, and opportunities the organization makes available to employees?

Can key people in the organization describe how the organization determines employee satisfaction?

Does the organization have a process for determining employee satisfaction?

Does the organization's process for determining employee satisfaction include specific metrics, frequency, and the key factors for all employees by category or type as appropriate?

Does the organization monitor trends in key indicators of well-being and satisfaction?

Does the organization evaluation of trends in the key indicators of well-being and satisfaction include the following?

- Satisfaction
- Safety
- Absenteeism
- Turnover
- Turnover rate for customer-contact personnel
- Grievances
- Strikes
- Worker compensation

Does the organization explain and report important adverse results, if any, in the trends in the key indicators of well-being and satisfaction?

Does the organization for such adverse results describe how root causes were determined and corrected and/or give current status?

Does the organization for such adverse results compare results on the most significant indicators with those of industry averages, industry leaders, key benchmarks, and local/regional averages as appropriate?

5.0 Management of process quality

5.1 Design and introduction of quality products and services

Does the organization have a disciplined system for the design and introduction of quality products and services?

Can key people in the organization describe how designs of products, services, and processes are developed so that the following can be accomplished?

- Customer requirements are translated into product and service design requirements.
- All product and service quality requirements are addressed early in the overall design process by appropriate company units.
- Designs are coordinated and integrated to include all phases of production and delivery.
- Key process performance characteristics are selected based on customer requirements.
- Appropriate performance levels are determined.
- Measurement systems are developed to track performance for each of the performance characteristics.

Does the organization review and validate designs, taking into account the following key factors?

- Product and service performance
- Process capability and future requirements
- Supplier capability and future requirements

Does the organization have a process to improve its designs and design processes so that new product and service introductions and product and service modifications progressively improve in quality and cycle time?

5.2 Process management: Product and service production and delivery processes

Does the organization have a production and delivery process?

Does the organization maintain the quality of production and delivery processes in accord with the product and service design requirements?

Does the production and delivery process consider the following?

- The key processes and their requirements
- Key indicators of quality and operational performance
- How quality and operational performance are determined and maintained, including types and frequencies of in-process and end-of-process measurements used

Does the production and delivery process include methods for the following?

- Indicating significant (out-of-control) variations in processes or outputs
- Determining root causes
- Taking and verifying corrective actions

Can everyone in the production and delivery organization describe how the process is improved to achieve better quality, cycle time, and overall operational performance?

Do all the organizations involved in the production and delivery process participate in process-improvement activities?

Does the production and delivery improvement process improvement use the following?

- Process analysis/simplification
- Benchmarking information
- Process research and testing
- Alternative technology
- Information from customers of the processes, within and outside the company
- Challenge goals

5.3 Process management: Business processes and support services

Can everyone in the organization identify the organization's key business processes?

Can everyone in the organization identify the organization's key support services?

Do the organization's key business processes and support services meet customer and/or company quality and operational performance requirements?

Do the organization's business processes and support services include the following?

- The key processes and their requirements
- Key indicators of quality and performance
- How quality and performance are determined and maintained, including types and frequencies of in-process and end-of-process measurements used

Do the organization's business processes and support services include methods for the following?

- Indicating significant (out-of-control) variations in processes or outputs
- Determining root causes
- Taking and verifying corrective actions

Does the organization's business processes and support services improvement process improvement use the following?

- Process analysis/simplification
- Benchmarking information
- Process research and testing
- Alternative technology
- Information from customers of the processes, within and outside the company
- Challenge goals

5.4 Supplier quality

Does the organization have a supplier quality process?

Does the organization promote partnerships with supplier?

Can everyone in the organization describe how the organization defines and communicates its quality requirements to suppliers?

Can key people in the organization provide a brief summary of the principal quality requirements of key suppliers?

Does the organization have established key indicators that it uses to evaluate supplier quality?

Does the organization have methods to ensure that its quality requirements are met by suppliers?

Does the organization have regular communication with suppliers to describe the results of suppliers meeting quality requirements and other relevant performance information?

Does the organization have a continuous improvement effort for procurement activities?

Can everyone involved in the organization's procurement process describe how the organization evaluates and improves its own procurement activities?

Does the organization seek and use feedback from suppliers in its procurement-improvement process?

Does the organization have current plans and actions to improve suppliers' abilities to meet key quality and response time requirements?

5.5 Quality assessment

Does the organization have a quality-assessment process?

Does the organization's quality-assessment process assess the following?

- Systems, processes, and practices
- Products and services

Can each person in the organization describe how their process is assessed, to include (1) what is measured, (2) how often is it measured, and (3) how measurement quality and adequacy of documentation of processes and practices are ensured?

Does the organization use assessment findings to improve products and services, systems, processes, practices, and supplier requirements?

Does the organization have a monitoring process that verifies that assessment findings lead to actions and that actions are effective?

6.0 Quality and operation results

6.1 Product and service quality results

Does the organization monitor trends and current levels for all key measures of product and service quality?

Does the organization evaluate current quality level comparisons with principal competitors in the company's key markets, industry averages, industry leaders, and appropriate benchmarks?

6.2 Company operational results

Does the organization monitor trends and current levels for key measures of organization operational performance?

Does the organization compare performance with that of competitors, industry averages, industry leaders, and key benchmarks?

6.3 Business process and support service results

Does the organization monitor trends and current levels for key measures of business processes and support services?

Does the organization compare performance of business processes and support services with that of competitors, industry averages, industry leaders, and key benchmarks?

6.4 Supplier quality results

Does the organization monitor trends and current levels for the most important indicators of supplier quality?

Does the organization compare supplier quality levels with those of appropriately selected companies/organizations and/or benchmarks?

7.0 Customer focus and satisfaction

7.1 Customer expectations: Current and future

Does the organization have a process for determining *current* and near-term requirements and expectations of customers?

Does the customer expectations determination process include the following?

1. How customer groups and/or market segments are determined, including how customers of competitors and other potential customers are considered
2. The process for collecting information, including what information is sought, frequency and methods of collection, and how objectivity and validity are ensured
3. The process for determining specific product and service features and the relative importance of these features to customer groups or segments
4. How other information such as complaints, gains and losses of customers, and product/service performance is cross-compared to support the determination
5. Future requirements and expectation of customers

Does the customer expectations determination process for future requirements and expectations of customers include the following?

1. The time horizon for the determination
2. How important technological, competitive, societal, economic, and demographic factors that may bear on customer requirements, expectations, or alternatives are considered
3. How customers of competitors and other potential customers are considered
4. How key product and service features and the relative importance of these features are projected
5. How changing or emerging market segments are addressed and their implications on new product/service lines as well as on current products and services are considered

Does the organization continuously evaluate and improve its processes for determining customer requirements and expectations?

Does the customer requirements and expectations improvement process consider the following?

- New market opportunities
- Extension of the time horizon for the determination

7.2 Customer relationship management

Does the organization foster developing customer relationships?

Does the organization have a system for identifying and monitoring the most important processes and transactions that its people have with customers?

Can everyone in the organization summarize the key requirements for maintaining and building relationships with customers (external and internal)?

Does the organization derive key quality indicators from customer relationship requirements?

Do the organization's service standards that address the key quality indicators include the following?

- How service standards requirements are deployed to customer-contact employees and to other company units that provide support for customer-contact employees
- How the overall service standards system is tracked

Does the organization provide information and easy access to enable customers to seek assistance, to comment, and to complain?

Can customer-contact people in the organization describe their types of contact with customers and how they maintain easy access to customers for assistance, comments, or complaints?

Does the organization have a process to follow up with customers on products, services, and recent transactions to seek feedback and to help build relationships?

Does the organization have an established system that addresses for customer-contact employees the following?

- Selection factors
- Career path
- Deployment of special training, including knowledge of products and services, listening to customers, soliciting comments from customers, how to anticipate and handle problems or failures ("recovery"), skills in customer retention, and how to manage expectations
- Empowerment and decision making

- Satisfaction determination
- Recognition and reward
- Turnover

Does the organization have a process that ensures that formal and informal complaints and feedback received by all company units are aggregated for overall evaluation and use throughout the company?

Does the organization ensure that complaints and problems are resolved promptly and effectively?

Does the organization set priorities for improvement projects based on analysis of complaints, including types and frequencies of complaints and relationships to customer's repurchase intentions?

Does the organization have a continuous customer relationship improvement process that evaluates and improves its customer relationship management strategies and practices?

Does the customer relationship improvement process include the following?

- How the company seeks opportunities to enhance relationships with all customers or with key customers
- How evaluations lead to improvements in service standards, access, customer-contact employee training, and technology support

Can everyone in the organization describe how customer information is used in the improvement process?

7.3 Commitment to customers

Does the organization make commitments to customers to promote trust and confidence in its products/services and to satisfy customers when product/service failures occur?

Can all customer-contact employees describe the normal commitments made to customers?

Can all customer-contact employees describe how the commitments made to customers accomplish the following?

1. Address the principal concerns of customers
2. Be free from conditions that might weaken customers' trust and confidence
3. Be communicated to customers clearly and simply

Does the organization have a continuous improvement process to evaluate and improve its commitments and the customers' understanding of them to avoid gaps between expectations and delivery?

Does the organization's commitment to the customer improvement process include the following?

- How information/feedback from customers is used
- How product/service performance improvement data are used
- How competitors' commitments are considered

7.4 Customer satisfaction determination

Does the organization have a system to determine customer satisfaction?

Does the organization's customer satisfaction determination system include the following?

- A brief description of methods, processes, and measurement scales used, frequency of determination, and how objectivity and validity are ensured
- Consideration of significant differences, if any, in these satisfaction methods, processes, and measurement scales for different customer groups or segments
- How customer satisfaction measurements capture key information that reflects customers' likely market behavior, such as repurchase intentions

Does the organization have a process to monitor customer satisfaction relative to that for competitors?

Does the organization's process to monitor customer satisfaction relative to that for competitors include the following?

1. Company-based comparative studies
2. Comparative studies or evaluations made by independent organizations and/or customers
3. Objectivity and validity of studies

Does the organization have a continuous improvement process to evaluate and improve its overall processes, measurement, and measurement scales for determining customer satisfaction and customer satisfaction relative to that for competitors?

Does the organization's customer satisfaction improvement process include the following?

- How other indicators (such as gains and losses of customers) are used
- How customer dissatisfaction indicators (such as complaints) are used in this improvement process
- How competitors' commitments are considered

7.5 Customer satisfaction results

Does the organization monitor trends in indicators of customer satisfaction segmented by customer group as appropriate?

Are the trends supported by objective information and/or data from customers demonstrating current or recent (past 3 years) satisfaction with the company's products/services?

Does the organization monitor trends in indicators of customer dissatisfaction?

Do the customer satisfaction results address the most relevant indicators for the organization's products/services?

7.6 Customer satisfaction comparison

Does the organization monitor trends in indicators of customer satisfaction relative to competitors segmented by customer group as appropriate?

Are trends supported by objective information and/or data from independent organizations, including customers?

Does the information and/or data that support the trend include survey results, competitive awards, recognition, and ratings?

Does the organization monitor trends in gaining and losing customers or customer accounts to competitors?

Does the organization monitor trends in gaining or losing market share to competitors?

Individual Assessment

The purpose of this assessment is for the team leader and/or team members to perform a self-analysis. This survey forms the basis for specific emphasis on individual development actions.

Ability to work in teams

1. Do you spend time working on team projects?
2. Do you choose to be part of collective and shared effort instead of preferring to work on individual tasks?
3. Do you participate effectively in team meetings?

Customer orientation

4. Do you consider the "user" of your work as a "customer"?
5. Do you put yourself in the shoes of your downstream customers?

6. Do you adjust what you do according to your customers?
7. Do you focus on doing it right the first time?
8. Do you assume no one can fix your work as well as you?
9. Do you feel you "own" your work?
10. Do you let your suppliers know your requirements clearly?

Ability to see "work flows" and the "big picture"

11. Do you consider the whole process in which you work?
12. Do you design and create your work process?
13. Do you review your work process with your immediate customer?
14. Do you ensure that your work process adds value to the ultimate customer?
15. Do you see how your work process fits into the overall system?
16. Do you see the "big-picture" issues of the organization?
17. Do you ensure that your work process role fits your role in the "big picture" of the organization?
18. Do you make improvements in your work process?
19. Do you suggest improvements to the "big-picture" issues of the organization?

Ability to do high-quality work

20. Do you value doing your work well and right the first time?
21. Do you take pride in meeting your own expectations for the work?
22. Do you solicit your customers' views of what high-quality work is and adjust your standards accordingly?
23. Do you produce work that is thoroughly researched and accurate according to the customer?

Ability to do a high quantity of high-quality work

24. Do you produce outputs and work products regularly?
25. Do you meet your customers' expectations for turnaround?
26. Do you produce a high volume of the right work?

Use of resources and time

27. Are you an effective and efficient user of time and resources?

28. Are you aware of the passing of time and expenditures of money in the work you do?
29. Are you cost conscious?
30. Are you schedule conscious?
31. Do you use a computer to assist you in the managing of tasks, time, and budgets?

Communications

32. Do you speak clearly?
33. Do you make effective presentations to groups?
34. Do you keep your communications to the point?
35. Do you communicate specifically to the audience?
36. Do you prepare clear and concise written communications?
37. Do you listen effectively?
38. Do you actively listen to your customer(s)?
39. Do you provide feedback when warranted?

Interpersonal relationships

40. Do you interact successfully with a wide range of people?
41. Do you focus on the process and not the person?
42. Do you maintain control and composure in conflict situations?

Conceptual skills

43. Do you think critically about the issues and look beyond superficial symptoms to discover underlying causes?
44. Do you have a good model in you mind about how the overall process should go before you pursue a course of action?
45. Do you have an understanding of the customers' needs and expectations so that you can meet them?

Problem-solving skills

46. Are you effective at identifying the problem and addressing it?
47. Do you simplify problems in order to solve them?
48. Do you gain consensus on issues and solutions before taking action?
49. Are you comfortable working plans and programs to solve problems?

50. Do you stay with implementation to see the problem through resolution?

Job knowledge

51. Do you keep current in your area of work?
52. Do you effectively translate your technical knowledge to guidelines for others?
53. Do you actively relate your job knowledge to requirements of the customer?
54. Do you pursue the latest technical knowledge in your field?

Organization of work

55. Do you set clear objectives?
56. Do you stay focused on objectives until they are complete?
57. Do you work several tasks at once?
58. Do you prioritize tasks?
59. Do you work one task at a time?
60. Do you use a computer to assist you in organizing your work?

Personal initiative

61. Do you take the initiative to change processes and procedures that do not work?
62. Do you willingly take on jobs and tasks that are not part of your job?
63. Do you communicate your ideas for continuous improvement, even if they suggest more work for you?

Coaching and mentoring

64. Do other workers seek you out for advice on work-related issues?
65. Do you take enough interest in the growth of others to listen to their issues and provide guidance?
66. Do you act as a resource person?
67. Are you accessible to others?

Technical and professional competence

68. Do peers seek your advice on technical issues?

69. Do you take the opportunity for self-development and professional development activities?
70. Do you design your own measures for your technical effectiveness?

Team Assessment: Teamwork

The purpose of this assessment is for the team leader and/or team members to perform a self-analysis of their teamwork. This survey forms the focus for team development.

This section assists you with the assessment of teamwork within your team. This analysis is based on the characteristics of successful teams.

Trust

- Is the level of trust among team members sufficient to allow open and honest communication without tension?
- Can team members promote an innovative or creative idea?
- Do team members tell each other how they are feeling?
- Does the team share as much information as possible within the team?
- Does the team share information with outside teams, support, suppliers, and customers?
- Do all team members make an input into the decision-making process?
- Do team members keep team integrity and confidences?
- Do team member freely admit mistakes?
- Do team members avoid blaming, scapegoats, and fault-finding?
- Do team members demonstrate respect for each other's opinions?
- Do team members demonstrate respect for other people's opinions?
- Do team members confront issues directly with people rather than avoid or go around them?
- Do team members credit the proper individual(s)?
- Do team members strive to always maintain everyone's self-esteem?

Effective communication, especially listening

- Does everyone have a chance to express ideas?
- Do team members actively listen to each other?

- Is the message communicated clarified as much as possible?
- Do team members observe body language?
- Do team members make their points short and simple?
- Do team members take time to understand others' points of view?
- Are others' feelings considered when communicating?
- Do team members involve themselves in the message being conveyed?
- Do team members generally comprehend most messages being communicated?
- Do team members pay attention to the messages of others?
- Do team members talk judiciously?
- Is listening emphasized by all team members?
- Do team members listen actively while others convey their message?
- Is summarizing and paraphrasing used frequently to develop understanding?
- Do team members empathize with others' views?
- Do team members nurture active listening skills?
- Does the team foster an environment conducive to sharing feedback?
- Does the team encourage feedback as a matter of routine?
- Does the team have guidelines for providing feedback?
- Are all unclear communications openly discussed by the team?
- Is direct feedback given to team members when appropriate?
- Do team members ask questions to get feedback?
- Do team members consider the "real" feelings of other team members when giving feedback?
- Is feedback focused on the issue and not made personal?

Attitude, positive "can do"

- Do team members maintain a positive, "can do" approach in all team activities?
- Do team members see opportunities even in negative situations?
- Do team members display a willingness to take risks?
- Do all team members maintain a positive outlook even when faced with adversity?

- Do team members work toward a small success to overcome negative attitudes?
- Does the team explore the root causes of any negative attitude?

Motivation

- Are team members actively participating?
- Do team members demonstrate the self-confidence and self-esteem to actively participant on the team?
- Do team members direct all their energy toward the team's mission?
- Does everyone on the team truly believe the team goal is shared?
- Are individual team members' contributions valued by other team members?
- Is satisfying individual team members' needs considered during team activities?
- Does everyone on the team know the team's rewards and recognition system?
- Does the team's rewards and recognition system recognize intrinsic rewards such as a feeling of accomplishment, opportunity for personal growth, improvement of self-esteem, and a sense of belonging?
- Does the team's rewards and recognition system provide fair compensation, an opportunity for advancement, and competitive benefits?
- Does the team's rewards and recognition system credit actual performance?
- Do team members provide praise when it is deserved?
- Does the organization empower the team to provide recognition and rewards outside the normal system for extraordinary performance?
- Does the team celebrate successes together?
- Does the team allow fun?

"We" mentality

- Do team members demonstrate a togetherness in words and actions?
- Do all team members feel a sense of belonging to the team?
- Are important decisions based on consensus?
- Do team members cooperate rather than compete?

- Do team members orient toward the mission rather than the person?
- Do team members avoid making issues personal?
- Does the team focus on negotiating win/win solutions?
- Does the team take an organizationwide perspective?
- Does the team recognize conflict as natural?
- Do team members recognize the limits of arguing?
- Do team members empathize with others' views?
- Are all team members viewed as equal regardless of perceived status differences?
- Does the team examine all sides of every issue?
- Does the team support constructive relationships?
- Does the team recognize the strengths of individual differences?
- Does the team use individual differences as an advantage to achieve the mission?
- Does the team know that too much agreement can be negative?
- Does the team appoint a "devil's advocate" when everyone agrees too readily?
- Do team members maintain team integrity?
- Do team members avoid gossiping, fault-finding, blaming, and back-stabbing?
- Does the team recognize that every team goes through four stages of team development?
- Does the team know its current stage of team development?

Ownership

- Do team members take the initiative to solve problems and/or improve their processes as a natural course of action?
- Do team members demonstrate constructive team behaviors to set the example?
- Are team members' roles and responsibilities defined?
- Do team members know their roles and responsibilities?
- Does each team member know the roles and responsibilities of all the other team members?
- Are individual contributions recognized by the team?

- Do team members take pride in the accomplishments of the team?
- Do all team members have the authority, responsibility, and resources needed to perform?
- Do all team members view themselves as "owners"?
- Do all team members understand the nature of their process with inputs from suppliers and outputs to customers?
- Do all team members know their suppliers?
- Do all team members know all their customers?
- Do all team members know all their customers' needs and expectations?
- Do all team members see themselves as suppliers of deliverables to others?
- Do all team members view their team as providing value added to a deliverable?
- Do all team members do whatever it takes personally to satisfy a customer?
- Does the team seek partnerships with suppliers?
- Does the team have an ongoing relationship with customers?
- Do all team members consider themselves marketers for the team's deliverable?
- Are team members empowered to perform and make improvements as necessary?
- Does the team's organization foster a constant learning environment?
- Are all team members technically competent?
- Can all team members describe and diagram their process?
- Do team members understand the business aspects of their process, i.e., budgets, plans, competition, return on investment, etc.?
- Does the team stress optimal life-cycle costs?
- Do team members concentrate on prevention of defects and designing in quality?
- Does the team use a disciplined approach to problem solving and process improvement?
- Does the team know through metrics if the deliverable is achieving total customer satisfaction?
- Does the team know the relationship of inputs to the output of the process?

- Does the team know the critical issues or opportunities?
- Does the team use data statistical analysis to analyze critical issues or opportunities?
- Does the team process improvements on critical issues and opportunities?

Respect, consideration of others

- Do team members respect individual differences?
- Are people's differences managed by the team to the team's advantage?
- Are constructive relationships being developed and maintained?
- Do team members feel that their individual self-worth is important?
- Do all team members treat others as they would want to be treated?
- Does the team seek outside assistance from others as necessary?
- Does the team coordinate activities with other teams and functions?
- Does the team actively seek to develop long-term relationships with customers, suppliers, other teams, and functional areas?
- Does the team have the respect of others in the organization?
- Does the team attend to the other's viewpoints?

Keeping focused

- Does the team know exactly what it needs to accomplish to be successful?
- Does the team have a clear focus?
- Does the team remain targeted on its focus?
- Does the team focus on the situation, issue, or behavior and not make it personal?
- Does the team have a vision for the future where they want to go?
- Does the team have a common reason for action?
- Is the focus oriented to specific customer expectations?
- Does the focus target excellence?
- Is the vision set by the leaders in the organization?

- Do the leaders in the organization set the example for the focus?
- Is the vision communicated so that everyone understands?
- Are sufficient resources committed to pursue the vision?
- Does the team have a mission statement?
- Is the mission statement customer-driven?
- Is the mission understood, clear, and achievable?
- Does the team have consensus on the mission statement?
- Does the mission statement provide the purpose for the team?
- Does the mission statement set the common direction for the team?
- Does the mission statement set the expected results?
- Does the mission statement involve all team members?
- Did the team identify goals after a thorough analysis?
- Are the goals geared to a specific result?
- Can the goals be observed by measurement or metrics?
- Can the goals be achieved by the team?
- Do the goals provide a challenge for the team?
- Are the goals limited to a specific time period?
- Are all goals set by the team?

The following assessment applies to the customer-driven quality improvement team.

- Was the mission originated by manager or top leader?
- Was the mission negotiated and clarified by the team?
- Does the mission include the magnitude of improvement expected?
- Does the mission include the beginning process or perceived problem?
- Does the mission state the boundaries for the team?
- Does the mission state the authority for the team?
- Does the mission identify resources?

Abbreviations and Acronyms

ACWP	Actual cost of work performed
AON	Activity on node
AI	Artificial intelligence
ALAP	As late as possible
ASAP	As soon as possible
B&P	Bid and proposal
BAC	Budget at completion
BCWP	Budgeted cost of work performed
BCWS	Budgeted cost of work scheduled
CAD	Computer-aided design
CAE	Computer-aided engineering
CALS	Computer-aided acquisition and logistics support
CAM	Computer-aided manufacturing
CCB	Change control board
CCN	Contract change notice
CDPM	Customer-driven project management
CDR	Critical design review
CDRL	Contract data requirements list
CDT	Customer-driven team
CE	Concurrent engineering
CIM	Computer-integrated manufacturing
CIS	Continuous improvement system
COPQ	Cost of poor quality
COQ	Cost of quality
CPFF	Cost plus fixed fee
CPIF	Cost plus incentive fee
CPM	Critical-path method
CQI	Continuous quality improvement
CS	Computer system
C/SCSC	Cost/schedule control system criteria
C/SSR	Cost/schedule status report
CWBS	Contract work breakdown structure
DAB	Defense acquisition board
DCIMI	Defense Council on Integrity and Management Improvement
DDR	Detailed design review
DoD	Department of Defense
DOE	Design of experiments
DR	Design review

DRL Data requirement list
DS Design specification
DTC Design to cost
DTLCC Design to life-cycle cost
EAC Estimate at completion
ECN Engineering change notice
ECP Engineering change proposal
EF Earliest finish
ES Earliest start
ETC Estimate to complete
FF Finish to finish
FFP Firm fixed price
FOT&E Follow-on operational test and evaluation
FP Fixed price
FQR Formal qualification review
FQT Formal qualification testing
FS Finish to start
G&A General and administrative
GFE Government-furnished equipment
HW Hardware
ILS Integrated logistics support
I/O Input/output
IQUE In-plant quality engineering
IS Information system
JIT Just-in-time
LCC Life-cycle cost
LCL Lower control limit
LF Latest finish
LOE Level of effort
LS Latest start
MBO Management by objectives
MIS Management information system
MRPII Manufacturing resource planning II
MTBF Mean time between failures
NPRDC Navy Personnel Research and Development Center
OBS Organizational breakdown structure
ODC Other direct cost
OFI Opportunities for improvement
OSD Office of Secretary of Defense
PCA Physical configuration audit
PDCA Plan, do, check, act
PDR Preliminary design review
PERT Program evaluation and review techniques
PM Project (or program) manager (or management)
PMIS Project management information system
PMS Project management system
PO Purchase order
PWA Project work authorization
QA Quality assurance
QFD Quality function deployment
QPD Quality policy deployment
RD Robust design
R&D Research and development
RFP Request for proposal
RFQ Request for quotation
R&M Reliability and maintainability

SDRL	Subcontractor data requirement list
SF	Start to finish
SOW	Statement of work
SPC	Statistical process control
SPECS	Specifications
SQI	Service quality indicator
SRR	Systems requirements review
SS	Start to start
STQM	Strategic total quality management
STQMP	Strategic total quality management plan
SW	Software
TCS	Total customer service
TEMP	Test and evaluation master plan
TF	Total float
TIL	Total integrated logistics
T&M	Time and material
TPM	Total production maintenance
TQI	Total quality improvement
TQL	Total quality leadership
TQM	Total quality management
T&V	Test and verification
UCL	Upper control limit
VE	Value engineering
V&V	Verification and validation
WBS	Work breakdown structure
WO	Work order
WP	Work package

Glossary

Activities The steps of a process.

Agenda A plan for the conduct of a meeting.

Analogy comparisons Use of criteria from similar past programs to provide criteria for new programs.

Appraisal costs Costs associated with inspecting a product to ensure that it meets the customer's needs and expectation.

Bar chart A chart for comparing many events or items.

Benchmarking A method of measuring your organization against those of recognized leaders or best of class.

Best of class One of a group of similar organizations whose overall performance, effectiveness, efficiency, and adaptability are superior to all others.

Bottleneck A process, facility, function, department, person, or the like that delays performance.

Brainstorming Technique that encourages the collective thinking power of a group to create ideas.

Budget A plan which includes an estimate of future costs and revenues related to expected activities.

Calculated risk taking The taking of risks where possible consequences are well-considered and evaluated with the potential rewards being greater than acceptable and affordable losses. Calculated risk taking is systematic, reasonable, informed risk taking.

Capacity The total output with current resources.

Cause The reason for an action or condition.

Cause-and-effect analysis Technique for helping a group examine underlying causes; fishbone.

Chance An unknown or unconsidered force.

Chart A graphic picture of data that highlights important trends and significant relationships.

Checksheet A list made to collect data.

Coach Person who acts as a guide for organization or individual development.

Collaborate Working jointly with others.

Commitment Personal resolve to do something.

Common cause Normal variation in an established process.

Communication Technique for exchanging information.

Competition Anyone or anything competing for total customer satisfaction.

Computer-aided design Automated system to assist in design process.

Computer-aided engineering Automated system to assist in engineering process.

Computer-aided manufacturing Automated system to assist process design for manufacturing.

Computer-integrated manufacturing The integration of CAD/CAM for all design and manufacturing processes.

Computer systems Items such as hardware, software, firmware, robotics, expert systems, and artificial intelligence.

Concurrent engineering Systematic approach to the integrated concurrent design of products and their related processes, including manufacture and support. This approach is intended to cause the developers, from the outset, to consider all elements of the product life cycle from conception through disposal, including quality, cost, schedule, and user requirements.

Consensus The agreement reached by everyone that all team members can understand and support.

Constant A quantity which has a fixed value.

Constraint Something that puts a limitation on the maximization or minimization of a goal.

Continuous improvement The never-ending pursuit of excellence.

Continuous improvement system A disciplined methodology to achieve the goal of commitment to excellence by continually improving all processes.

Contract A document that defines a project. It usually consists of the deliverable specification, statement of work, and data requirements.

Control chart A chart that shows process performance in relation to control limits.

Cooperate Acting together with others.

Corrective action An action to correct an unwanted condition.

Correlation The relationship between two items.

Cost of poor quality Term for techniques that focus on minimizing the cost of nonconformance.

Cost of quality Term for a technique used to identify cost of conformance and nonconformance. This includes the costs of prevention, appraisal, and failure.

Criteria A standard or guidelines on which a decision can be based.

Critical path A set of tasks that is the longest route between the start and completion of a project.

Critical-path method (CPM) A network planning technique used for planning and controlling the activities in a project.

Critical-path network Those interconnected tasks in the project network which because of their duration will take the longest time to complete, thus determining the total duration of the project.

Critical task A task whose delay will cause the total project to be delayed.

Culture A prevailing pattern of activities, interactions, norms, sentiments, beliefs, attitudes, values, and products in an organization; the shared experience of a group.

Customer Everyone affected by the product and/or service. The customer can be the ultimate user of the product and/or service, known as an *external customer*. The customer also can be the next person or process in the organization, known as an *internal customer*.

Customer-driven The customer or customer's voice is the primary focus. The customer leads the way. Customer satisfaction becomes the focus of all efforts. This provides the constancy of purpose vital to success.

Customer-driven team Any team in the CDPM structure led by the customer or customer's voice.

Customer-driven project lead team The team in a CDPM structure with overall responsibility for the entire project.

Customer-driven project management A management philosophy, a set of guiding principles, a methodology, tools, and techniques stressing customer-driven deliverables including products and services. It focuses on the performance and improvement of a project and the design and delivery of a deliverable. It applies the proven techniques of project management, continuous improvement, measurement, people involvement, and technology. It integrates project management, total quality management, and a customer-driven structure.

Customer-driven project management life cycle The time from an agreement between a customer and supplier that a certain deliverable is needed until it is no longer fulfilling the customer's needs and expectations. Typically, it follows the life cycle of concept, definition, production, operations, continuous improvement, and eventually closeout.

Customer-driven project management methodology The disciplined, structured process for ensuring that a deliverable totally satisfies the customer.

Customer-driven project team Any team in the CDPM structure responsible for a project.

Customer-driven quality-improvement team A team in the CDPM structure empowered to make quality improvement its stated mission.

Customer-driven process team A team in the CDPM structure empowered to perform and improve its work process.

Customer logistics All elements of logistics focusing on customer satisfaction.

Customer satisfaction In CDPM, quality.

Customer/supplier analysis Techniques that provide insight into the customer's needs and expectations and involve an organization's suppliers in the development of an organization's requirements and its suppliers' conformance to them.

Customer/supplier strategy team Top-level team in an organization responsible for the development and implementation of the organization's business strategy.

Customer and supplier project steering team A top-level project management team that oversees many projects.

Customer's voice A customer's advocate that provides the views of the customer to drive the deliverable and the project for total customer satisfaction.

Cycle time The time from the beginning of the process to the end of the process.

Data Information or a set of facts.

Data statistical analysis Tools for collecting, sorting, charting, and analyzing data to make decisions.

Decision-making The process of making a selection.

Decision tree A method of analysis which evaluates alternative decisions in a treelike structure to estimate values and probabilities of outcomes.

Defect Any state of nonconformance to requirements.

Deliverable The output of a process provided to a customer. The deliverable can be a product or service or a combination of a product and a service.

Dependency A relationship between tasks that sets the sequence of the tasks.

Design of experiments Traditional design of experiments is an experimental tool used to establish both parametric relationships and a product/process model in the early (applied research) stages of the design process.

Design for manufacturing/producibility A design to ensure the deliverable can be produced effectively and efficiently.

Design phases The design of a product or process, according to Taguchi, has three phases: systems design, parameter design, and tolerance design.

Detailed process diagram A flowchart consisting of symbols and words that completely describes a process.

Detection Identification of nonconformance after the fact.

Deviation Any nonconformance to a standard or requirement.

Deviation The difference between a number and the mean of a set of numbers or between a forecast value and the actual value.

Disciplined continuous-improvement methodology The continuous-improvement system.

Driving forces Those forces which are pushing toward the achievement of a goal.

Duration The time from the start of a task or constraint to the end.

Earned value A calculation made during project implementation which evaluates the amount of acceptable project work that has actually been completed on schedule and within budget and which meets customer satisfaction.

Effectiveness A characteristic used to describe a process in which the process output conforms to requirements.

Effects A problem or defect or outcome that occurs on the specific job to which each group or team is assigned.

Efficiency A characteristic used to describe a process that produces the required output at a perceived minimum cost.

Empowerment The power of people to do whatever is necessary to do the job and improve the system within their defined authority, responsibility, and resources.

Engineering change proposal (ECP) A formal document used to make engineering changes in an existing document.

External customer The ultimate user of a product and/or service.

Extrinsic rewards Rewards given by other people.

Facilitator One who assists the team in developing teamwork and applying the CDPM tools and techniques. One who makes it easier to perform.

Feasibility study An analysis designed to establish the practicality of a given project.

Fishbone See *Cause-and-effect analysis.*

Float The amount of time a task can slip before it affects a project's end date.

Flow diagram A drawing combined with words used for defining a process. This tool provides an indication of problem areas, unnecessary loops and complexity, non-value-added tasks, and areas where simplification of a process is possible.

Focus setting Technique used to focus on a specific outcome.

Force-field analysis Technique that helps a group describe the forces at work in a given situation.

Forward scheduling A scheduling technique where the scheduler proceeds from a known start date and computes the completion date proceeding from the first operation to the last.

Frequency distribution A table that indicates the frequency with which data fall into each of any number of classes of the variable.

Functional organization An organization responsible for a major organizational function such as marketing, engineering, sales, design, manufacturing, and distribution.

Functional team A team consisting of representatives from only one functional area.

Gantt chart A graphic representation of the project network placed in the context of a calendar to relate tasks to real time.

Goal The specific desired outcome.

Guideline A suggested practice that is not mandatory in programs intended to comply with a standard.

Hierarchical nature of a process The various levels of a process.

Histogram A chart that shows frequency of data in column form. A graph showing frequency distribution.

Histogram (project) A graphic bar chart comparing the resource needs of various project tasks with each other and with available resources over time.

"House of quality" Quality functional deployment (QFD) planning chart.

Human resources The people in the organization.

Improvement methodology A method for making improvements in an organization.

In control A process within the upper and lower control limits.

Individual involvement Involvement of each person in the output of an organization.

Information system Automated system used throughout an organization to review, analyze, and take corrective action.

In-plant quality-evaluation program Method by which in-plant government people evaluate contractor controls over product quality.

Input What is needed to do a job.

Input/output analysis Technique for identifying interdependency problems.

Institutionalize To make an integral part of the organization's way of life.

Internal customer Next person or process in an organization.

Intrinsic reward Reward that is an integral part of the system. These rewards are within the individual person.

Invention A creation of something for the first time.

Just-in-time Method of having the right material just in time to be used in an operation.

Lead team A team that oversees several other teams.

Leadership The guiding of a group of people to accomplish a common goal.

Leveling The process of allocating resources to a project to optimize use of resources.

Life-cycle cost The total cost of a system or item over its full life. This includes the cost of acquisition, ownership, and disposal.

Line chart A chart that describes and compares quantifiable information.

Listening Technique for receiving and understanding information.

Logistics The aspect of military science dealing with procurement, maintenance, and transportation of military materials, facilities, and personnel.

Loss function A function that examines the costs associated with any variation from the target value of a quality characteristic.

Lower control limit The lower control limit of the process usually minus three sigma standard deviations of the statistic.

Maintenance The repair of an item.

Malcolm Baldrige National Quality Award An annual award created by public law to recognize U.S. companies that excel in quality achievement and quality management to help improve quality and productivity.

Management The leadership of an organization charged with optimizing resources. This means getting the most out of both technology and people. The target is on managing the project and leading the people to a deliverable that totally satisfies the customer.

Management functions Include planning, organizing, staffing, directing, controlling, and coordinating.

Manufacturing resource planning II System for planning and controlling a manufacturing company's operation.

Matrix organization An organization where resources are shared between both functional management and project management.

Mean The average of a group of data.

Mean time between failures (MTBF) The average time between successive failures of a given product.

Measurement The act or process of measuring to compare results to requirements. A quantitative estimate of performance.

Median The middle value of a set of measured values when the items are arranged in order of magnitude.

Meeting Technique of bringing a group together to work for a common goal.

Mentor A person assigned as management interface support for a team. A counselor for a team or individual.

Metric Meaningful measures focusing on total customer satisfaction and targeting improvement action.

Mission The intended result. The basic organizational view of the role and function of the organization in satisfying customer expectations.

Mistake-proofing Technique for avoiding simple human error at work; *poka-yoke.*

Mode The most common or frequent value in a group of values.

Multifunctional team A team consisting of representatives from more than one function.

Noise A factor that disturbs the function of a process or function.

Nominal group technique Technique similar to brainstorming that provides structured discussion and decision making.

Non-value-added Term used to describe a process, activity, or tasks that does not provide any value to a deliverable.

Optimization Achieving the best possible solution.

Out-of-control process A process for which the outcome is unpredictable.

Output The results of a process.

"Owner" The person who can change a process without further approval.

"Ownership" The power to have control over a process or project or activity or task.

Paradigm An expected pattern or view.

Parameter (or robust) design This stage focuses on making the product performance (or process output) insensitive to variation by moving toward the best target values of quality characteristics.

Parametric design The design phase where the sensitivity to noise is reduced.

Pareto's principle That a large percentage of the results are caused by a small percentage of the causes. For instance, 80 percent of results are caused by 20 percent of causes.

People involvement Individual and group activities involving people.

Performance A term used to describe both the work product itself and general process characteristics. The broad performance characteristics that are of interest to management are quality (effectiveness), cost, and schedule (efficiency). Performance is the highly effective common measurement that links the quality of the work product to efficiency and productivity.

Pie chart A chart in circular form that is divided to show the relationship between items and the whole.

Plan A specified course of action designed to attain a stated objective.

Poka-yoke See *Mistake-proofing.*

Policy A statement of principles and beliefs or a settled course adopted to guide the overall management of affairs in support of a stated aim or goal. It is mostly related to fundamental conduct and usually defines a general framework within which other business and management actions are carried out.

Population A complete collection of items (product observations, data) about certain characteristics of which conclusions and decisions are to be made for purposes of process assessment and quality improvement.

Presentation Tool for providing information, gaining approval, or requesting action.

Prevention A future-oriented approach to quality management that achieves quality improvement through corrective action on the process.

Prevention costs These are the costs associated with actions taken to plan the product or process to ensure that defects do not occur.

Priority Relative importance.

Problem A question or situation proposed for solution. The result of not conforming to requirements or, in other words, a potential task resulting from the existence of defects.

Process A series of activities that takes an input, modifies the input (work takes place/value is added), and produces an output.

Process analysis Tool used to improve the process and reduce process time.

Process capability Long-term performance level after the process has been brought under control.

Process control In statistics, the set of activities employed to detect and remove special causes of variation in order to maintain or restore stability.

Process design The development of the process.

Process diagram A tool for defining the process by using graphic symbols.

Process improvement The set of activities employed to detect and remove common causes of variation in order to improve process capability to optimum target. Process improvement leads to quality improvement.

Process-improvement team A team of associates with representative skills and functions to work specific process(es).

Process management Management approach comprising quality management and process optimization.

Process optimization The major aspect of process management that concerns itself with the efficiency and productivity of the process, i.e., economic factors.

Process "owner" A designated person within the process who has authority to manage the process and is responsible for its overall performance.

Process performance A measure of how effectively and efficiently a process satisfies customer requirements.

Process review An objective assessment of how well the methodology has been applied to your process. Emphasizes the potential for long-term process results rather than the actual results achieved.

Product An output of a process provided to a customer (internal/external), includes goods, systems, equipment, hardware, software, services, and information.

Product design The development of a product.

Production process The manufacturing of a product.

Productivity The value added by a process divided by the value of the labor and capital consumed.

Project Any series of activities that has a specific end objective. Almost all activities in an organization can be defined in terms of a project.

Project baseline A calculation of the planned allocation of resources for a project before it begins and used as a frame of reference for periodic evaluations of earned value.

Project management The process of planning and controlling the coordinated work of a project team to produce a product or deliverable for the customer within defined cost, time, and quality constraints.

Project schedule A plan for carrying out tasks in sequence on the basis of their interdependency.

Project team leader A customer or customer's voice that drives a project.

Project team facilitator A person who ensures that the customer-driven team creates and maintains teamwork.

Project manager A leader in the project supplier's organization who coordinates all project activities within the organization.

Project management system A business system to analyze, plan, implement, and evaluate a project to ensure that the deliverable satisfies the customer.

Project management information system A system that provides the project information support for reporting and decision making.

Project network (or network schedule or precedence diagram) A graphic representation of all tasks required to complete a project.

Project objective A narrative statement of the project deliverable and its basic dimensions and a description of how the deliverable will improve quality and customer satisfaction.

Project plan The combination of project objective, work breakdown structure, project network, budget, Gantt chart, and scope of work.

Pull The customer drives the deliverable from the marketplace.

Push The marketplace creates the demand for the deliverable for the customer.

Quality In customer-driven project management, total customer satisfaction.

Quality (DoD) Conformance to a set of customer requirements that, if met, result in a product or service that is fit for its intended use.

Quality (product) Conformance to requirements.

Quality (Taguchi) The (measure of degree of) loss a product causes after being shipped, other than any losses caused by its intrinsic functions.

Quality function deployment A disciplined approach for transforming customer requirements, the voice of the customer, into product development requirements at each phase of product development.

Quality-improvement team A group of individuals charged with the task of planning and implementing quality improvement.

Quantitative methods Use of measurements.

Random Having no predictable pattern.

Random sample A selection of observations taken from all the observations of a phenomenon where each observation has an equal possibility of selection.

Random variation A fluctuation in data which is due to random or uncertain occurrences.

Range The difference between the maximum and the minimum values of data.

Recognition Special attention paid to an individual or group.

Reengineering Redesigning starting from scratch.

Reliability The probability that an item can perform its intended function for a specified interval under stated conditions.

Requirement A formal statement of a need and the expected manner in which it is to be met.

Requirements Expectations for a product or service. The "it" in "do it right the first time." Specific and measurable customer needs with an associated performance standard.

Restraining forces Forces that keep a situation from improving.

Reward Rewards are external or internal. External rewards are controlled by other people; they are pay, promotion, and benefits. Internal rewards are part of the task or individual; they are items like challenge, feeling of accomplishment, feeling of belonging, and sense of pride.

Risk management A technique for continually assessing the risk in each task of the project not only in terms of time and cost but also in terms of technical feasibility of the task.

Robust design Designing a product having minimal quality losses.

Root cause Underlying reason for nonconformance within a process. When the root cause is removed or corrected, the nonconformance will be eliminated.

Rules of conduct Rules that provide guidance for the team's conduct.

Sample A finite number of items taken from a population.

Sampling The collection of some, not all, of the data.

Scatter chart A chart that depicts the relationship between two or more factors.

Scope of work A broad narrative discussion of the project objective, tasks, and network, placing limits on the project and defining how the work will progress.

Service A range of customer satisfiers. A deliverable that is not a tangible product.

Selection grid Tool for comparing each problem, opportunity, or alternative against all others.

Selection matrix Technique for rating problems, opportunities, or alternatives based on specific criteria.

Seven basic tools of quality The seven tools are Pareto charts, cause-and-effect diagrams, process flow diagram, checksheets, the histograms, scatter diagrams, and control charts.

Skew The degree of nonsymmetry shown by frequency distribution.

Sigma A common designation for the standard deviation.

Simulation Technique of observing and manipulating an artificial mechanism (model) that represents a real-world process that, for technical or economical reasons, is not suitable or available for direct experimentation.

Slack Same as float.

Solution The answer to a problem.

Sorting Arranging information into some order, such as in classes or categories.

Special/assignable causes Abnormal causes of variation in a process.

Specification A document containing a detailed description or enumeration of particulars. Formal description of a work product and the intended manner of providing it (the provider's view of the work product).

Stages of team development The phases every team goes through from orientation, dissatisfaction, and resolution, to production.

Standard deviation A parameter describing the spread of the process output. The positive square root of the variance.

Statistic Any parameter that can be determined on the basis of the quantitative characteristics of a sample.

Statistical control Term used to describe a process from which all special causes of variation have been removed and only common causes remain. Such a process is also said to be *stable.*

Statistical estimation The analysis of a sample parameter in order to predict the values of the corresponding population parameter.

Statistical methods The application of the theory of probability to problems of variation.

Statistical process control Statistical tool for monitoring and controlling a process to maintain and possibly improve quality.

Statistics The branch of applied mathematics that describes and analyzes empirical observations for the purpose of predicting certain events in order to make decisions in the face of uncertainty.

Steering group An executive-level steering committee.

Strategic customer-driven project management planning The overall method for creating and maintaining CDPM.

Strategy A broad course of action chosen from a number of alternatives to accomplish a stated goal in the face of uncertainty.

Stratification Arranging data into classes.

Suboptimization A term describing a solution that is best from a narrow point of view but not from an overall or higher point of view.

Subprocesses The internal processes that make up a process.

Supplier An individual, organization, or firm that provides inputs to a process. The supplier can be internal or external to a company, firm, or organization.

Supplier/customer analysis Technique used to obtain and exchange information for conveying needs and requirements to suppliers and mutually determining needs and expectations of customers.

Supplier logistics All the logistics inputs into an organization.

Support system A system within an organization that guides the organization through the CDPM process.

Synergy A team of people working together in a cooperative effort.

System Many processes together to accomplish a specific function.

System improvement A method that focuses on the development or redesign of systems.

Systems (parts) or concept design This phase arrives at the design architecture (size, shape, materials, number of parts, etc.) by looking at the best available technology. Note that this phase is commonly known as the *parts design phase* in American terminology.

Taguchi approach Techniques for reducing variation of product or process performance to minimize loss.

Task One of a number of actions required to complete an activity.

Task list The list of individual tasks determined necessary to produce all the components of the work breakdown structure.

Team Group of people working together toward a common goal.

Teamwork Shared responsibility for the completion of a common task or problem.

Time-sensitive diagram The precedence diagram compared against a calendar to relate all tasks to actual dates.

Tolerance design This design stage focuses on setting tight tolerances to reduce variation in performance. Because it is the phase most responsible for adding costs, it is essential to reduce the need for setting tight tolerances by successfully producing robust products and processes in the parameter design phase.

Top-down process diagram A chart of the major steps and substeps in a process.

Top-level process diagram A diagram of the entire, overall process.

Total Term used to describe the involvement of everyone and everything in a continuous-improvement effort. Everyone is committed to "one" common organizational purpose as expressed in the vision and mission. They are also empowered to act to make it a reality. Besides people, everything in the organization, including systems, processes, activities, tasks, equipment, and information, must be aligned toward the same purpose.

Total float The maximum amount of time a path can slip before it affects the project's end date. It is the time difference between the latest allowable date and the expected completion date.

Total integrated logistics The integration of all the logistics elements involved in the inputs to the organization, all the processes within the organization, and the outputs of the organization to ensure total customer supportability at an optimal life-cycle cost.

Total production maintenance System for involving the total organization in maintenance activities.

Total quality management/leadership (TQM/TQL) A leadership philosophy and set of guiding principles stressing continuous improvement through people involvement, usually in teams, and a disciplined, structured methodology emphasizing process measurement and focusing on total customer satisfaction.

Total quality management action planning The specific roadmap for the overall TQM effort that establishes a clear focus for an organization.

TQM Total quality management.

TQM environment An internal organizational environment of openness, honesty, trust, communication, involvement, "ownership," pride of workmanship, individuality, innovation, creativity, and personal commitment to be the "best."

TQM philosophy The overall, general concepts for a continuously improving organization.

TQM principles The essential, fundamental rules required to achieve victory.

TQM process The process that transforms all the inputs into the organization into a product and/or service that satisfies a customer.

TQM umbrella The integration of all the fundamental management techniques, existing improvement efforts, and technical tools under a disciplined approach focused on continuous improvement.

Trainers People in an organization who provide skills training.

Training Gives the skills to accomplish actions.

Upper control limit The upper control limit of a process usually plus three sigma standard deviation of the statistic.

Values The principles that guide the conduct of an organization.

Variable A data item that takes on values within some range with a certain frequency or pattern.

Variance In quality management terminology, any nonconformance to specifications. In statistics, it is the square of the standard deviation.

VICTORY-C TQM model A systematic, integrated, consistent, organization-wide model consisting of all the elements required for victory focused on total customer satisfaction.

VICTORY elements *V*ision and leadership, *I*nvolvement of everyone and everything, *C*ontinuous improvement of all systems and processes, *T*raining and education, *O*wnership, *R*ewards and recognition, and *Y*ears of support and commitment from management focused on customer satisfaction.

Vision Where the organization wants to go.

Vote Technique to determine majority opinion.

What-if analysis The process of evaluating alternative strategies.

Work breakdown structure (WBS) An organizational chart of the basic components of the project deliverable derived from the process of defining customer needs through quality function deployment and used to estimate costs and control project implementation.

Bibliography

Aldag, Raymond J., and Timothy M. Stearns. *Management.* Cincinnati: South-Western Publishing Co., 1987.

Amsden, Robert T., Howard E. Butler, and Davida M. Amsden. *Statistical Process control, Simplified.* White Plains, N.Y.: Quality Resources, 1989.

Aubrey, Charles A., II, and Patricia K. Felkins. *Teamwork: Involving People in Quality and Productivity Improvement.* Milwaukee: Wisc.: Quality Press, 1988.

Benedetto, Richard F., and Beverly Jones Benedetto. *Management Concepts for the 90s: Matrix and Project Management.* Dubuque, Iowa: Kendall/Hunt, 1989.

Block, Peter. *The Empowered Manager: Positive Political Skills at Work.* San Francisco: Jossey-Bass, 1991.

Brassard, Michael. *The Memory Jogger Plus.* Methuen, Mass.: GOAL/QPC, 1989.

Camp, Robert C. *Benchmarking.* Milwaukee, Wisc.: ASQC Quality Press, 1989.

Champy, James, and Michael Hammer. *Reengineering the Corporation: A Manifesto for Business Revolution.* New York: Harper Collins, 1991.

Chandler, Arthur D., Jr. *Strategy and Structure.* Cambridge, Mass.: MIT Press, 1962.

Cleland, David I., James M. Gallagher, and Ronald S. Whitehead. *Military Project Management.* New York: McGraw-Hill, 1993.

Cleland, David I., and William King. *Project Management Handbook.* New York: Van Nostrand Reinhold, 1983.

Cleland, David I. *Project Management: Strategic Design and Implementation.* Blue Ridge Summit, Pa.: Tab Books, 1990.

Ciampa, Dan. *Total Quality: A User's Guide for Implementation.* Reading, Mass.: Addison-Wesley, 1992.

Clavell, James (ed.). *Art of War.* New York: Dell, 1983.

Crosby, Philip B. *Quality Is Free.* New York: McGraw-Hill, 1979.

Crosby, Philip B. *Running Things.* New York: McGraw-Hill, 1986.

Davidow, William H., and Bro. Uttal. *Total Customer Service.* New York: Harper & Row, 1989.

Deming, W. Edwards. *Out of the Crisis.* Cambridge, Mass.: MIT Center for Advanced Engineering Study, 1982.

Department of Defense. CSDL-R-2161: *Findings of the U.S. Department of Defense Technology Assessment Team on Japanese Manufacturing Technology, Washington: Final Report, June 1989.*

Department of Defense. DoD 4245.7M: Transition from Development to Production: Solving the Risk Equation. Washington, September 1985.

Department of Defense, HQ AFSD/FMC: *Metrics Handbook.* Washington: 1991.

Department of Defense. *Total Quality Management Guide,* Vols. I and II. Washington: Final Draft, Feb. 15, 1990.

Department Systems Management College. *Risk Management: Concepts and Guidance.* Ft. Belvoir, Va.: 1990.

Drucker, Peter. *Management: Tasks, Responsibilities, and Practices.* New York: Harper & Row, 1974.

Drucker, Peter F. *The New Realities.* New York: Harper & Row, 1989.

Feigenbaum, Armand V. *Total Quality Control Handbook.* New York: McGraw-Hill, 1983.
Garvin, David. "How the Baldrige Award Really Works," *Harvard Business Review,* November-December 1991, p. 94.
Goldratt, Eliyahu M., and Jeff Cox. *The Goal.* Croton-on-Hudson, N.Y.: North River Press, 1986.
Harrington, H. James. *Business Process Improvement.* New York: McGraw-Hill, 1991.
Harrington, H. James. *The Improvement Process.* New York: McGraw-Hill, 1987.
Hersey, Paul, and Kenneth H. Blanchard. *Management of Human Behavior: Utilizing Human Resources.* Englewood Cliffs, N.J.: Prentice-Hall, 1988.
Hurdeski, Michael. *Computer Integrated Manufacturing.* Blue Ridge Summit, Pa.: Tab. Books, 1988.
Imai, Masaaki. *Kaizen.* New York: Random House, 1986.
Institute for Defense Analysis: *IDA Report R-338: The Role of Concurrent Engineering in Weapons System Acquisition.* Alexandria, Va.: December 1988.
International Standards. *ISO 9001, Quality Systems: Model for Quality Assurance in Design/Development, Production, Installation and Servicing.* Geneva: International Organization for Standardization.
International Standards. *ISO 9002, Quality Systems: Model for Quality Assurance in Production and Installation.* Geneva: International Organization for Standardization.
International Standards. *ISO 9003, Quality Systems: Model for Quality Assurance in Final Inspection and Test.* Geneva: International Organization for Standardization.
Ishikawa, Kaoru. *Guide to Quality Control.* Tokyo: Asian Productivity Organization, 1982.
Jones, James V. *Integrated Logistics Support Handbook.* Blue Ridge Summit, Pa.: Tab Books, 1987.
Juran, Joseph M. *Juran on Planning for Quality.* New York: Free Press, 1988.
Juran, Joseph, M., and Frank M. Gryna, Jr. *Quality Planning and Analysis.* New York: McGraw-Hill, 1986.
Kayser, Thomas A. *Mining Group Gold: How to Cash in on the Collaborative Brain Power of a Group.* El Segundo: Serif Publishing, 1990.
Kerzner, Harold. *Project Management: A Systems Approach to Planning, Scheduling, and Controlling.* New York: Van Nostrand, 1992.
Kezsbom, Deborah S., Donald L. Schilling, and Katherine A. Edward. *Dynamic Project Management: A Practical Guide for Managers and Engineers.* New York: Wiley, 1989.
King, Bob. *Better Designs in Half the Time, Implementing QFD Quality Function Deployment in America.* Methuen, Mass.: GOAL/QPC, 1989.
Lacy, James. A. *Systems Engineering Management.* New York: McGraw-Hill, 1992.
Lareau, William. *American Samurai.* Clinton, N.J.: New Win Publishing, 1991.
Levine, Harvey A. *Project Management Using Microcomputers.* New York: McGraw-Hill, 1986.
Levitt, Theodore. *The Marketing Imagination.* New York: Free Press, 1980.
Lubben, Richard T. *Just-in-Time Manufacturing.* New York: McGraw-Hill, 1988.
Mansir, Brian E., and Nicholas R. Schacht. Report IR806R1: *Continuous Improvement Process.* Bethesda, Md.: Logistics Management Institute, August 1989.
McGill, Michael E. *American Business and the Quick Fix.* New York: Holt & Company, 1988.
Nakajima, Sceiichi. *Total Productive Maintenance.* Cambridge, Mass.: Productivity Press, 1988.
Office of the Assistant Secretary of Defense. *Total Quality Management: An Education and Training Strategy for Total Quality Management in the Department of Defense.* Washington, July 1989.
Office of the Assistant Secretary of Defense for Production and Logistics. *The Role of Concurrent Engineering in Weapons System Acquisition.* Washington, December 1988.
Office of Deputy Assistant Secretary of Defense for TQM. *Total Quality Management: A Guide for Implementation* (DoD Guide 5000.51G). Washington, 1989.

Peters, Thomas J. *Thriving on Chaos.* New York: Knopf, 1987.
Peters, Thomas, J. *Liberation Management.* New York: Alfred A. Knopf, 1992.
Peters, Thomas J., and Robert H. Waterman, Jr. *In Search of Excellence.* New York: Harper & Row, 1982.
Pyzdek, Thomas. *Pyzdek's Guide to SPC: Volumes One and Two, and Workbooks.* Tucson, Quality Publishing.
Ross, Phillip J. *Taguchi Techniques of Quality Engineering.* New York: McGraw-Hill, 1988.
Saylor, James H. *TQM Field Manual.* New York: McGraw-Hill, 1992.
Scherkenbach, William W. *The Deming Route to Quality and Productivity.* Washington: Cee Press, 1988.
Scholtes, Peter R. *The Team Handbook.* Madison, Wisc.: Joiner Associates, 1988.
Schonberger, Richard J. *World Class Manufacturing.* New York: Free Press, 1986.
Senge, Peter. *The Fifth Discipline: The Art and Practice of the Learning Organization.* New York: Doubleday, 1990.
Shores, A. Richard. *Survival of the Fittest.* Milwaukee, Wisc.: ASQC Quality Press, 1988.
Symantec. *Timeline: The Corporate Choice for Project Management and Presentations, User Manual.* Cupertino, Calif.: Symantec Corporation, 1991.
Taguchi, Genichi. *Introduction to Quality Engineering.* Tokyo: Asian Productivity Organization, 1986.
Tenner, Arthur, and Irving DeToro. *Total Quality Management: Three Steps to Continuous Improvement.* Reading, Mass.: Addison-Wesley, 1992.
Townsend, Patrick L. *Commit to Quality.* New York: Wiley, 1986.
Walker, Robert A. *Project Management.* College Park: University of Maryland University College Press, 1992.
Wallace, Thomas E. *MRPII: Making It Happen.* Essex Junction,: Oliver Wright, 1985.
Walton, Mary. *Deming Management at Work.* New York: Perigee Books, 1990.

Index

ABOUT THE AUTHORS

BRUCE T. BARKLEY is Manager, Consulting Services, Maryland Center for Quality and Productivity. He was formerly Director of Total Quality Management and Associate Director and Instructor in the technology management program at the University of Maryland. He is the recipient of the 1990 "Excellence in Teaching" award. Mr. Barkley has several years' consulting and training experience in both the public and private sectors, including 25 years in the federal government for two cabinet level agencies.

JAMES H. SAYLOR owns The Business Coach. He has over 25 years' experience in organizational development, quality management, project management, integrated logistics support, and training. He has coached, trained, and facilitated many public and private organizations in achieving their specific victory through application of continuous improvement efforts. Jim is the author of the widely acclaimed *TQM Field Manual* (McGraw-Hill, 1992).